T0201401

Business Risk Management

Business Risk Management

Models and Analysis

Edward J. Anderson

The University of Sydney Business School, Australia

WILEY

Library of Congress Cataloging-in-Publication Data
Anderson, E. J. (Edward J.), 1954-
  Business risk management : models and analysis / Edward Anderson, PhD.
      pages cm
  Includes bibliographical references and index.
  ISBN 978-1-118-34946-5 (hardback)
  1. Risk management. I. Title.
  HD61.A529 2014
  658.15'5 – dc23

                                                        2013028911

A catalogue record for this book is available from the British Library.

ISBN: 978-1-118-34946-5

Set in 10/12pt Times by Laserwords Private Limited, Chennai, India
Printed and bound in Malaysia by Vivar Printing Sdn Bhd

1   2014

To my wife, Margery, and my children:
Christian, Toby, Felicity, Marcus, Imogen, Verity and Clemency.

Contents

---

*Sections marked by an asterisk may be skipped by readers requiring a less detailed discussion of the subject.

Preface

What does this book try to do?

Managers operate in a world full of risk and uncertainty and all managers need to manage the risks that they face. In this book I introduce a number of different areas that I think are important in understanding risk and in making good decisions when the future is uncertain. This is a book aimed at all students who want to learn about risk management in a business environment.

The best way to achieve a clear understanding of risk is to use quantitative tools and probability models, and this book is unashamedly quantitative in its emphasis. However, that does not mean the use of advanced mathematics: the material is carefully chosen to be accessible to those without a strong mathematical background.

The book is aimed at either postgraduate or senior undergraduate students. It would be suitable for MBA students taking an elective course on Business Risk Management. This text is for a course aimed at all business students rather than those specializing in finance. The book could also be used for self-study by a manager who wishes to improve their understanding of this important area.

Risk management is an area where a manager's instinct may run counter to the results of a careful analysis. This book explores the critical issues for managers who need to understand both how to make wise decisions in risky environments and how people respond to risk.

There are many different types of risk and there are existing textbooks that look at specific kinds of risk: for example, environmental risk, engineering risk, political risk (particularly for companies operating in an international environment), or health and safety risks. These books give advice on evaluating specific types of risk, whether that be pollution issues or food safety, and they are aimed at students who will work in specific industries. Their focus is on understanding particular aspects of the business environment and how these generate risk; on the other hand, my focus is on the decisions that managers must take.

This textbook is unusual in providing a comprehensive treatment of risk management from a quantitative perspective, while being aimed at general business students rather than finance specialists. In fact, many of the topics that I discuss can only be found in more advanced monographs or research papers.

In writing this book I wanted to bring together a great range of material, and to include some modern advanced approaches alongside the fundamentals. So I discuss the basic probability ideas needed to understand the principle of diversification, but at the same time I include an introduction to the treatment of heavy tails through extreme value theory. I discuss the fundamental ideas of utility theory, but I also give an extensive discussion of Prospect Theory which describes how people actually make decisions on risk. I introduce Monte Carlo methods for making good decisions in a risky environment, but I also discuss modern ideas of robust optimization. To bring all these topics together is an ambitious aim, but I hope that this book will demonstrate that it is natural to teach this material together.

It is my belief that some important topics that have traditionally been seen as the realm of finance specialists need to be made accessible to those with a more general business focus. Thus, we will cover some of the classic financial risk areas, such as the Basel framework of market, credit and operational risk; the use of value at risk in practice; credit scoring; and real options. We do all this without requiring any advanced financial mathematics.

The book has been developed from teaching material used in courses at both advanced undergraduate and master's level at the University of Sydney Business School. These are full semester courses (13 weeks) but the design of the book would enable a selection of chapters to be taught in a shorter course.

What is the structure of this book?

The first chapter is introductory: it sets out my understanding of the essence of risk management and covers the framework for the rest of the book.

The next three chapters deal with the analysis of risk. Chapter 2 works through some fundamental ideas about risks that depend on events and risks that depend on values. It introduces the important idea of diversification of risk and looks in detail at how this can fail when diversification takes place over a portfolio where different elements tend to move in tandem. This leads up to a brief discussion of copulas as a way to model dependence. Chapter 3 moves from the theory of Chapter 2 to the more practical topic of value at risk. Anyone working in this area needs to know what this is and how it is calculated; as well as understanding both the strengths and the weaknesses of value at risk as a measure of risk. This chapter also discusses expected shortfall as an alternative to value at risk. Chapter 4 takes us deeper into the essential problems of risk management that involve the tails of a probability distribution. The chapter introduces heavy-tailed distributions and shows how extreme value theory can be used to help us estimate risk from data that inevitably do not contain many extreme values.

The next four chapters are concerned with making decisions in a risky environment. The fundamental insight here is that we need to think not only of how much profit or loss is made, but also how those different outcomes affect us, either as individuals or as a firm. This leads to the idea of a utility function that we want to maximize. Chapter 5 gives a thorough treatment of Expected Utility Theory, which is a powerful normative description of how we should take decisions. It turns out, however, that individual decision makers do not keep to the 'rules' of Expected Utility Theory. Chapter 6 describes the way that choices are made in risky environments by real people. Prospect Theory can be a helpful predictor of these decisions and I describe this in detail. Chapter 7 looks at the difficulties of making the right decision in complex problems, particularly where the situation evolves over time. We show how such problems can be formulated and solved and explain how to use Monte Carlo simulation in finding solutions. One of the problems with these methods is that they require a complete description of the probability distributions involved. In practice, this can involve more guesswork than actual knowledge. Chapter 8 discusses a modern approach, termed 'robust optimization', to overcome this problem by specifying a range of possible values rather than a complete distribution.

The last two chapters of the book have a different emphasis. Chapter 9 describes the important topic of real options. This switches the focus from the negative events to the positive ones. It is enormously valuable for managers to understand the concept of an option value: and how this implies that more variability will lead to a higher value for the project. In a sense, this is an example of how risk can be good. The final chapter returns to the Basel distinction between three different kinds of risk: market risk, credit risk and operational risk. After Chapter 1 our emphasis has been mainly on market risk, but in Chapter 10 we discuss credit risk. We look at credit scoring approaches both at the firm level, where agencies like Standard & Poor's dominate, and also at the consumer level, where credit scoring can determine the terms of a loan.

How can this book be used?

An important question in teaching quantitative risk management is how much mathematical maturity one should assume. This book is aimed at students who have taken an introductory statistics course or quantitative methods course, but do not otherwise have much mathematical background. I have included an appendix that gives a reminder of the probability theory that will be used. The idea of finding the area under the tail of a distribution function to calculate a probability is quite fundamental for risk management and so some knowledge of elementary calculus will be helpful, but I have limited the material in which calculus is used. There is no need for knowledge of matrix algebra. However, it should not be thought that this implies a superficial treatment of the material. This text requires students to come to grips with advanced concepts and students taught from this material in Sydney have found it challenging. To make it easier to use this textbook for a more elementary course, I have starred certain subsections that can be omitted by those who want to understand the important ideas without too much of the theoretical detail.

Excel spreadsheets are used throughout to illustrate the material and for some exercises. There is no requirement for any other special purpose software. The excel spreadsheets mentioned can be found in the companion website to the book: http://www.wiley.com/go/business_risk_management

Throughout the text I will discuss small examples set in fictitious companies. The exercises too are often based around decision problems faced by imaginary companies. I believe that the best way to come to grips with this sort of material is to spend time working through the problems (while resisting the temptation to look too quickly at the answer provided). I have provided a substantial number of end-of-chapter exercises. The answers to the even-numbered exercises are given in Appendix B and full worked solutions are available for instructors (see the instructions in the companion website).

Early versions of this manuscript were used in my classes on Business Risk Management at the University of Sydney in both 2011 and 2012. I would like to thank everyone who took those classes for their comments and questions which have helped me in improving the presentation, and I would particularly like to thank Heying Shi who managed to uncover the greatest number of mistakes.

Eddie Anderson
Sydney

1

What is risk management?

The biggest fraud of all time
A number of banks have succeeded in losing huge sums of money in their trading operations, but Société Générale ('SocGen') has the distinction of losing the largest amount of money as the result of a fraud. This took place in 2007, but was uncovered in January 2008. SocGen is one of the largest banks in Europe and the size of the fraud itself is staggering; SocGen estimated that it lost 4.9 billion Euros as a result of unwinding the positions that had been entered into. With a smaller firm this could well have caused the bank's collapse, as happened to Barings in 1995, but SocGen is large enough to weather the storm. The employee responsible was Jérôme Kerviel, who did not profit personally (or at least only through his bonus payments being increased). In effect, he was taking enormous unauthorized gambles with his employer's money. For a while these gambles came off, but in the end they went very badly wrong.

In America the news broke on January 24, 2008, when the *New York Times* reported as follows:

> 'Société Générale, one of the largest banks in Europe, was thrown into turmoil Thursday after it revealed that a rogue employee had executed a series of "elaborate, fictitious transactions" that cost the company more than $7 billion US, the biggest loss ever recorded in the financial industry by a single trader.
>
> Before the discovery of the fraud, Société Générale had been preparing to announce pretax profit for 2007 of €5.5 billion, a figure that Bouton (the Société Générale chairman) said would have shown the company's "capacity to absorb a very grave crisis." Instead, Bouton – who is forgoing his salary through June as a sign of taking responsibility – said the "unprecedented" magnitude of the loss had prompted it to seek

Business Risk Management: Models and Analysis, First Edition. Edward J. Anderson.
© 2014 John Wiley & Sons, Ltd. Published 2014 by John Wiley & Sons, Ltd.
Companion website: www.wiley.com/go/business_risk_management

about €5.5 billion in new capital to shore up its finances, a move that secures the bank against collapse.

Société Générale said it had no indication whatsoever that the trader – who joined the company in 2000 and worked for several years in the bank's French risk-management office before being moved to its Delta One trading desk in Paris – "had taken massive fraudulent directional positions in 2007 and 2008 far beyond his limited authority." The bank added: "Aided by his in-depth knowledge of the control procedures resulting from his former employment in the middle-office, he managed to conceal these positions through a scheme of elaborate fictitious transactions."

When the fraud was unveiled, Bouton said, it was "imperative that the enormous position that he had built, and hidden, be closed out as rapidly as possible." The timing could hardly have been worse. Société Générale was forced to begin unwinding the trades on Monday "under conditions of extreme market volatility," Bouton said, as global stock markets plunged amid mounting fears of an economic recession in the United States.'

A story like this inevitably prompts the question: How could this have happened? Later in this chapter we will give more details about what went wrong. SocGen was a victim of an enormous fraud but the defense lawyers at Kerviel's trial argued that the company itself was primarily responsible. Whatever degree of blame is assigned to SocGen, it clearly paid a heavy price. It is easy to be wise after the event, but good business risk management calls on us to be wise beforehand. Later in this chapter we will discuss the things that can be learnt from this episode (and that need to be applied in a much wider sphere than just the world of banks and traders.)

1.1   Introduction

In essence, *risk management* is about managing effectively in a risky and uncertain world. Banks and financial services companies have developed some of the key ideas in the area of risk management, but it is clearly vital for any manager. All of us, every day, operate in a world where the future is uncertain.

When we look out into the future there is a myriad of possibilities: there can be no comprehension of this in its totality. So our first step is to simplify in a way that enables us to make choices amidst all the uncertainty. The task of finding a way to simplify and comprehend what the future might hold is conceptually challenging and different individuals will do this in different ways. One approach is to set out to build, or imagine, a set of different possible futures, each of which is a description of what might happen. In this way we will end up with a range of possible future scenarios that are all believable, but have different likelihoods.

Though it is obviously impossible to describe every possibility in the future, at least having a set of possibilities will help us in planning.

One way to construct a scenario is to think of chains of linked events: if one thing happens then another may follow. For example, if there is a typhoon in Hong Kong, then the shipment of raw materials is likely to be late, and if this happens then we will need to buy enough to deal with our immediate needs from a local supplier, and so on. This creates a *causal chain*.

A causal chain may, in reality, be a more complicated network of linked events. But in any case it is often helpful to identify a particular *risk event* within the chain that may or may not occur. Then we can consider both the probability of the risk event occurring and also the consequences and costs if it does. In the example of the typhoon in Hong Kong, we need to bear in mind both the probability of the typhoon and the costs involved in finding an alternative temporary source.

Risk management is about seeking better outcomes, and so it is critical to identify different risk events and to understand both their causes and consequences. Usually risk in this context refers to something that has a negative effect, so that our interest in the causes of negative risk events is to reduce their probability or, better still, eliminate them altogether. We are concerned about the consequences of risk events so that we can act beforehand in a way that reduces the costs if a negative risk event does occur. The open-ended nature of this exercise makes it important to concentrate on the most important causal pathways – we can think of this as identifying *risk drivers*.

At the same time as looking at actions specifically designed to reduce risk, we may need to think about the risk consequences of management decisions that we make. For example, we may be considering moving to an overseas supplier who is able to deliver goods at a lower price but with a longer lead time, so that orders will need to be placed earlier: then we need to ask what extra risks are involved in making this change. In later chapters we will give much more attention to the problems of making good decisions in a risky environment.

Risk management involves planning and acting before the risk event. This is proactive rather than reactive management. We don't just wait and see what happens, with the hope that we can manage our way through the consequences; instead we work out in advance what might happen and what the consequences are likely to be. Then we plan what we should do to reduce the probability of the risk event and to deal with the consequences if it occurs.

Sometimes the risk event is not in our control; for example, we might be dealing with changes in exchange rates or government regulation – usually this is called an *external risk*. On other occasions we can exercise some control over the risk events, such as employee availability, supply and operations issues. These are called *internal risks*. The same distinction between what we can and cannot control occurs with consequences too. Sometimes we can take actions to limit negative consequences (like installing sprinklers for a fire), but at other times there are limits to what we can do and we might choose to insure against the event directly (e.g. purchasing fire insurance).

We will use the term risk management to refer to the entire process:

- *Understanding risk:* both its drivers and its consequences.

- *Risk mitigation:* reducing or eliminating the probability of risk events as well as reducing the severity of their impact.

- *Risk sharing:* the use of insurance or similar arrangement so that some of the risk is transferred to another party, or shared between two parties in some contractual arrangement.

The risk framework we are discussing makes it sound as though all risk is bad, but this is misleading in two ways. First we can use the same approach to consider good outcomes as well as bad ones. This would lead us to try to understand the most important causal chains, with the aim of maximizing the probability of a positive chance event, and of optimizing the benefits if this event does occur. Second we need to recognize that sometimes the more risky course of action is ultimately the wiser one. Managers are schizophrenic about risk. Most see risk taking as part of a manager's role, but there is a tendency to judge whether a decision about risk was good or bad simply by looking at the results. Though it is rarely put in these terms, the idea seems to be that it is fine to take risks provided that nothing actually goes badly wrong! Occasionally managers might talk of 'controlled risk' by which they mean a course of action in which there may be negative consequences but these are of small probability and the size of the cost is tolerable.

In their discussion of the agile enterprise, Rice and Franks (2010) say, 'While uncertainty impacts risk, it does not necessarily make business perilous. In fact, risk is critical to any business – for nothing can improve without change – and change requires risk.' Much the same point was made by Prussian Marshall Helmuth von Moltke in the mid-1800s: 'First weigh the considerations, then take the risks.'

Our discussion so far may have implied an ability to list all the risks and discuss the probability that an individual risk event occurs. But often there is no way to identify all the possible outcomes, let alone enter into a calculation of the probability of their occurrence. Some people use the term *uncertainty* (rather than risk) to refer to this idea. Frank Knight was an economist who was amongst the first to distinguish clearly between these two concepts and he used 'risk' to refer to situations where the probabilities involved are computable. In many real environments there may be a total absence of information about, or awareness of, some potentially significant event. In a much-parodied speech made at a press briefing on February 12, 2002, former US Defense Secretary Donald Rumsfeld said:

'There are known knowns. These are things we know that we know. There are known unknowns. That is to say, there are things that we now know we don't know. But there are also unknown unknowns. These are things we do not know we don't know.'

In Chapter 8 we will return to the question of how we should behave in situations with uncertainty, when we need to make decisions without being able to assign probabilities to different events.

1.2 Identifying and documenting risk

Many companies set up a formal *risk register* to document risks. This enables them to have a single point at which information is gathered together and it encourages a careful assessment of risk probabilities and likely responses to risk events.

A carefully documented risk management plan has a number of advantages. There is first of all a benefit in making it more likely that risk will be managed appropriately, with major risks identified and appropriate measures taken. Secondly there is an advantage in defining the responsibility for managing and responding to particular categories of risk. It is all too easy to find yourself in a company in which something goes wrong and no person or department admits to being the responsible party.

Moreover, a risk management plan allows stakeholders to approve the risk management approach and helps to demonstrate that the company has exercised an appropriate level of diligence in the event that things do go wrong.

There are really three steps in setting up a risk register:

1. *Identify the important risk events.* The first step is to make some kind of list of different risks that may occur, and in doing this a systematic process for identifying risk can be helpful. A good starting point is to think about the context for the activity: the objectives; the external influences; the stages that are gone through. The next step is to go through each element of the activity asking what might happen that could cause external factors to change, or that could affect the achievement of any objective.

2. *Understand the causes of the risk events.* Risk does not occur in a vacuum. Having identified a set of risk events, the next step is to come to grips with the factors that are involved in causing the risk events. In order to understand what can be done to avoid these risks, we should ask the following questions, for each risk:

   • How are these events likely to occur?

   • How probable are these events?

   • What controls currently exist to make this risk less likely?

   • What might stop the controls from working?

3. *Assess the consequences of the risk events.* The final step is to understand what may happen as a result of these risk events. The aim is to find ways to reduce the bad effects. For each risk we will want to know:

   • Which stakeholders might be involved or affected? For example, does it affect the return on share capital for shareholders? Does it affect the assurance of payment for suppliers? Does it affect the security that is offered to our creditors? Does it affect the assurance of future employment for our employees?

- How damaging is this risk?
- What controls currently exist to make this risk less damaging?
- What might stop the controls from working?

At the end of this process we will be in a better position to build the risk register. This will indicate, for each risk identified:

- its causes and impacts;
- the likelihood of this risk event;
- the controls that exist to deal with this risk;
- an assessment of the consequences.

Because the risk register will contain a great many different risks, it is important to focus on the most important ones. We want to construct some sort of priority rating – giving the overall level of risk. This then provides a tool so that management can focus on the most important risk events and then determine a risk treatment plan to reduce the level of risk. The most important risks are those with serious consequences that are relatively likely to occur. We need to combine the likelihood and the impact and Figure 1.1 shows the type of diagram that is often used to do this, with risk levels labeled L = Low; M = Medium; H = High; and E = Extreme.

This type of diagram of risk levels is sometimes called a *heat map*, and often red is used for the extreme risk boxes; orange for the high risks; and yellow for the medium risks. It is a common tool and is recommended in most risk management standards. It should be seen as an important first step in drawing

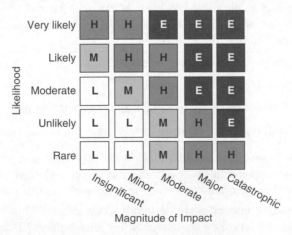

Figure 1.1   Calculating risk level from likelihood and impact.

up a risk management plan, prior to making a much fuller investigation of some specific risks, but nevertheless there are some significant challenges associated with the use of this approach.

One problem is related to the use of a scale based on words like 'likely' or 'rare': these terms will mean very different things to different people. Some people will use a term like 'likely' to mean a more than two thirds chance of occurring (this is the specific meaning that is ascribed in the IPCC climate change report). But in a risk management context, quite small probabilities over the course of a year may seem to merit the phrase 'likely'.

The use of vague terms in a scale of this sort will make misunderstandings far more likely. Douglas Hubbard describes an occasion when he asked a manager 'What does this mean when you say this risk is "very likely"?' and was told that it meant there was about a 20% chance of it happening. Someone else in the room was surprised by the small probability, but the first manager responded, 'Well this is a very high impact event and 20% is too likely for that kind of impact.' Hubbard describes the situation as 'a roomful of people who looked at each other as if they were just realizing that, after several tedious workshops of evaluating risks, they had been speaking different languages all along.' This story illustrates how important it is to be absolutely clear about what is meant when discussing probabilities or likelihoods in risk management.

The heat map method is clearly a rough and ready tool for the identification of the most important risks. But its greatest value is in providing a common framework in which a group of people can pool their knowledge. Far too often the methodology fails to work as well as it might, simply because there has not been any prior agreement as to what the terms mean. A critical point is to have a common view of the time frame or horizon over which risks are assessed. Suppose that there is a 20% probability of a particular risk event occurring in the next year, but the group charged with risk management is using an implicit 10-year time horizon. This would certainly allow them to assess the risk as very likely, since, if each year is independent of the last and the probability does not vary, then the probability that the event does not occur over 10 years is $0.8^{10} = 0.107$. So there is a roughly 90% chance that the event *does* occur at some point over a 10-year period.

More or less the same argument applies to the terms used to identify the magnitude of the impact. It will not be practicable to give an exact dollar figure associated with losses, just as there is little point in trying to ascribe exact probabilities to risk events. But it is worthwhile having a discussion on what a 'minor' or a 'moderate' impact really means. For example, we might initiate a conversation about the evaluation we would give for the impact of an event that led to an immediate 5% drop in the company share price.

1.3 Fallacies and traps in risk management

In this introductory chapter it is appropriate to give some 'health warnings' about the practice of risk management. These are ideas about risk management that can be misleading or dangerous.

It is worth beginning with the observation that society at large is increasingly intolerant of risk which has no obvious owner – no one who is responsible and who can be sued in the event of a bad outcome. Increasingly it is no longer acceptable to say 'bad things happen' and we are inclined to view any bad event as someone's fault. This is associated with much management activity that could be characterized as 'covering one's back'. The important thing is no longer the risk itself but the demonstration that appropriate action has been taken so that the risk of legal liability is removed. The discussion of risk registers in the previous section demonstrates exactly this divergence between what is done because it brings real advantage, and what is done simply for legal reasons. Michael Power makes the case that greater and greater attention is placed on what might be called *secondary risk management*, with the sole aim of deflecting risk away from the organization or the individuals within it. It is fundamentally wrong to spend more time ensuring that we cannot be sued than we do in trying to reduce the dangers involved in our business. But in addition to questions of morality, a focus on secondary risk management means we never face up to the question of what is an appropriate level of risk, and we may end up losing the ability to make sound judgments on appropriate risks: the most fundamental requirement for risk management professionals.

Another trap we may fall into is the feeling that good risk management requires a scenario-based understanding of all the risks that may arise. Often this is impossible, and trying to do so will distract attention from effective manage-ment of important risks. As Stulz (2009) argues, there are two ways to avoid this trap. First there is the use of statistical tools (which we will deal with in much more detail in later chapters).

> 'Contrary to what many people may believe, you can manage risks without knowing exactly what they are – meaning that most of what you'd call unknown risks can in fact be captured in statistical risk management models. Think about how you measure stock price risk. ... As long as the historical volatility and mean are a good proxy for the future behavior of stock returns, you will capture the relevant risk characteristics of the stock through your estimation of the statistical distribution of its returns. You do not need to know why the stock return is +10% in one period and −15% in another.'

The second way to avoid getting bogged down in an unending set of almost unknowable risks is to recognize that important risks are those that make a difference to management decisions. Some risks are simply so low in probability that a manager would not change her behavior even if this risk was brought to her attention. This is like the risk of being hit by an asteroid – it must have some small probability of occurring but it does not change our decisions.

A final word of caution relates to the use of historical statistical information to project forward. We may find a long period in which something appears to be varying according to a specific probability distribution, only to have this change quite suddenly. An example with a particular relevance for the author is in the

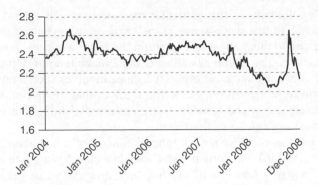

Figure 1.2   Australian dollars to one British pound 2004–2008.

exchange rate between the Australian dollar and the British pound. The graph in Figure 1.2 shows what happened to this exchange rate over a five-year period from 2004 to 2008.

The weekly data here have a mean of 2.38 Australian dollars per pound and the standard deviation is 0.133. Fifteen months later, in March 2010, the rate had fallen to 1.65 (and continued to fall after that date). Now, if weekly exchange rate data followed a normal distribution then the chance of observing a value as low as 1.65 (more than five standard deviations below the mean) would be completely negligible. Obviously the foreign exchange markets do not behave in quite the way that this superficial historical analysis suggests. Looking over a longer period and considering also other foreign exchange rates would suggest that the relatively low variance over the five-year period taken as a base was unusual. In this case the fallout from the global financial crisis quickly led to exchange rate values that reflect historically very high levels for the Australian dollar and a low level for the British pound.

We may be faced with the task of estimating the risk of certain events on the basis of statistical data but without the benefit of a very long view and with no opportunity to compare any related data. In this situation all that we might have to guide us is a set of data like Figure 1.2. Understanding how hard it is in a foreign exchange context to say what the probabilities are of certain outcomes should help us to be cautious when faced with the same kind of task in a different context.

1.4   Why safety is different

This book is about business risk management and is aimed at those who will have management responsibility. There are significant differences between how we may behave as managers and how we behave in matters of our personal safety. Every day as we grow up, and throughout our adult lives, we make decisions which involve personal risk. The child who decides to try jumping off the playground swing is weighing up the risk of getting hurt against the excitement involved. And the driver who overtakes a slower vehicle on the road is weighing up the risks of that particular road environment against the time or frustration saved. In that sense we are all risk experts; it's what we do every day.

It is tempting to think about safety within the framework we have laid out of different risk events, each with a likelihood and a magnitude of impact. With this approach we could say that a car trip to the shops involves such a tiny likelihood of being involved in a collision with a drunk driver that the overall level of risk is easily outweighed by the benefits. But there are two important reasons why thinking in this way can be misleading.

First we need to consider not only the likelihood of a bad event, but also its consequences. And if I am worried about someone else driving into me, then the consequence might be the loss of my life. Just how does that get weighed up against the inconvenience of not using a car? Most of us would simply be unable to put a monetary value on our own lives, and no matter how small the chance of our being killed in a car crash, the balance will tilt against driving the car if we make the value of our life high enough. But yet we still drive our cars and do all sorts of other things that carry an element of personal risk.

A second problem with treating safety issues in the same way as other risks is that the chance of an accident is critically determined by the degree of care taken by the individual concerned. The probability of dying in a car crash on the way to the shops is mostly determined by how carefully I drive. This makes my decision on driving a car different to a decision on traveling by air, where once on board I have no control over the level of risk. However, there are many situations where being careful will dramatically reduce the risk to our personal safety. Paradoxically, the more dangerous we perceive the activity to be then the more careful we are. The risks from climbing a ladder may end up being greater than from using a chain saw if we believe that the ladder is basically safe, but that the chain saw is extremely dangerous.

A better way to consider personal safety is to think of each of us as having an in-built 'risk thermostat' that measures our own comfort level with different levels of risk. As we go about our lives there comes a time with certain activities when we start to feel uncomfortable with the risk we are taking; this happens when the amount of risk starts to exceed our own risk thermostat setting. The risk we will tolerate varies according to our own personalities, our age, our experience of life, etc. But if the level of risk is below this personal thermostat setting then there is very little that holds us back from increasing the risk. So, if driving seems relatively safe then we will not limit our driving to occasions when the benefits are sufficiently large. John Adams points out that some people will actively seek risk so that they return to the risk thermostat setting which they prefer. So, in discussing the lives that might be saved if motorcycling was banned, he points out that, 'If it could be assumed that all the banned motorcyclists would sit at home drinking tea, one could simply subtract motorcycle accident fatalities from the total annual road accident death toll. But at least some frustrated motorcyclists would buy old bangers and try to drive them in a way that pumped as much adrenaline as their motorcycling'.

These are important issues and need to faced by businesses in which health and safety are big concerns, such as mining. If the aim is to get as close as possible to eliminating accidents in the workplace, then it is vital to pay attention to the workplace culture, which can have a role in resetting the risk thermostat of our employees to a lower level.

1.5 The Basel framework

The Basel Accords refer to recommendations made by the Basel Committee on Banking Supervision about banking regulations. The second of these accords (Basel II) was first published in 2004 and defines three different types of risk for banks – but the framework is quite general and can apply to any business.

Market risk. Market risk focuses on the uncertainties that are inherent in market prices which can go up or down. Market risk applies to any uncertainty where the value is dependent on prices that cannot be predicted fully in advance. For example, we might build a plant to extract gold from a low-yield resource, but there is a risk that the gold price will drop and our plant will no longer be profitable. This is an example of a *commodity risk*. Other types of market risk are *equity risk* (related to stock prices and their volatility); *interest rate risk*; and *currency risk* (related to foreign exchange rates and their volatility).

Credit risk. Any business will be involved in many different contractual arrangements. If the counterparty to the contract does not deliver what is promised then legal means can be used to extract what is owed. But this assumes that the counterparty still has funds available. Credit risk is the risk of a counterparty to a contract going out of business. For example, a business might deliver products to its customers and have 30-day payment terms. If the customer goes out of business there may be no way of getting back more than a small percentage of what is owed. In its most direct form, the contract is a loan made to another party and credit risk is about not being repaid due to bankruptcy.

Operational risk. Operational risk is about something going badly wrong. This category of risk includes many of the examples we have discussed so far that are associated with negative risk events. Operational risk is defined as arising from failures in internal processes, people or systems, or due to external events.

Since we are interested in more general risk management concerns, not just risk for banks, it is helpful to add a fourth category to the three discussed by Basel II.

Business risk. Business risk relates to those parts of our business value proposition where there is considerable uncertainty. For example, there may be a risk associated with changes in costs, or changes in customer demand, or changes in the security of supply of raw materials. Business risk is like market risk but does not relate directly to prices.

Both market risk and credit risk are, to some extent, entered into deliberately as a result of calculation. Market risk is expected, and we can make calculations on the basis of the likelihood of different market outcomes. Business risk also often has this characteristic: for example, most businesses will have a clear idea of what will happen under different scenarios for customer demand. Credit risk is always present, and in many cases we assess credit risk explicitly through credit ratings. But operational risk is different: it is not entered into in the expectation of reward. It is inherent and is, in a sense, the unexpected risk in our business. It may well fit into the 'unknown unknown' description in the quotation from Rumsfeld that we gave earlier. Usually operational risk involves low-probability and high-severity events and this makes it particularly challenging to deal with.

1.6   Hold or hedge?

When dealing with market or business risk a manager is often faced with an ongoing risk, so that it recurs from day to day or month to month. In this case there is the need to take strategic decisions related to these risks.

An example of a recurring risk occurs with airlines who face ongoing uncertainty related to the price of fuel (which can only be partially offset by adding fuel surcharges). The question that managers face is: when to hold on to that risk, when to insure or hedge it, and when to attack the risk so that it is reduced?

A financial hedge is possible when we can buy some financial instrument to lessen the risk of market movements. For example, a power utility company might trade in futures for gas prices. If the utility is buying gas and selling electricity then it is exposed to a market risk if the price of gas rises and it is not able to raise the price of electricity to the same extent. By holding a futures contract on the gas price, the company can obtain a benefit when the price of gas increases: if the utility knows how much gas it will purchase then the net effect will be to fix the gas price for the period of the contract and eliminate this form of market risk. Even if the utility cannot exactly predict the amount of gas it will burn, there will still be the opportunity to hedge the majority of its potential losses from gas price rises.

Sometimes we have an operational hedge which achieves the same thing as a financial hedge through the way that our operations are organized. For example, we may be concerned about currency risk if our costs are primarily in US dollars but our sales are in the Euro zone. Thus, if the Euro's value falls sharply relative to the US dollar, then we may find our income insufficient to meet our manufacturing expenses even though our sales have remained strong. An option is to buy a futures contract which has the effect of locking in an exchange rate. However, another 'operational hedge' could be achieved by moving some of our manufacturing activity into a country in the Euro zone, so that more of our costs occur in the same currency as the majority of our sales.

In holding on to a risk the company deliberately decides to accept the variation in profit which results. This may be the best option when a company has sufficient financial resources, and when it has aspects of its operations that will limit the consequences of the risk. For example, a vertically integrated power utility company that sets the price of electricity for its customers may decide not to fully hedge the risks associated with rises in the cost of gas if there are opportunities to quickly change the price of the electricity that it sells in order to cover increased costs of generation.

1.7   Learning from a disaster

We began this chapter with the remarkable story of Jérôme Kerviel's massive fraud at Société Générale, which fits into the category of operational risk. Now we return to this example with the aim of seeing what can be learnt. To understand what happened we will start by giving some background information on the world of bank trading. A bank, or any company involved in trading in a financial marketplace, will usually divide its activities into three areas. First the traders themselves: these are the people who decide what trades to make and when to make them (the 'front office'). Second, a risk management area responsible for monitoring the traders' activity measuring and modeling risk levels etc. (the 'middle office'). And finally an area responsible for carrying out the trades, making the required payments and dealing with the paperwork (the 'back office').

The trading activities are organized into *desks*: groups of traders working with a particular type of asset. The Kerviel story takes place in SocGen's Delta One desk in Paris. Delta One trading refers to buying and selling straightforward derivatives that do not involve any options. Options are derivatives which give 'the right but not the opportunity' to make a purchase or sale. The trading of options gives a return that depends non-linearly on whatever is the underlying security (we explain more about this in Chapter 9), but trading activities for a Delta One desk are simpler than this – the returns just depend directly on what happens to the underlying security. In fact, the delta in the terminology refers to the first derivative of the return as a function of the underlying security, and 'Delta One' is shorthand for 'delta equals one,' implying this direct relationship.

For example, a trade might involve buying a future on the DAX, which is the main index for the German stock market and comprises the 30 largest and most actively traded German companies. Futures can be purchased in relation to different dates (the end of each quarter) and are essentially a prediction of what the index will be at that date. One can also buy futures in the individual stocks that make up the index and by creating a portfolio of these futures in the proportions given by the weights in the DAX index, one would mimic the behavior of the future for the index as a whole. However, over time the weights in the DAX index are updated (in an automatic way based on market capitalization), so holding the portfolio of futures on individual stocks would lead to a small divergence from the DAX future over a period of time.

The original purpose of a Delta One trading desk is to carry out trades for the bank's clients, but around that purpose has grown up a large amount of proprietary trading where the bank intends to make money on its own account. One approach is for a trader to make a bet on the convergence of two prices that (in the trader's view) should be closer than they are. If the current price of portfolio A is greater than that of portfolio B and the two prices will come back together before long, then there will be an opportunity to profit by buying B and selling A, and then reversing this transaction when the prices move closer together. Since both portfolios are made up of derivatives, the 'buying' and 'selling' here need not involve ownership of the underlying securities, just financial contracts based on their prices. This type of trading, which intends to take advantage of a mis-pricing in the market, is called an arbitrage trade, and since trades of one sort are offset by trades in the opposite direction, the risk involved should, in theory, be very low.

Many of these trading activities take advantage of quite small opportunities for profit (in percentage terms) and therefore, in order to make it worthwhile, they require large sums of money to be involved. Kerviel was supposed to act as an arbitrageur, looking for small differences in price between different stock index futures. In theory this means that trades in one direction are offset by balancing trades in the other direction. But Kerviel was making fictitious trades: reporting trades that did not occur. This enabled him to hold one half of the combined position but not the other. The result of the fictitious trade is to change an arbitrage opportunity with large nominal value but relatively small risk into a simple (very large) bet on the movement of the futures price.

When Kerviel started on this process in 2006 things went reasonably well–his bets came off and the bank profited. Traders are allowed to make some speculative trades of this sort, but there is a strict limit on the amount of risk they take on: Kerviel breached those limits repeatedly (and spectacularly). Over time the amounts involved in these speculations became greater and greater, and things still went well. During 2007 there were some ups and downs in the way that these bets turned out, but by the end of the year Kerviel was well ahead. He has claimed that his speculation made 1.5 billion Euros in profits for SocGen during 2007. None of this money made its way to him personally; he would only have profited through earning a large bonus that year.

In January 2008, however, his good fortune was reversed when some large bets went very wrong. The senior managers at the bank finally discovered what was happening on January 18th 2008. There were enormous open positions and SocGen decided that it had no option but to close off those positions and take the losses, whatever these turned out to be. The timing was bad and the market was in any case tumbling; the net result was that SocGen lost 4.9 billion Euros. The news appeared on January 24th. The sums of money involved are enormous and a smaller bank would certainly have been bankrupted by these losses, but SocGen is very large and some other parts of its operation had been going well.

Nevertheless, the bank was forced into seeking an additional 5.5 billion Euros in new capital as a result of the losses.

Banks such as SocGen have elaborate mechanisms to ensure that they do not fall into this kind of situation. Outstanding positions are checked on a daily basis, but each evening Kerviel, working late into the night, would book offsetting fictitious transactions, without any counterparties, and in this way ensure that his open positions looked as if they were appropriately hedged. Investigations after the event revealed more than a thousand fake trades; there is no doubt that these should have been picked up.

Kerviel, who was 31 when the scandal broke, was tried in June 2010. He acknowledged what he had done in booking fake trades, but he argued that his superiors had been aware of what he was doing and had deliberately turned a blind eye. He said 'It wasn't me who invented these techniques – others did it, too.' Finally, in October 2010, Kerviel was found guilty of breach of trust, forging documents and entering false data into computers; he was sentenced to three years in prison and ordered to repay SocGen's entire trading loss of 4.9 billion Euros. The judge held Kerviel solely responsible for the loss and said that his crimes had threatened the bank's existence. The case came to appeal in October 2012 and the original verdict was upheld. There is, of course, no possibility of Kerviel ever repaying this vast sum, but SocGen's lawyers have said that they will pursue him for any earnings he makes by selling his story.

1.7.1 What went wrong?

There is no doubt that what happened at SocGen came about because of a combination of factors. First there was Kerviel himself, who had some knowledge of the risk management practices of the middle office through previously having worked in this area. It seems that he kept some access appropriate to this, even when he became a trader. This is exactly what happened with Nick Leeson at Barings – another famous example of a trader causing enormous losses at a bank. Kerviel was someone whose whole world was the trading room and, over the course of a year or so, he was drawn into making larger and larger bets with the company's money. There remains a mystery about what might have been his motivation. In his appeal he offered no real explanation, simply describing his actions as 'moronic', but maintaining that he was someone trying 'to do his job as well as possible, to make money for the bank'.

A second factor was the immediate supervision at the Delta One desk. Whether or not one accepts Kerviel's claims that his bosses knew what was going on, they certainly should have known and done something about it. It is hard at this point to determine what is negligence and what is tacit endorsement. Eric Cordelle, who was Kerviel's direct superior, was only appointed head of the Delta One desk in April 2007, and did not have any trading experience. He was sacked for incompetence immediately after the fraud was discovered. He claims that during this period his team was seriously understaffed and he had insufficient time to look closely at the activities of individual traders.

A third important factor is the general approach to risk management being taken at SocGen in this period. It is easy to take a relaxed attitude to the risk of losses when everything seems to be going well. During 2007 there was an enormous increase in trading activity on the Delta One desk and large profits were being made. The internal reports produced by SocGen following the scandal were clear that there had been major deficiencies in the monitoring of risks by the desk. The report by PriceWaterhouseCoopers on the fraud stated that: 'The surge in Delta One trading volumes and profits was accompanied by the emergence of unauthorized practices, with limits regularly exceeded and results smoothed or transferred between traders.' Moreover, 'there was a lack of an appropriate awareness of the risk of fraud.'

In fact there were several things which should have alerted the company to a problem:

- there was a huge jump in earnings for Kerviel's desk in 2007;

- there were questions which were asked about Kerviel's trades from Eurex, the German derivatives exchange, who were concerned about the huge positions that Kerviel had built up;

- there was an unusually high level of cash flow associated with Kerviel's trading;

- Kerviel did not take a vacation for more than a few days at a time – despite a policy enforcing annual leave;

- there was a breach of Kerviel's market risk limit on one position.

We can draw some important general lessons from this case. I list five of these below.

1. *Company culture is more important than the procedures.* The organizational culture in SocGen gave precedence to the money-making side of the business (trading) over the risk management side (middle office), and this is very common. Whether or not procedures are followed carefully will always depend on cultural factors, and the wrong sort of risk culture is one of the biggest factors leading to firms making really disastrous decisions.

2. *Good times breed risky behavior.* In the SocGen case the fact that Kerviel's part of the operation was doing well made it easy to be lax in the care with which procedures were carried out. It may be true that the reverse of this statement is also true: in bad times taking risks may seem the only way through, but whether wise or not these are at least a conscious choice. Risks that managers enter into unconsciously seem to generate the largest disasters.

3. *Companies often fail to learn from experience.* One example occurs when managers ignore risks in similar companies, such as we see in the uncanny

resemblance between SocGen and Barings. But it can also be surprisingly hard to learn from our own mistakes in a corporate setting. Often a scapegoat is found and moved on, without a close look at what happened and why. Dwelling on mistakes is a difficult thing to do and will inevitably be perceived as threatening, and perhaps that is why a careful analysis of bad outcomes is often ducked.

4. *Controls need to be acted upon.* On many occasions risks have been considered and controls put in place to avoid them. The problem occurs when the controls that are in place are ignored in practice. SocGen had a clear policy on taking leave (as is standard in the industry) but failed to act upon it.

5. *There must be adequate management oversight.* Inadequate supervision is a key ingredient in poor operational risk management. In the SocGen case, Kerviel's supervisor had inadequate experience and failed to do his job. More generally, risks will escalate when a single person or a small group can make decisions that end with large losses, either through fraud or simple error. Companies need to have systems that avoid this through having effective oversight of individuals by managers, who need to supervise their employees sufficiently closely to ensure that individuals do what they are supposed to do.

This book is mostly concerned with the quantitative tools that managers can use in order to deal with risk and uncertainty. It is impossible to put into a single book everything that a manager might need to know about risk. In fact, the most important aspects of risk management in practice are things that managers learn through experience better than they learn in an MBA class. But paying attention to the five key observations above will be worthwhile for anyone involved in risk management, and may end up being more important than all the quantitative methods we are going to explore later in this book.

It is hard to overstate the importance of the culture within an organization: this will determine how carefully risks are considered; how reflective managers are about risk issues; and whether or not risk policies are followed in practice. A culture that is not frightened by risk (where employees are prepared to discuss risk openly and consider the appropriate level of risk) is more likely to avoid disasters than a culture that is paranoid about risk (where employees are uncomfortable in admitting that risks have been not been eliminated entirely). It seems that when we are frightened of risk we are more likely to ignore it, or hide it, than to take steps to reduce it.

Notes

This chapter is rather different than the rest of the book: besides setting the scene for what follows, it also avoids doing much in the way of quantification. I have tried to distill some important lessons rather than give a set of models to be used. I have found the book by Douglas Hubbard one of the best resources for understanding the basics of risk management applied in a broad business context. His book covers not only some of the material in this chapter but also has useful things to say about a number of topics we cover in later chapters (such as the question of how risky decisions are actually made, which we cover in Chapter 6).

A good summary of approaches which can be used to generate scenarios and think about causal chains as well as the business responses can be found in Miller and Waller (2003). The discussion on why we need to think differently about safety issues is taken from the influential book by John Adams, who is a particular expert on road safety.

The material on the Société Générale fraud has been drawn from a number of newspaper articles: Société Générale loses $7 billion in trading fraud, *New York Times*, January 24, 2008; Bank Outlines How Trader Hid His Activities, *New York Times*, January 28, 2008; A Société Générale Trader Remains a Mystery as His Criminal Trial Ends, *New York Times*, June 25, 2010.; Rogue Trader Jerome Kerviel 'I Was Merely a Small Cog in the Machine' *Der Spiegel* Online, November 16, 2010.

We have said rather little about company culture and its bearing on risk, but this is by no means a commentary on the importance of this aspect of risk, which probably deserves a whole book to itself (some references on this are Bozeman and Kingsley, 1998; Flin *et al.*, 2000; Jeffcot *et al.*, 2006 as well as the papers in the book edited by Hutter and Power, 2005).

References

Adams, J. (1995) *Risk*. UCL Press.

Bozeman, B. and Kingsley, G. (1998) Risk Culture in Public and Private Organizations. *Public Administration Review*, **58**, 109–118.

Flin, R., Mearns, K., O'Connor, P. and Bryden, R. (2000) Measuring safety climate: identifying the common features. *Safety Science*, **34**, 177–192.

Hubbard, D. (2009) *The Failure of Risk Management*. John Wiley & Sons.

Hutter, B. and Power, M. (2005) *Organizational Encounters with Risk*. Cambridge University Press.

Jeffcott, S., Pidgeon, N., Weyman, A. and Walls, J. (2006) Risk, trust, and safety culture in UK train operating companies. *Risk Analysis*, **26**, 1105–1121.

Miller, K. and Waller, G. (2003) Scenarios, real options and integrated risk management. *Long Range Planning*, **36**, 93–107.

Power, M. (2004) The risk management of everything. *Journal of Risk Finance*, **5**, 58–65

Rice, J. and Franks, S. (2010) Risk Management: The agile enterprise. *Analytics Magazine*, INFORMS.

Ritchie, B. and Brindley, C. (2007) Supply chain risk management and performance. *International Journal of Operations & Production Management*, **27**, 303–322.

Stulz, R. (2009) Six ways companies mismanage risk. *Harvard Business Review*, **87** (3), 86–94

Exercises

1.1 Supply risk for valves

DynoRam makes hydraulic rams for the mining industry in Australia. It obtains a valve component from a supplier called Sytoc in Singapore. The valves cost 250 Singapore dollars each and the company uses between 450 and 500 of these each year. There are minor differences between valves, with a total of 25 different types being used by DynoRam. Sytoc delivers the valves by air freight, typically about 48 hours after the order is placed. Deliveries take place up to 10 times a month depending on the production schedule at DynoRam. Because of the size of the order, Sytoc has agreed a low price on condition that a minimum of 30 valves are ordered each month. On the 10th of each month (or the next working day) DynoRam pays in advance for the minimum of 30 valves to be used during that month and also pays for any additional valves (above 30) used during the previous month.

(a) Give one example of market risk, credit risk, operational risk and business risk that could apply for DynoRam in relation to the Sytoc arrangement.

(b) For cach of the risks identified in part (a) suggest a management action which would have the effect either of reducing the probability of the risk event or minimizing the adverse consequences.

1.2 Connaught

The following is an excerpt from a newspaper report of July 21, 2010 appearing in the UK *Daily Telegraph*.

> 'Troubled housing group Connaught has been driven deeper into crisis after it discovered a senior executive sold hundreds of thousands of pounds worth of shares ahead of last month's shock profit warning.
>
> The company which lost more than 60% of its value in just three trading days in June, and saw its chief executive and finance director resign, has launched an internal investigation into the breach of city rules... Selling shares with insider information when a company is about to disclose a price-sensitive statement is a clear breach of FSA rules.
>
> Connaught, which specializes in repairing and maintaining low cost (government owned) housing, has fallen a total of 68% since it gave a warning that a number of public sector clients had postponed capital expenditure, which would result in an 80 million pound fall in expected revenue this year.

The group said that it had been hit by deferred local authority contracts which would knock 13m pounds off this year's profits and 16m pounds from next year's. It also scaled back the size of its order book from the 2.9 billion pounds it said it was worth in April to 2.5 billion.

The profit warning also sparked renewed concerns about how Connaught accounts for its long-term repair and maintenance contracts. Concerns first surfaced late last year with city analysts questioning whether the company was being prudent when recognizing the revenue from, and costs of, its long term contracts.

The company vehemently defended its accounting practices at the time and continues to do so. Chairman Sir Roy Gardner has tried to steady the company since his arrival earlier this year.'

(a) How would you describe the 'profits warning' risk event: is it brought about by market risk, credit risk, operational risk or business risk?

(b) From the newspaper report can you make any deductions about risk management strategies the management of the company could have taken in advance of this in order to reduce the loss to shareholders?

1.3 Bad news stories

Go through the business section of a newspaper and find a 'bad news' story, where a company has lost money.

(a) Can you identify the type of risk event involved: market risk, credit risk, operational risk or business risk?

(b) Look at the report with the aim of understanding the risk management issues in relation to what happened. Was there a failure to anticipate the risk event? Or a failure in the responses to the event?

1.4 Product form for heat map

Suppose that the risk level is calculated as the expected loss and that the likelihoods are converted into probabilities over a 20-year period as follows: 'very likely' $= 0.9$; 'likely' $= 0.7$; 'moderate' $= 0.4$; 'unlikely' $= 0.2$; and 'rare' $= 0.1$. Find a set of dollar losses associated with the five different magnitudes of impact such that the expected losses are ordered in the right way for Figure 1.1: in other words, so that the expected losses for a risk level of low are always lower than the expected losses for a risk level of medium, and these are lower than the expected losses for a risk level of high, which in turn are lower than the expected losses for a risk level of extreme. You should set the lowest level of loss ('insignificant') as $10\,000$.

1.5   Publication of NHS reform risk register

Risk registers may take various forms, but the information they contain can sometimes be extremely sensitive. In 2012 the UK government discussed whether or not to release the full risk register that had been created for the highly controversial reform of the National Health Service. The health secretary, Andrew Lansley, told parliament in May 2012 that only an edited version of this document would be made available, on the principle that civil servants should be able to use 'direct language and frank assessments' when giving advice to ministers. Lansley argued that if this advice were to be released routinely then 'future risk registers [would] become anodyne documents of little use.' The net result would be that 'Potential risks would be more likely to develop without adequate mitigation. That would be detrimental to good government and very much against the public interest.' Using this example as an illustration, discuss whether there is a tension between realistic assessment of risk and the openness that may be implicit in a risk register.

2

The structure of risk

Did the global financial crisis signal a failure in diversification?
The idea of diversification is simple but vital in managing investments. Invest everything in one stock and you may be unlucky with a bad result for that particular stock; invest in 50 stocks and no single bad event will be able to trip you up. That is the advantage of buying shares in a mutual fund (or investment trust) – the investor is automatically diversifying through holding a whole range of different stocks. But sometimes this principle of diversification seems to fail: if you had $100 000 invested in the average US mutual fund at the beginning of 2008, you would have lost $39 500 during that year. That was a year in which all stocks did badly. The US bear market that began in October 2007 ran till March 2009 and in that period US equity markets fell by 57%.

Diversification amongst different US stocks did not help in 2008. In fact, the only way to avoid losses was not to diversify, but to invest in one of the small number of stocks that did much better than the market (93% of all US equities lost money in 2008 but not all: for example, McDonalds and Wal-Mart were exceptions). So, if diversification is the answer then the right strategy in that year would include investing in areas outside of equities. If you were a risk-averse investor then it would have made sense to diversify into many different asset classes with the intention of avoiding large losses. Unfortunately, not only did stocks fall, but commodities, real estate and emerging markets fell as well. In fact, virtually every asset class did badly and, as a result, even well-diversified portfolios saw massive losses. This is exactly what diversification is supposed to avoid. So the difficult question that needs to be faced is follows: if it is the case that in really bad times everything goes down at once, what is the point of diversification in the first place?

In order to understand the benefits of diversification and avoid being led astray by models that leave out critical parts of the picture, we need to go back to first principles.

Business Risk Management: Models and Analysis, First Edition. Edward J. Anderson.
© 2014 John Wiley & Sons, Ltd. Published 2014 by John Wiley & Sons, Ltd.
Companion website: www.wiley.com/go/business_risk_management

2.1   Introduction to probability and risk

The origin of our ideas of probability can be traced back to games of chance. There have always been high stakes gamblers and whenever there is a large amount of money involved there is also a powerful motivation for developing tools to predict the chance of winning or losing. The intellectual history of the ideas of probability and risk is a long one, but one early reference occurs in a book by Luca Paccioli that appeared in 1494 with the title *Summa de Arithmetic, Geometria et Proportionalita*. This is a book on mathematics but it includes an influential description of double entry bookkeeping, as well as the following problem:

> 'A and B are playing a fair game of *balla*. They agree to continue until one has won six rounds. The game actually stops when A has won five and B three. How should the stakes be divided?'

How would we answer this problem? Just one more win for A (Adam) will seal his victory, but B (Ben) still has a chance: if he can win the next three games then he will be the overall winner. A fair division of the stakes needs to reflect the relative likelihood of one or other player winning. But at the time the question was posed there was no language that could be used to capture these ideas of likelihood. Perhaps it is not surprising that this problem ('the problem of the points') would not be solved till many years later.

In fact the problem was discussed in the letters between two famous mathematicians in the 1650s, Blaise Pascal and Pierre de Fermat. Both these men were brilliant: Pascal was a child prodigy who worked on everything from calculating machines and hydraulics to barometers, but renounced his scientific work at the age of 31 after a mystical experience; while Fermat was a mathematical genius who did far more than leave us the puzzle of 'Fermat's last theorem'. Pascal describes the right form of division of the initial stakes in the problem of the points by saying, 'the rule determining that which will belong to them will be proportional to that which they had the right to expect from fortune'.

Rather than leave the problem of the points hanging in the air, we can quickly sketch how Pascal and Fermat approached this issue. The key is to think about what may happen over the next three rounds. The game will certainly finish then (after 11 rounds in all) if it has not finished already. But it does no harm to suppose that all three rounds are played, with a total of eight possible outcomes, e.g. one of the outcomes is first Ben wins, then Ben wins again, then finally Adam wins. We can write down the possible sequences AAA, AAB, ABA, ABB, BAA, BAB, BBA, BBB. Then the question of how to determine a fair distribution can be resolved by appealing to the assumption that the underlying game is fair. Each time that Adam plays Ben there is an equal chance that either of them will win (making each of these eight possible outcomes equally likely). But since only one of the eight, BBB, has Ben winning the stake, a fair division is to divide the stake with a proportion 1/8 going to Ben and 7/8 to Adam. Pascal and Fermat discussed how this calculation could be effectively carried out for any possible starting position and any number of rounds of play.

Notice that, even without knowing the exact rules of the game of balla, we may object that Adam's superior performance so far provides evidence that he is the stronger player and, if that is so, then even giving Ben one eighth of the stake is too generous.

Jakob Bernoulli had these issues in mind when he wrote *Ars Conjectandi (The Art of Conjecture)* which was published posthumously in 1713 (eight years after his death). At this stage some more modern ideas of probability were beginning to emerge. Bernoulli was conscious of the need to use the past as an indicator of the likelihood of future events and in particular drew attention to the way that (if nothing changes in the underlying circumstances) an increasing number of observations leads to increasing certainty regarding the actual probability of something occurring. He wrote:

> 'Because it should be assumed that each phenomenon can occur and not occur in the same number of cases in which, under similar circumstances, it was previously observed to happen and not to happen. Actually, if, for example, it was formerly noted that, from among the observed three hundred men of the same age and complexion as Titius now is and has, two hundred died after ten years with the others still remaining alive, we may conclude with sufficient confidence that Titius also has twice as many cases for paying his debt to nature during the next ten years than for crossing this border. Again, if someone will ... be very often present at a game of two participants and observe how many times either was the winner, he will thus discover the ratio of the number of cases in which the same event will probably happen or not also in the future under circumstances similar to those previously existing.'

There are clearly limitations in this approach. There are inevitable variations between circumstances in the future and those in the past – so how do we know that Titius really faces the same probability of an early death as the 300 men 'of the same age and complexion'? There has always been a dispute about the extent to which we can rely on calculations based on what has happened in the past to guide us in our future actions. When dealing with everyday decisions we are much more likely to be guided by a more subjective understanding of what the future holds. Peter Bernstein talks of the 'tension between those who assert that the best decisions are based on quantification and numbers, determined by the patterns of the past, and those who base their decisions on more subjective degrees of belief about the uncertain future.'

In this book we will focus on the quantitative tools for modeling and managing risk. But the student who wants to apply these tools needs to be aware constantly of their limitations. We have now reached the point in the historical story at which the mathematical theory of probability takes off, but rather than talk about the various contributions of de Moivre, Bayes, Laplace, Gauss, D'Alembert, Poisson and others we will move on to the specifics of risk. A short discussion of all the probability theory we will need in this book is given in Appendix A: Tutorial on probability theory.

2.2   The structure of risk

Our contention is that probability and probability distributions are the right tools to use in understanding risk. We will begin by thinking about the structure of the risk that we face and see how this is reflected in the probabilities involved.

The first distinction to make is between *event risk* and *quantity risk*. Event risk has a simple yes/no form: What is the risk that a particular company goes bankrupt? What is the risk that a new drug fails to pass its safety checks? Quantity risk relates to a value which can vary (a *random variable* in the parlance of probability theory). Most often the value is measured in monetary terms. This is a type of risk where there is no simple yes/no result: What is the risk of large losses in an investment project? What is the risk of a high cost in a construction project? Quantity risks can always be converted to event risks by adding some sort of hurdle: for example, rather than asking about losses in general, we may ask about the risk of losing more than $500 000.

2.2.1   Intersection and union risk

Sometimes event risk involves a number of separate things failing at once. For example, we may consider the risk of a failure in power supply to a hospital as the risk that there is a power cut *and* the emergency generator fails. We call this an *intersection risk*, since it relates to the intersection of the two events: 'mains power fails' and 'emergency power fails'.

On the other hand, we may need to analyze risks where there are a number of different failure paths, each of which leads to the same outcome. If we consider the risk of a failure in a rocket launch then any one of a number of different things can go wrong in the last few seconds before takeoff, and each will produce the same end result. We call this a *union risk*, since the probability of failure is the probability that one or more of the events takes place.

The basic tool to visualize these situations is a Venn diagram, where each event is represented as a set in the diagram, and the overlap between sets A and B represents the event that both A and B occur. This is shown for three risk events A, B and C in Figure 2.1. The intersection risk is the probability of the event described by the intersection of A, B and C. In the diagram, X is the set around everything and represents all possible events.

We say that two events are independent if one of them occurring makes no difference to the likelihood of the other occurring. This means that the probability

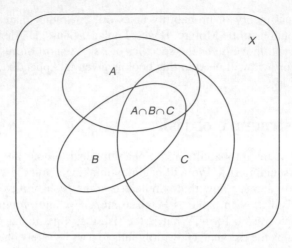

Figure 2.1   Venn diagram showing three different risk events.

of both A and B occurring is given by the product of the two individual probabilities (see Appendix A: Tutorial on probability theory for more details about this). This allows us to calculate the intersection risk for independent events.

Returning to the hospital power supply example, suppose we know that the probability of a power cut on any given day is 0.0005 and the probability that the hospital emergency power fails on any given attempt to start the generator is 0.002. If the two events are independent (as seems likely in this example), then the probability of a power supply failure at the hospital on any given day is $0.0005 \times 0.002 = 0.000001$.

Now we want to consider union risk, and as an example of this consider the probability of a catastrophic rocket launch failure during takeoff. Suppose that the three main causes of failure are as follows: A = failure in fuel ignition system; B = failure of the first stage separation from the main rocket; and C = failure in the guidance and control systems. Suppose that the probabilities are as follows $\Pr(A) = 0.001$, $\Pr(B) = 0.0002$, and $\Pr(C) = 0.003$. What is the overall probability of failure if the events A, B and C are all independent?

The probability that we want is the probability that one or other of A, B and C occur. We want to find the entire area covered by the union of the sets A, B and C in the Venn diagram. This is given by the formula

$$\Pr(A) + \Pr(B) + \Pr(C) - \Pr(A \cap B) - \Pr(B \cap C) - \Pr(A \cap C)$$
$$+ \Pr(A \cap B \cap C).$$

This is called the *inclusion–exclusion formula* – the idea is that if we just add up the probabilities of A, B and C we will double count the intersections, so the second three terms correct for this, but then anything in the intersection of all three sets will have been counted three times initially and then been taken away three times, so the final term restores the balance to make all the components of the Venn diagram end up being counted just once. With the probabilities given and using the product form for the probability of the intersection of independent events, we end up with a probability of launch failure given by:

$$10^{-3} + 2 \times 10^{-4} + 3 \times 10^{-3} - 2 \times 10^{-7} - 6 \times 10^{-7} - 3 \times 10^{-6} + 6 \times 10^{-10}$$
$$= 0.004196.$$

This example shows how, when there are small risk probabilities and independent events, we can more or less ignore the extra terms after the sum $\Pr(A) + \Pr(B) + \Pr(C)$. In this example, simply adding the three probabilities gives the value 0.0042.

It is obvious that there is an enormous difference between the end result of a union risk where (approximately) probabilities get added, and an intersection risk where probabilities get multiplied.

There is an important trick which we will introduce here and that is often helpful in calculating risk probabilities. We will write A^C for the *complement* of A; that is the event that A does not occur. Thus:

$$\Pr(A^C) = 1 - \Pr(A).$$

It is helpful to look at the Venn diagram of Figure 2.1 where the event A^C corresponds to the set X with A removed. Notice that $A^C \cap B^C = (A \cup B)^C$. This is clear by looking at the diagram, and we can put this into words by saying that 'the intersection of complements is the complement of the union'. From the point of view of calculating risk probabilities, the benefit of this rearrangement arises from the way that we can multiply the probabilities of independent events to get the probability of their intersection. If A and B are independent then so are A^C and B^C. Thus we have the following chain of implications for independent events A and B:

$$\Pr(A \cup B) = 1 - \Pr((A \cup B)^C)$$
$$= 1 - \Pr(A^C \cap B^C)$$
$$= 1 - \Pr(A^C) \Pr(B^C)$$
$$= 1 - (1 - \Pr(A))(1 - \Pr(B)).$$

This might not seem like much of an advance, but the same trick can be used for any number of independent events and this is helpful in avoiding the complications of the inclusion–exclusion formula. For example, in dealing with the probability of failure in the rocket launch example we have

$$\Pr(A \cup B \cup C) = 1 - (1 - \Pr(A))(1 - \Pr(B))(1 - \Pr(C))$$
$$= 1 - 0.999 \times 0.9998 \times 0.997 = 0.004196.$$

So we get to the same answer as we obtained before using inclusion–exclusion.

2.2.2   Maximum of random variables

Now we consider quantity risks, involving random variables rather than events. Again we need to start with an understanding of the structure of the risk and the way that different random variables are combined. We first look at a situation in which the risk we want to measure is determined by the largest of a set of random variables.

Example 2.1   IBM stock price losses over 20 days

Suppose that we want to find the probability that the price of IBM shares drops by more than 10% in one day at some point over the next four weeks, given that the probability of a loss of more than 10% in a single day is 0.01 (so we expect that it will happen on one trading day in 100), and the behavior on successive days is independent.

There are 20 trading days and so this is a question about the probability that the largest single-day drop in value over a 20-day period is more than 10%. To calculate the answer we want, we begin by defining the event:

$$A = \text{the daily loss in IBM stock price is less than 10\%.}$$

Note that A includes all the days with price rises too. The probability of a loss of more than 10% is simply $1 - \Pr(A)$, and this is the number we are told is 0.01. Hence, we deduce that $\Pr(A) = 0.99$. With independence, the probability that on both day 1 and day 2 the loss is less than 10% is the intersection probability:

$$\Pr(A) \times \Pr(A) = 0.99^2 = 0.9801.$$

The probability that the daily loss is less than 10% on all 20 days is just

$$\Pr(A)^{20} = 0.99^{20} = 0.8179.$$

But if this doesn't happen then there is at least one day when the loss is more than 10%, which is exactly the probability we want to find. So the answer we need is $1 - \Pr(A)^{20} = 0.1821$.

The task of determining the risk that the largest of a set of random variables is higher than a certain value is just like the analysis for the union risk: it is the probability that one or more of these random variables is greater than a certain value. The analysis we have given here is an example of the use of the complement trick to convert this problem into a problem of evaluating an intersection risk. We can rewrite the analysis of the paragraph above in a slightly more formal way as follows. We define the event:

$$B_i = \text{IBM stock drops by more than 10\% on day } i.$$

So, B_i is the complement of $A_i = $ IBM stock drops by less than 10% on day i. A_i is the same as the event A, but we add a subscript to indicate the day in question. We have $\Pr(B_i) = 1 - \Pr(A_i)$. Then, from our complement trick we know that

$$\Pr(B_1 \cup B_2) = 1 - \Pr(A_1 \cap A_2),$$

and a similar expression holds when three or more days are considered. So the number we want to find is

$$\Pr(B_1 \cup B_2 \cup \ldots \cup B_{20}) = 1 - \Pr(A_1 \cap A_2 \cap \ldots \cap A_{20}).$$

Now if all the A_i have the same probability and are independent (as we are assuming here) then this becomes $1 - \Pr(A)^{20}$.　　　　□

We can convert a discussion about the probabilities of events into a discussion of cumulative distribution functions or CDFs. Remember that we define the CDF for a random variable X as $F_X(z) = \Pr(X \leq z)$. Now consider a random variable U defined as the maximum of two other random variables X and Y. Thus, $U = \max(X, Y)$. To find the CDF for U we need to find the probability that the maximum of X and Y is less than a given value z. This is just the probability that both X and Y are less than z, so

$$F_U(z) = \Pr(X \leq z \text{ and } Y \leq z).$$

Hence, when X and Y are independent,

$$F_U(z) = \Pr(X \leq z) \times \Pr(Y \leq z) = F_X(z) \times F_Y(z).$$

The same idea can be used with more than two random variables.

When we are dealing with random variables having the same distribution, the formula becomes simpler. Suppose that $X_1, X_2, \ldots, X_N$ are identically distributed and independent random variables, all with the same CDF given by $F_X(\cdot)$. Then the CDF for the random variable $U = \max(X_1, X_2, \ldots, X_N)$ is given by

$$F_U(z) = \left(F_X(z)\right)^N.$$

One of the most common questions we need to answer is not about the largest (or smallest) of several different random variables, but instead relates to the risk arising when random variables are added. Hence we are concerned with $X + Y$, rather than with $\max(X, Y)$. In our example above we asked about the probability that an IBM share price falls by more than 10% in a single day during a 20-day period. But we are just as likely to be interested in the total change in price over the 20-day period, and to calculate this we need to add together the successive price movements over those 20 days. The fundamental insight here is that extreme events in one day's movement are quite likely to be canceled out by movements on other days. As a result we can say that, unless price movements are strongly positively correlated, the risk for the sum of many individual elements is less than the sum of the individual risks. In the next section we explore this idea in more detail.

2.3   Portfolios and diversification

We began this chapter by talking about diversification in share portfolios and now we return to this theme. The essential risk idea can be captured with the advice: 'Don't put all your eggs in one basket'. If there is the option to do so, then it is better to spread risk so that different risk events act on different parts of an entire portfolio of activities. In a stock market context, investing in a single share will carry the risk that all one's money is lost if that firm goes bankrupt. Splitting an investment between a portfolio of many different shares automatically reduces the probability of this very extreme result. The final result for the investor is the sum of the results obtained for each share in the portfolio (weighted according to the amount invested). Adding these together ensures that a bad result in one part of the portfolio is likely to be balanced by a good (or less bad) result in another part of the portfolio.

2.3.1   Adding random variables

We will start by looking in more detail at what happens when random variables are added together. If we consider the sum of two random variables X and Y, each representing a loss, then we can ask: What is the probability that the sum of the two is greater than a given value? To answer this question, we take $U = X + Y$ and consider $\Pr(U \geq z) = 1 - F_U(z)$. This is not an easy calculation to do in general, since we need to balance the value of X with the value of Y.

Example 2.2   Combination of two discrete random variables.

To illustrate this we look at an example where X and Y can each take values between 1 and 5 with the probabilities given in Table 2.1.

We can calculate the probability of $U = X + Y$ being 8 or more by considering the three possibilities: $X = 3$ and $Y = 5$; $X = 4$ and $Y \geq 4$; and $X = 5$ and $Y \geq 3$. When X and Y are independent, this shows that the probability of $U \geq 8$ is given by

$$0.2 \times 0.1 + 0.2 \times (0.1 + 0.1) + 0.2 \times (0.3 + 0.1 + 0.1) = 0.16.$$

Table 2.1   Probability of different values for X and Y.

Value	Probability for X	Probability for Y
1	0.1	0.2
2	0.3	0.3
3	0.2	0.3
4	0.2	0.1
5	0.2	0.1

At first sight the probability here is smaller than we might expect. There is a probability of 0.4 that $X \geq 4$ and a probability of 0.2 that $Y \geq 4$. Yet the probability that $X + Y \geq 8$ is smaller than both these figures. This is a simple example of the way that adding independent random variables tends to reduce overall risk levels. □

The same kind of calculation can be made for more general random variables taking integer values $1, 2, \ldots, M$, where we write $p_k = \Pr(X = k)$ and $q_k = \Pr(Y = k)$. Then

$$\Pr(X + Y \geq z) = p_{z-M}(q_M) + p_{z-M+1}(q_{M-1} + q_M)$$
$$+ \ldots + p_M(q_{z-M} + \ldots + q_M).$$

We need $2M \geq z > M$ for this formula to hold (so that the subscript $z - M$ is in the range $1, 2, \ldots, M$).

We can translate this formula into an integral form for continuous random variables. Suppose that X and Y are independent and the random variable X has density function f_X and CDF F_X, while the random variable Y has density function f_Y and CDF F_Y. To start with we suppose that both random variables take values in the range $[0, M]$. As before, we take $U = X + Y$. Then

$$1 - F_U(z) = \Pr(X + Y \geq z) = \int_{z-M}^{M} f_X(x)(1 - F_Y(z - x))dx. \qquad (2.1)$$

Since f_X is a probability density and integrates to 1, we have

$$\int_{z-M}^{M} f_X(x)dx = 1 - \int_{0}^{z-M} f_X(x)dx = 1 - F_X(z - M).$$

So, Equation (2.1) can be written

$$1 - F_U(z) = 1 - F_X(z - M) - \int_{z-M}^{M} f_X(x)F_Y(z - x)dx,$$

and so

$$F_U(z) = F_X(z - M) + \int_{z-M}^{M} f_X(x)F_Y(z - x)dx.$$

This is intuitively reasonable: the first term is the probability that X takes a value so low that $X + Y$ is guaranteed to be less than z.

There is an equivalent formula that applies when the variables do not have finite ranges. This is like taking M infinitely large and we get

$$F_U(z) = \int_{-\infty}^{\infty} f_X(x)F_Y(z - x)dx.$$

The integral here is called a *convolution* between the functions f_X and F_Y.

Worked Example 2.3   Failure and repair combination

Suppose that the time to the next breakdown of a piece of machinery is distributed as an exponential distribution with mean 10 days. When the item fails it will require repair, which will take anywhere between 0 and 20 days and the repair time is equally likely to take any value in this range. Suppose that the repair time is independent of the failure time. What is the probability that the item has failed and already been repaired within 30 days? (You will need to use the fact that the exponential distribution with mean λ has a CDF $F(x) = 1 - e^{-x/\lambda}$.)

Solution

We let Y = failure time. Then the CDF for Y is given by $F_Y(x) = 1 - e^{-x/10}$. We let X = repair time. Then the density for X is constant and given by $f_X(x) = 1/20$ for x in the range 0 to 20. Thus, if U is the time till the repair of the first failure, we have $U = X + Y$ and we want to find the probability that $U \leq 30$. So we need to evaluate:

$$F_U(30) = \int_{-\infty}^{\infty} f_X(x) F_Y(30 - x) dx.$$

Because $f_X(x)$ is zero unless x is in the range 0 to 20, we can take the integral over this range and we get

$$F_U(30) = \int_0^{20} f_X(x) F_Y(30 - x) dx$$

$$= \frac{1}{20} \int_0^{20} \left(1 - e^{-(30-x)/10}\right) dx.$$

Now $e^{-(30-x)/10} = e^{-3} e^{x/10}$ and $e^{x/10}$ integrates to $10 e^{x/10}$, so

$$F_U(30) = \frac{1}{20} \left[x - 10 e^{-3} e^{x/10}\right]_0^{20}$$

$$= \frac{1}{20} \left(20 - 10 e^{-3} e^2 + 10 e^{-3}\right)$$

$$= 1 - \frac{e^{-1}}{2} + \frac{e^{-3}}{2} = 0.84095.$$

This is the probability we need: there is about a 16% chance that after 30 days the item has either not yet failed or is still being repaired.   □

It is often difficult to find mathematical expressions for the convolution integrals that appear when adding distributions together. The equivalent formula if we want to consider the sum of more than two variables is even harder. The usual way to deal with these difficulties is to work with the *moment generating function* of the distribution rather than the distribution itself. But to start to discuss moment generating functions will take us too far away from our main aim in this book. Instead we will introduce a different approach to evaluating risk in a portfolio; we will give up something in terms of exact probabilities, but we will make a big gain in terms of ease of evaluation.

Instead of looking at specific probabilities we look instead at the spread of values, as measured by the standard deviation (or the variance). Again we consider two independent random variables. Remember that when X and Y are independent we can add their variances, i.e. for independent X and Y, the variance of $X + Y$ is the sum of the variances of X and Y, which we can write as

$$\text{var}(X + Y) = \text{var}(X) + \text{var}(Y).$$

The standard deviation of a random variable X, which we write as σ_X, is just the square root of the variance, $\text{var}(X)$, so when X and Y are independent,

$$\sigma_{X+Y} = \sqrt{\text{var}(X) + \text{var}(Y)} = \sqrt{\sigma_X^2 + \sigma_Y^2}.$$

We can extend this formula to any number of random variables. The simplest case of all is where we have a set of random variables $X_1, X_2, \ldots, X_N$ which are all independent and also all have the same standard deviation, so we can write

$$\sigma_X = \sigma_{X_1} = \sigma_{X_2} = \ldots = \sigma_{X_N}.$$

Then

$$\sigma_{X_1+X_2+\ldots+X_N} = \sqrt{\sigma_{X_1}^2 + \sigma_{X_2}^2 + \ldots + \sigma_{X_N}^2} = \sqrt{N\sigma_X^2} = \left(\sqrt{N}\right)\sigma_X.$$

This formula will obviously apply when all the variables have the same distribution (automatically making their standard deviations equal). For example, the individual random variables might be the demand for some product in successive weeks, when we have no reason to expect changes in the average demand over time. Then the standard deviation of the total demand over, say, 10 weeks is just given by $\sqrt{10}\sigma$ where σ is the standard deviation over a single week, provided that demand in successive weeks is independent.

The key point to remember is:

> The standard deviation of the sum of N identical independent random variables is square root N times the standard deviation of one of the random variables.

2.3.2   Portfolios with minimum variance

Now consider a situation where a portfolio is constructed from investing an amount w_i in a particular investment opportunity i, where $i = 1, 2, \ldots, N$. We let

X_i be the random variable giving the value of investment i at the end of the year. So the value of the portfolio is

$$Z = w_1 X_1 + w_2 X_2 + \ldots + w_N X_N.$$

We want to find the variance of Z, and again for simplicity we will suppose not only that all the X_i are independent, but also that they all have the same variance, σ_X^2 (so σ_X is the standard deviation of X_i).

When a random variable is multiplied by w, the standard deviation is multiplied by w and the variance is multiplied by w^2. So the variance of the value of the entire portfolio is

$$\operatorname{var}(Z) = \operatorname{var}(w_1 X_1) + \operatorname{var}(w_2 X_2) + \ldots + \operatorname{var}(w_N X_N)$$
$$= w_1^2 \sigma_{X_1}^2 + w_2^2 \sigma_{X_2}^2 + \ldots + w_N^2 \sigma_{X_N}^2 = \sigma_X^2 \left(w_1^2 + w_2^2 + \ldots + w_N^2 \right).$$

If we have a total amount W to invest and we split our investment equally (after all, each investment opportunity has the same variance), then each $w_i = W/N$ and

$$\operatorname{var}(Z) = \sigma_X^2 N (W/N)^2 = (1/N)\sigma_X^2 W^2.$$

We may want to minimize the standard deviation of the value of the portfolio when the individual investments have different standard deviations. This will be a good idea if there is no difference between the investments in terms of their average performance. Perhaps the first thought we have is to put all of our money into the best of the investment opportunities; in other words, put everything into the single investment that has the smallest standard deviation. It will certainly be sensible to invest more of our total wealth in investments with small standard deviations, but the principle of diversification means that we can do better by spreading our investment across more than one investment opportunity.

To illustrate the principle we can consider investing a total amount W in one of two stocks which are independent of each other. We will suppose that investing \$1 in stock 1 gives a final value which is a random variable with mean μ and standard deviation σ_1. On the other hand, investing \$1 in stock 2 gives the same average final value μ, but with a standard deviation σ_2. So, whatever investment choice is made, the expected final value is μW. Then the problem of minimizing the standard deviation can be written as an optimization problem

$$\text{minimize} \quad \sqrt{w_1^2 \sigma_1^2 + w_2^2 \sigma_2^2}$$
$$\text{subject to} \quad w_1 + w_2 = W,$$
$$w_1 \geq 0, w_2 \geq 0.$$

In this case, with just two investments, the problem has a simple geometrical interpretation, since the expression $\sqrt{w_1^2 \sigma_1^2 + w_2^2 \sigma_2^2}$ gives the distance from a point with coordinates $(w_1 \sigma_1, w_2 \sigma_2)$ to the origin. Moreover, the constraints

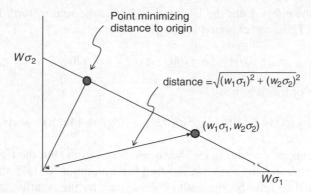

Figure 2.2   Choice of investment amounts to minimize standard deviation of return.

imply that this point lies somewhere on the straight line between $(W\sigma_1, 0)$ and $(0, W\sigma_2)$. These two endpoints correspond to what happens if we invest only in one or other of the two options. All this is illustrated in Figure 2.2 for a case with $\sigma_1 = 2\sigma_2$.

In the case shown in the figure (with $\sigma_1 = 2\sigma_2$) we can find the best choice of weights simply by substituting $w_2 = W - w_1$, which means that the objective is to minimize

$$\sqrt{w_1^2 4\sigma_2^2 + (W - w_1)^2 \sigma_2^2}.$$

We can use calculus to find the minimum of this expression, which occurs when $w_1 = W/5$ and $w_2 = 4W/5$. The resulting standard deviation is

$$\sqrt{\frac{4}{25} W^2 \sigma_2^2 + \frac{16}{25} W^2 \sigma_2^2} = \frac{\sqrt{20}}{5} W\sigma_2.$$

Worked Example 2.4   Minimizing variance with two investments

Andy has \$100 000 to invest for three years and believes that investment in US equities will deliver the same average returns as investment in an emerging market fund. He wants to split his investment between a mutual fund investing in US stocks, which he believes will, on average, deliver him \$120 000 after three years with a standard deviation of \$4000; and an emerging market fund that he believes will also deliver \$120 000 after three years, but with a standard deviation of \$12 000. Assuming the returns in the two funds are independent, how should he split his investment to minimize his risk?

Solution

We work in \$1000s. Suppose Andy invests an amount x in the US fund and $100 - x$ in the emerging market fund. His return after three years is $xU + (100 - x)V$, where U is the return from the US fund and V is the return from the emerging market fund. Thus, the expected return is

$$xE(U) + (100 - x)E(V) = 120.$$

The variance of this return is

$$x^2\text{var}(U) + (100 - x)^2\,\text{var}(V) = 16x^2 + 144(100 - x)^2.$$

We want to choose x to minimize the square root of this, but the right choice of x will also minimize the variance. To find the minimum we take the derivative and set it equal to zero. So the optimal x is given by the solution to

$$32x - 288(100 - x) = 0.$$

Thus, $x = 28\,800/320 = 90$. Andy should invest \$90\,000 in the US fund and the remaining \$10\,000 in the emerging market fund.   □

We can also ask what happens with a large number of investment opportunities, so that the number N goes to infinity. We begin by thinking about the case when all N stocks have the same standard deviation. We have already shown that the standard deviation of the overall return when all the individual stocks have standard deviation σ_X is given by

$$\sigma_X\sqrt{w_1^2 + w_2^2 + \ldots + w_N^2}.$$

This expression is minimized by splitting the investment of W equally, so that $w_i = W/N$ for $i = 1, 2 \ldots N$, giving a standard deviation of

$$\sigma_X\sqrt{(W/N)^2 + (W/N)^2 + \ldots + (W/N)^2} = \sigma_X\frac{W}{N}\sqrt{N} = \frac{\sigma_X W}{\sqrt{N}}.$$

Hence, in the case of independent investments, as the number of different investments goes to infinity and the amount invested in each gets smaller and smaller, the overall standard deviation goes to zero. And so the risk is also reduced to zero.

We can establish that the same behavior occurs in the more general situation, where stocks have different standard deviations (see Exercise 2.6). If none of the standard deviations is more than $\sigma_{\max}$, then we can create a portfolio with standard deviation less than $W\sigma_{\max}/\sqrt{N}$. Again this expression approaches zero as N gets larger and larger. Thus, we have established that there is really no upper bound to the benefits of diversification. Provided we can find new investment opportunities which are independent of our existing portfolio, and there is no extra cost to investing in these, then we always reduce the risk by adding these extra investments into our portfolio and rebalancing accordingly.

2.3.3 Optimal portfolio theory

Now we will look at the case when different potential investments have different expected profits as well as different variances. This is the foundation of what is often called simply *portfolio theory*. When there are differences in expected profit for individual investments, there will also be differences in the expected profit for a portfolio and so we can no longer simply find the portfolio which achieves the minimum standard deviation, we need to also consider the expected return of the portfolio. This will mean a trade-off: greater diversification will lead to less risk but will inevitably involve more of the lower return investments, and along with this a reduction in the overall expected return.

To illustrate this idea, suppose that we have three potential investments: A, B and C. We can explore the result of putting different weights on different components within the portfolio, and end up with a set of possible trade-offs between risk and return. Suppose that the expected profit from a \$10 000 investment and the standard deviations for A, B and C are as follows:

		Expected profit	Standard deviation
A	$R_A = \$1000$	$\sigma_A = 100$
B	$R_B = \$950$	$\sigma_B = 80$
C	$R_C = \$900$	$\sigma_C = 85$

At first sight it may seem that investment C will not be used, since it is dominated by investment B, which has a higher expected profit and at the same time a lower risk (in the sense of a less variable return). But we will see that the advantage of having one more investment in the portfolio may outweigh the fact that it is an unattractive investment.

Consider the problem of finding the least risk way of achieving a given profit, R. This can be written as an optimization problem:

$$\text{minimize} \quad w_A^2 \sigma_A^2 + w_B^2 \sigma_B^2 + w_C^2 \sigma_C^2$$

$$\text{subject to} \quad w_A + w_B + w_C = W,$$
$$w_A R_A + w_B R_B + w_C R_C = R,$$
$$w_A \geq 0, w_B \geq 0, w_C \geq 0.$$

Here, w_A, w_B and w_C are the sums invested, so W, the total amount, is set to \$10 000; R_A, R_B and R_C are the expected profits obtained from investing \$10 000 in the different investments; and σ_A, σ_B and σ_C are the standard deviations of those profits. Notice that the objective we have chosen is to minimize the variance of the overall profit return rather than the standard deviation. But as the standard deviation is just the square root of the variance, whatever choice of weights minimizes one will also minimize the other.

It is possible to write down a complex formula for the optimal solution to this problem, but rather than do this we will just look at the numerical solution to the problem with the particular data given above. Figure 2.3 shows what

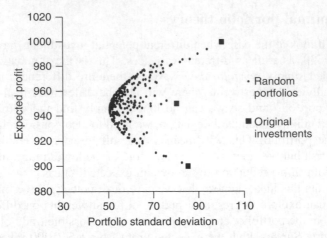

Figure 2.3   Profit versus standard deviation for random portfolios.

happens with a whole set of different random choices for the way that the total investment is split up. This figure shows quite clearly the different trade-offs that can be made: we can select a range of different overall standard deviations for the portfolio down to a minimum of around 50 at an expected profit of about $945.

We can look in more detail at the boundaries of the set of possible solutions. If we just consider a combination of two investments then, when we plot the expected profit against the standard deviation, we get a curved line joining the two points. In this example there are three such curved lines depending on which pair of the original investments we choose. These are the dashed lines in Figure 2.4. The lightly shaded area is the set of all possible results from different portfolios. The solid line is the boundary giving the minimum standard deviation that can be achieved at any given value for overall expected profit. For example, the dot at an expected profit of 960 and a standard deviation of 53.95 is the best possible at this profit level and is achieved by making $w_A = 0.3924$, $w_B = 0.4152$ and $w_C = 0.1924$.

2.3.4   When risk follows a normal distribution

Our discussion of portfolio risk so far has simply looked at the standard deviations for the overall profit (obtained from the sum of random variables). There is a critical assumption about independence of the different investments, but no assumption on the form of the distributions. When the distribution is known, then we can say more about the risks involved, and in particular we can calculate the probability of getting a result worse than some given benchmark level.

The most important distribution to look at is the normal distribution. Its importance stems from the way that it approximates the result of adding together a number of different random variables whatever their original distributions. This is

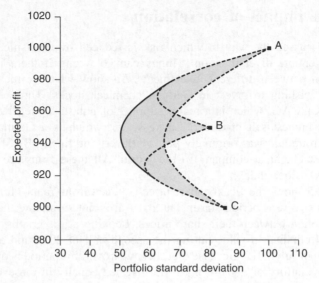

Figure 2.4   Boundary of the region that can be achieved through a portfolio of three investments.

the Central Limit Theorem discussed in Appendix A: Tutorial on probability theory. At the same time, a normal distribution is easy to work with because the sum of two or more random variables each with a normal distribution also has a normal distribution. If the distribution of profit follows a normal distribution, we can use tables or a spreadsheet to calculate any of the probabilities we might need.

Example 2.5   Probability calculation with two normal distributions

Consider an example where there are two independent investments, both having a normal distribution for the profits after one year. The first has an expected profit of $1000 with a standard deviation of $400 and the second has an expected profit of $600 with a standard deviation of $200. If we hold both these investments, what is the probability that we will lose money? Without information on the distribution of the profit, this probability is not determined, but with the knowledge that the profits follow a normal distribution it becomes easy to answer the question. The sum of the two returns is also a normal distribution with mean of $1000 + $600 = $1600 and, given that they are independent, the standard deviation is

$$\sqrt{400^2 + 200^2} = \sqrt{200\,000} = 447.21.$$

The probability of getting a value less than $0 can be obtained from tables (it's the probability of being more than z standard deviations from the mean, where $z = 1600/447.21 = 3.5777$) or, more simply, using the NORMDIST function in a spreadsheet. Specifically we have NORMDIST(0, 1600, 447.21, 1) = 0.00017329.   □

2.4   The impact of correlation

We have discussed the way in which risk is reduced by diversification, but it
has to be genuine diversification. Things can go wrong if there is too close
a connection between different investments. To show what we mean, consider
an investor wishing to invest in German chemical stocks. Three examples of
these are: K+S AG, which is a major supplier of potash for fertilizer and also
the world's biggest salt producer; Lanxess AG, which is a German specialty
chemicals group and was originally part of Bayer; and finally BASF SE, which
is the largest chemical company in the world. All these firms are part of the
German DAX stock index.

Figure 2.5 shows the weekly share price (sourced from Yahoo) for these three
stocks for a two-year period starting in 2010. It is not surprising that there is a
close connection between their share prices. Looking at the graphs, we can see
that they all climbed in value during the second half of 2011 and all fell quite
substantially in July 2012. Because their behavior over this period is quite similar,
the overall volatility of a share portfolio invested equally in these three stocks
would not be much less than investing in a single stock, so that diversification
would bring little benefit.

In fact, the main correlation here does not relate to the chemical industry,
but instead to overall movements in the German stock market, as shown by
Figure 2.6 which charts the behavior of the DAX Index over the same period.

In this section we want to look at the way that correlation between different
random variables will affect the behavior of their sum. This is a topic of great importance when assessing risk in practice but it is also best handled through some more
complex mathematical tools than we have used in the rest of this book (particularly
matrix algebra). So we will give a relatively broad brush treatment here.

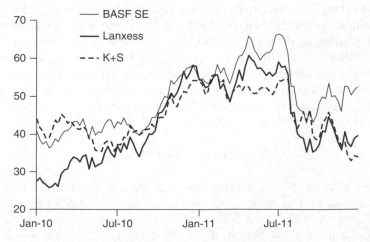

Figure 2.5   Three German chemical stocks: weekly prices in Euro.

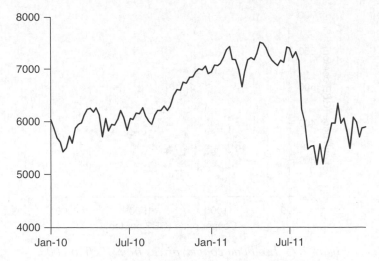

Figure 2.6   The DAX index over the period 2010–2011.

2.4.1   Using covariance in combining random variables

Figure 2.7 shows the monthly average prices over a 20-year period, starting in 1980, for lead and copper on the London metal exchange. It happens that the prices for these two commodities at the end of this period in December 1999 were not much greater than in January 1980. Suppose that we know that in a year's time we will need to purchase an amount of both lead and copper. The risk is related to the total purchase price. Following our previous argument, if the two commodity prices are independent then the risk associated with the combined purchase is reduced in comparison with purchasing just one of the commodities, since high prices for one commodity may well be balanced out by low prices for the other. But the scatter plot of Figure 2.7 shows that there is quite a high degree of correlation between these variables, and so the beneficial effect of diversification is reduced.

To make the discussion specific, suppose that we are interested in a combined price for 1 ton of copper and 2 tons of lead (which is cheaper). A good way to think about this is to recognize that the points which have the same value of $X + 2Y$ all lie on a straight line drawn in the (X, Y) plane. So if we looked at the monthly price for the purchase of 1 ton of copper and 2 tons of lead then the cost is the same at 1900 GBP if either (A) copper is 1500 GBP and lead is 200 GBP, or (B) copper is 500 GBP and lead is 700 GBP. And the same is true for any point on the straight line between these two points. Figure 2.8 has the price combinations (A) and (B) marked, and shows dashed lines for sets of price combinations that lead to the same overall price for this purchase. The effect is to project down onto the solid line all the different monthly combinations. The choice of solid line does not matter here: usually a projection means a mapping onto a line which is at right angles to the dashed contour lines of equal overall

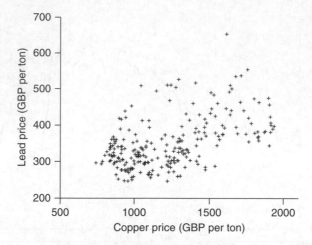

Figure 2.7   Lead and copper prices from 1980 to 2000.

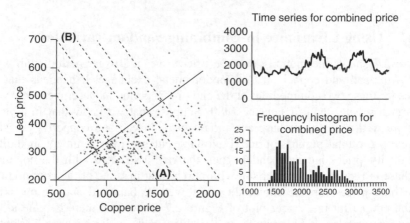

Figure 2.8   Price distribution for 1 ton of copper and 2 tons of lead.

price, but in this case some other choice of straight line just leads to a linear scaling of the results.

The right-hand side of Figure 2.8 shows how the price of 1 ton of copper and 2 tons of lead varies over time together with the frequency histogram for these prices. Notice how spread out this distribution is: the variance is high and the distribution itself does not have the nice shape of a normal distribution.

It is the covariance that measures the extent to which these two data series are correlated. Positive values of the covariance correspond to a positive correlation, and the covariance will be zero if the two variables are independent. Remember that the covariance between random variables X and Y is

$$\text{cov}(X, Y) = E(XY) - E(X)E(Y).$$

If we are interested in understanding the properties of a portfolio with weights w_X and w_Y then we can use the covariance between X and Y to get the variance of the portfolio (for more details about the formula here, see Appendix A: Tutorial on probability theory).

$$\text{var}(w_X X + w_Y Y) = w_X^2 \, \text{var}(X) + w_Y^2 \, \text{var}(Y) + 2 w_X w_Y \, \text{cov}(X, Y).$$

We can see how this works out for the lead and copper price example. The copper price has mean \$1231.16 and standard deviation \$312.40 (implying a variance of 97 590.80). The lead price has mean \$350.99 and standard deviation \$69.65 (implying a variance of 4851.55). The two sets of prices are positively correlated with a covariance 11 281.00. Thus, the formula implies that the variance of the combination 1 ton of copper and 2 tons of lead is

$$97\,590.80 + 4 \times 4\,851.55 + 4 \times 11\,281.00 = 162\,121.00,$$

giving a standard deviation of \$402.64. If the two prices had been independent, then the third term would not appear and the overall standard deviation would have been \$342.05.

2.4.2 Minimum variance portfolio with covariance

We can switch focus from thinking about the distribution of values for a particular combination purchase to deciding how we might invest if we had a choice of portfolios over assets with a correlated behavior. Suppose that the initial prices are X_0 and Y_0 and we will sell after one year at prices which are X and Y. How should we split our available investment sum W? The decision here will depend on the relationship between the purchase prices X_0 and Y_0 and the expected values for X and Y. In order to eliminate the question of different relative returns, let us assume that μ_X, the mean value of X, is a certain multiple of X_0, and μ_Y, the mean value of Y, is the same multiple of Y_0, so $\mu_X = k X_0$ and $\mu_Y = k Y_0$. Thus, the expected result from this investment after one year is that the initial investment is multiplied by k, no matter how we split the investment between X and Y. In this situation it makes sense to invest in a way that minimizes the variance of the return. Given a purchase of w_X units of asset X then we have a remaining $W - w_X X_0$ to invest. This means that we can purchase w_Y units of asset Y where

$$w_Y = (W - w_X X_0)/Y_0.$$

Then the variance of the portfolio is

$$w_X^2 \, \text{var}(X) + (W - w_X X_0)^2 \, \text{var}(Y)/Y_0^2 + 2 w_X (W - w_X X_0) \, \text{cov}(X, Y)/Y_0,$$

which (using calculus) we can show is minimized when

$$2 w_X \, \text{var}(X) - 2 X_0 (W - w_X X_0) \, \text{var}(Y)/Y_0^2$$
$$+ \left(-2 X_0 w_X + 2(W - w_X X_0) \right) \text{cov}(X, Y)/Y_0 = 0.$$

Simplifying we get

$$w_X \text{var}(X) - X_0(W - w_X X_0) \text{var}(Y)/Y_0^2 + (W - 2w_X X_0) \text{cov}(X, Y)/Y_0 = 0,$$

and we can solve this to find the w_X that minimizes the variance:

$$w_X = \left(\frac{W}{Y_0}\right) \frac{(X_0/Y_0) \text{var}(Y) - \text{cov}(X, Y)}{\text{var}(X) + (X_0^2/Y_0^2) \text{var}(Y) - 2(X_0/Y_0) \text{cov}(X, Y)}.$$

Example 2.6   Optimal portfolio weights with covariance

We look again at the example of lead and copper prices. Suppose that we have $1000 to invest, and the current prices of lead and copper are at their average for the 20-year period shown in Figure 2.7, so X_0 (for copper) is $1231.16 and Y_0 (for lead) is $350.99. Let $\alpha = X_0/Y_0 = 3.51$. Then we get that the optimal weight of the portfolio in copper is given by

$$w_{\text{copper}} = \left(\frac{W}{Y_0}\right) \frac{\alpha \, \text{var}(\text{lead}) - \text{cov}(\text{copper}, \text{lead})}{\text{var}(\text{copper}) + \alpha^2 \, \text{var}(\text{lead}) - 2\alpha \, \text{cov}(\text{copper}, \text{lead})}$$

$$= \left(\frac{1000}{350.99}\right) \frac{3.51 \times 4851.55 - 11\,281.00}{97\,590.80 + (3.51)^2 \times 4851.55 - 2 \times 3.51 \times 11\,281.00}$$

$$= 0.2095,$$

$$w_{\text{lead}} = \frac{W - w_{\text{copper}} \mu_{\text{copper}}}{\mu_{\text{lead}}} = \frac{1000 - 0.2095 \times 1231.16}{350.99} = 2.1142,$$

with expenditure of $0.2095 \times 1231.16 = \$257.93$ on copper and $2.1142 \times 350.99 = \$742.06$ on lead (this is one cent less than $1000 which has disappeared in the rounding of these calculations).   □

The result of this calculation may give a negative value for one of the weights w_X or w_Y. This would imply a benefit from selling one commodity in order to buy more of the other. The process to do this may well be available in the marketplace: in the language of finance this amounts to going short on one commodity and long on another. However, we will not pursue the idea here.

We have seen how a positive correlation between two investments reduces the diversification benefits on risk if both investments are held. Exactly the same thing takes place with more than two investments.

2.4.3   The maximum of variables that are positively correlated

Now we consider the other scenario in which the maximum value of two random variables is of interest (rather than their sum) and we ask how a positive correlation will impact on this measure. The probability of both X and Y being

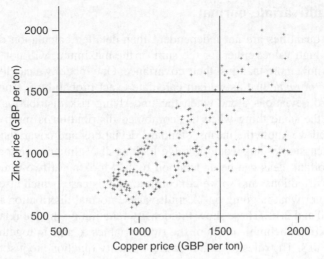

Figure 2.9   How does correlation affect the risk for the maximum of zinc and copper prices?

less than z is given by $F_X(z) \times F_Y(z)$ if the two variables are independent. The probability will be more than this if they are positively correlated, since a low value for one tends to happen at the same time as a low value for the other. Hence, if we define $U = \max(X, Y)$ then

$$F_U(z) = \Pr(\max(X, Y) < z)$$
$$> F_X(z) \times F_Y(z).$$

We can look at this in the other direction and say that the probability of U being more than a given value is reduced when X and Y are positively correlated.

To see what this looks like for a specific example, we consider the behavior of zinc and copper prices shown in Figure 2.9. Again this shows monthly prices over the 20-year period starting in January 1980. There are 51 occasions out of 240 in which the copper price is greater than \$1500 per ton and 24 occasions in which the zinc price is greater. So (approximately) $F_{copper}(1500) = 189/240 = 0.7875$ and $F_{zinc}(1500) = 216/240 = 0.9$. If these were independent then $F_U(1500) = 0.9 \times 0.7875 = 0.709$ and we would expect $0.709 \times 240 = 170$ occasions when the maximum of the two prices is below 1500 and 70 when it is greater than 1500. In fact there are only three occasions when the price of zinc is higher than 1500 and copper is not, meaning a total of $51 + 3 = 54$ occasions when the maximum of the two is greater than 1500 (much less than the 70 we might expect with independence). Thus, in this case the correlation between the two prices has *reduced* the risk that the maximum is very high, which is the opposite of what happens when looking at a sum (or average) of prices.

2.4.4   Multivariate normal

If different quantities are not independent then detailed calculation of the probability of a high value, either for the sum or the maximum, will not be possible unless we know more than just their covariance. One model we can look at is the *multivariate normal*. Just as we can calculate exact probabilities from the mean and standard deviations alone when the underlying risk distribution is normal, we can do the same thing when the combined distribution is multivariate normal provided we know the means, standard deviations and covariances. A more detailed discussion of this would involve looking at N-dimensional problems but all the important ideas can be understood by just looking at two-dimensional or *bivariate* distributions, and so we concentrate on this case, which also means we can easily plot what is going on. A multivariate normal distribution is shown in Figures 2.10 and 2.11. These show the density function giving the relative probabilities of different combinations of the two variables X and Y, together with the contours of this. The rules for a multivariate density function are just the same as for a univariate one – to calculate the probability of the (X, Y) pair being in any region of the (X, Y) plane, we just integrate the density function over that region.

The formula for a two-dimensional multivariate normal density function is

$$f(x, y) = K \exp\left(-\frac{1}{2(1 - \rho^2)}\left(\left(\frac{x - \mu_X}{\sigma_X}\right)^2 + \left(\frac{y - \mu_Y}{\sigma_Y}\right)^2 \right.\right.$$
$$\left.\left. -\frac{2\rho(x - \mu_X)(y - \mu_Y)}{\sigma_X\sigma_Y}\right)\right),$$

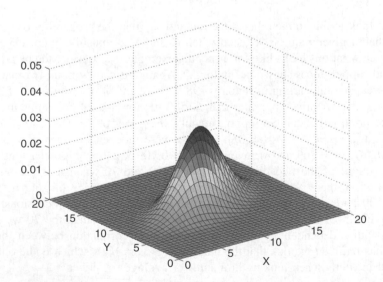

Figure 2.10   Density function for a bivariate normal distribution.

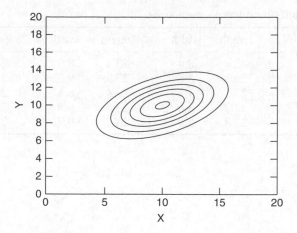

Figure 2.11   Contours of the density of the bivariate normal distribution.

where

$$K = \frac{1}{2\pi\sigma_X\sigma_Y\sqrt{1-\rho^2}} \text{ and } \rho = \frac{\text{cov}(X, Y)}{\sigma_X\sigma_Y}$$

(K is a normalizing constant and ρ is the correlation between X and Y). The figures show an example with $\mu_X = \mu_Y = 10$, $\sigma_X = 3$, $\sigma_Y = 2$ and $\rho = 0.5$. The contours in Figure 2.11 are all ellipses.

One of the most important properties of a multivariate normal is that any linear combination of the variables also has a multivariate normal distribution. It is easy to imagine that any vertical straight slice through the density function of Figure 2.10 would give a bell-shaped normal distribution curve. But this property is saying something a little different, since it involves the kind of projection down onto a straight line that we described in Figure 2.8.

Example 2.7   Children's shoes and linear combinations of values

Consider a children's shoe manufacturer interested in the distribution of children's foot sizes (widths and lengths). Table 2.2 gives the data collected for 20 fourth grade boys.[1]

Figure 2.12 shows a scatter plot of the data from Table 2.2. We find that the mean length is 23.105 (sample standard deviation = 1.217) and the mean width is 9.19 (sample standard deviation = 0.452). The covariance is 0.299. If we fit a multivariate normal then we expect that the distribution of any combination of height and width is also normal. For example, (Length) + 2 × (Width) should have a normal distribution with mean $23.105 + 2 \times 9.19 = 41.485$. Writing L and W for the two random variables, the variance of $L + 2W$ is

$$\text{var}(L) + 4\,\text{var}(W) + 4\,\text{cov}(L, W) = 1.217^2 + 4 \times 0.452^2 + 4 \times 0.299 = 3.494,$$

---

[1] Data are taken from kidsfeet.dat at http://www.amstat.org/publications/jse/jse_data_archive.htm

Table 2.2   Data on children's foot sizes.

length	width	length	width	length	width	length	width
20.9	8.8	22.4	8.4	23.1	8.9	24.1	9.6
21.6	9.0	22.4	8.6	23.2	9.8	24.1	9.1
21.9	9.3	22.5	9.7	23.4	8.8	25.0	9.8
22.0	9.2	22.7	8.6	23.5	9.5	25.1	9.4
22.2	8.9	22.8	8.9	23.7	9.7	25.5	9.8

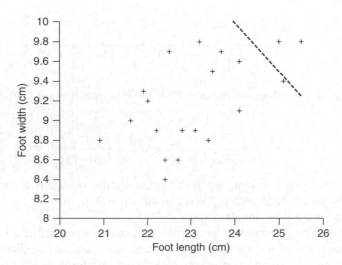

Figure 2.12   Data on foot sizes for 20 fourth grade boys.

giving a standard deviation of $\sqrt{3.494} = 1.869$. From this we can calculate the probability of getting different ranges of values for this linear combination. For example, suppose that we wish to estimate the probability that an individual has an $L + 2W$ value greater than 44cm. This corresponds to the dashed line in Figure 2.12. The z value is $(44 - 41.485)/1.869 = 1.346$. Under the normal assumption this would imply a probability of $1 - \Phi(1.346) = 1 - 0.9108 = 0.0892$ of achieving this value. Here, Φ is the cumulative distribution function for the standard $N(0, 1)$ normal distribution. Given 20 observations this would lead us to expect around 1.8 individuals with this characteristic. In fact, we observe two individuals – very much in line with our prediction.   □

The value of an explicit model is that it can help us to make predictions about the likelihood of events we have only occasionally (or never) observed. For example, in the children's shoes example we can estimate the probability of having an individual where the composite score is more than 45.5, which does not occur in this group of 20. However, we should be cautious in extrapolating beyond the data we have observed, and it is possible that the approximation of a multivariate normal fails as a predictor for more extreme results. We shall say more in Chapter 4 about modeling the tails of distributions.

Worked Example 2.8   One variable more than x times higher than the other

We return to the copper–lead example of Figure 2.7. We already know that for these data the copper price has mean \$1231.16 and standard deviation \$312.40; the lead price has mean \$350.99 and standard deviation \$69.65 and the two prices have a covariance 11 281.00. Suppose that the distribution is a multivariate normal. What is the probability that the copper price is more than five times as high as the lead price?

Solution

Write X for the copper price and Y for the lead price. We want to find the probability that $X > 5Y$. Thus, if we let $W = X - 5Y$ we want the probability that $W > 0$. Given the multivariate normal assumption, W is a normal random variable with mean and variance as follows:

$$E(W) = E(X) - 5E(Y)$$
$$= 1231.16 - 5 \times 350.99 = -523.8,$$
$$\mathrm{var}(W) = \mathrm{var}(X) + 25\,\mathrm{var}(Y) - 10\,\mathrm{cov}(X, Y)$$
$$= (312.40)^2 + 25(69.65)^2 - 11\,2810 = 106\,062.$$

This gives a standard deviation for W of $\sqrt{106\,062} = 325.7$. With this mean the probability of being greater than 0 is given by $1 - \Phi(523.8/325.7) = 0.0539$. In the data there are 7 out of 240 data points for which the inequality holds, giving an empirical probability of 0.029, rather less than the multivariate normal model predicts.   □

2.5   Using copulas to model multivariate distributions

On many occasions the multivariate normal is not a good approximation for the dependence between two or more variables and we need to look for a more flexible model. A good choice is to use copula models, which are a relatively modern development in probability theory. We will talk about a two-dimensional copula describing the joint behavior of variables X and Y, but the same ideas can be applied to multivariate models with three or more variables. The idea is to look at the distribution over values (x, y) expressed in terms of the positions of x and y within the distributions for X and Y respectively. This gives a way of distinguishing between what is happening as a result of the distribution of the underlying variables X and Y, and what is happening as a result of the dependence of one variable on the other. This means that we can, if we wish, apply the same copula model for different distributions of the underlying variables.

We begin by looking again at the data on lead and copper prices. Rather than look at the scatter plot of Figure 2.7 we will plot the points with their rank ordering for copper price on the horizontal axis and the rank ordering for lead price on the vertical axis. There are 240 points and we scale the ranks by dividing by 240, so that the scale is between 0 and 1. Thus, the lowest copper price has an x value of $1/240$ and the highest copper price translates to $240/240 = 1$ (and the same with lead prices). The result is shown in Figure 2.13, and this kind of scatter diagram is called an *empirical copula* plot.

Figure 2.13 shows some features of the data more clearly than the basic scatter plot of Figure 2.7. In this diagram we can see that the dependence between prices is stronger at the higher end than it is at the lower end. For example, we can see that a copper price that is in the top 10% corresponds to a lead price that is in the top 40% (roughly) of possible values. On the other hand, knowing that the copper price is in the bottom 10% of possible values implies much less: that the lead price is in the bottom 70% (roughly) of possible values.

The empirical copula scatter diagram is really a sample from the underlying *copula*, which is a way of describing the dependence between two variables. The *copula density* is the underlying function which gives the relative probabilities of different points occurring in the $(0, 1) \times (0, 1)$ square. We might expect this copula density to be smooth and well-behaved, but we will make deductions on the basis of a sample from it. It is helpful to think of a copula as part of a decomposition of a bivariate distribution. The idea is to take the overall distribution of the two values together and split this up by looking at the distribution of each value separately; often these are called the *marginal* distributions. The copula describes the way that the two marginal distributions are put together into the overall distribution.

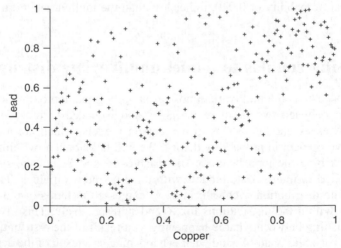

Figure 2.13   An empirical copula plot of copper and lead prices for 20 years from 1980.

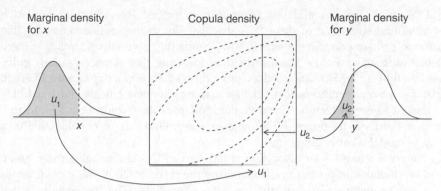

Figure 2.14   Decomposition using a copula. Sample for x first, then sample from the copula to get u_2, from which y is deduced.

If we have the two marginal distributions (say, for copper and lead prices) and the copula function that links them, then the process to generate a sample for the full bivariate distribution is illustrated in Figure 2.14. We take a sample, x, from the copper distribution according to its marginal distribution. Suppose this has a proportion u_1 of the distribution below it, then we look at the copula density on a vertical line through the point u_1 on the horizontal axis and use this density to sample a point u_2. This gives the proportion of lead prices below our required sample point and can be converted to a lead price y. The (x, y) pair then has the correct joint distribution.

The important thing about copulas is that they can indicate when dependence may increase. They are flexible enough to deal differently with the tails of the distribution than with the central parts. A good question to ask when dealing with multivariate data that include some correlation is whether the correlation will be greater or less at extreme values. This may be a question which is easier to ask than to answer, but giving it attention will at least ensure that we avoid some important risk management pitfalls. For example, if two variables are weakly correlated with a covariance close to 0, then we might be tempted to conclude that the diversification effect will imply a substantially reduced risk for their average value (e.g. a portfolio split equally between them) than either on its own. This is usually a correct deduction, but it will fail if the correlation between them is very low *except* at points where they take extremely large values, when they become closely correlated. In this case the risk of their average becoming very large will be close to the risk of an individual variable becoming large, and the diversification benefits we might expect will not occur.

The problem with making statements about correlation in the tails is that these are events we rarely see, and so there is unlikely to be much historical evidence to guide us. It is worth thinking about the underlying reasons for very large values in one or other of the variables. For example, consider the relationship between two stocks both traded in the New York stock exchange, let's say Amazon and Ford. They are both affected by market sentiment and by general economic conditions, and this will lead to correlation between their prices. But will this correlation be amplified when one of them falls sharply? For example, does knowing that Amazon's share price has fallen by 30% in one day give correspondingly more information about Ford's share price than knowing that Amazon's share price has fallen by 15%? Perhaps a sudden price drop of 30% in a day is most likely to arise from very specific factors (like an earnings announcement) and less likely to relate to general factors (like a market collapse) that would tend to lead to a drop in Ford's share price too. Whichever way round this is, copulas give us a way to model what is going on.

Now we return to our discussion at the start of this chapter about the failure of diversification to protect investors during the crisis of 2008. We can view this as a sudden increase in correlation as losses mounted. The implication is that the covariance dependence between different asset prices may not be symmetric between gains and losses. If Wall Street moves up by a large amount, the corresponding effect on the London stock market may be less than if Wall Street falls sharply. There is good empirical evidence to support exactly this behavior. A cynic might say that fear is an even more contagious emotion than greed, but whatever the mechanism, it is clear from a number of empirical studies that these asymmetries exist. The copula approach is a good way to make this explicit, for example through the use of a model for losses with upper tail dependence, but no lower tail dependence.

The next section gives a more detailed discussion of the mathematics that lies behind copula modeling.

2.5.1   *Details on copula modeling

A *copula density* $c(u_1, u_2)$ is a density defined on the unit square $0 \leq u_1 \leq 1$, $0 \leq u_2 \leq 1$, with the property that the resulting (marginal) distribution for u_1 is uniform on $[0, 1]$, and the distribution for u_2 is also uniform on $[0, 1]$. We can write these conditions as

$$\int_0^1 c(u, u_2)du = 1, \text{ for each } u_2 \text{ in } [0, 1],$$

$$\int_0^1 c(u_1, u)du = 1, \text{ for each } u_1 \text{ in } [0, 1].$$

The simplest way to make these conditions hold is to make the copula density uniform on the unit square, so that $c(u_1, u_2) = 1$ for every u_1 and u_2.

Before we talk about different examples of copula densities and how they relate to different kinds of dependence we need to show how to convert a copula density and the information on the underlying distributions into a distribution for the multivariate distribution as a whole. To do this we will change from densities (which are easier to visualize) into cumulative distribution functions (which are easier to work with). So we define a *copula* (really a cumulative copula distribution) as the distribution function $C(u_1, u_2)$ obtained from the copula density function $c(u_1, u_2)$. So $C(v_1, v_2)$ is the probability that both $u_1 \leq v_1$ and $u_2 \leq v_2$ when (u_1, u_2) has density function c, or, more formally,

$$C(v_1, v_2) = \int_0^{v_1} \left(\int_0^{v_2} c(u_1, u_2) du_1 \right) du_2.$$

Then the cumulative distribution for the multivariate distribution is obtained from the copula C and the underlying distribution functions F_X and F_Y for X and Y through the fundamental copula equation:

$$\Pr(X \leq x \text{ and } Y \leq y) = C(F_X(x), F_Y(y)). \tag{2.2}$$

In words, we take x and y and convert them into quantiles for the appropriate distributions, $F(x)$ and $F(y)$, and then use the copula function to determine the probability of being in the rectangle with (x, y) at its top-right corner.

To illustrate how this works, let's go back to the uniform copula density: $c(u_1, u_2) = 1$. This can be converted into a copula

$$C(v_1, v_2) = \int_0^{v_1} \left(\int_0^{v_2} du_1 \right) du_2 = v_1 v_2,$$

which is in product form. Then, from Equation (2.2), we have

$$\Pr(X \leq x \text{ and } Y \leq y) = F_X(x) F_Y(y) = \Pr(X \leq x) \Pr(Y \leq y).$$

This is exactly the formula for the probability if the two variables are independent. From this observation we get the result that a uniform copula density (or product form copula) is equivalent to the variables being independent.

We can also move from a copula back to its density by taking derivatives; more precisely we have

$$c(v_1, v_2) = \partial^2 C(v_1, v_2) / \partial v_1 \partial v_2.$$

Taking derivatives with respect to x and y of Equation (2.2) we obtain the formula

$$f(x, y) = \frac{\partial^2}{\partial x \partial y} C(F_X(x), F_Y(y)) = c(F_X(x), F_Y(y)) f_X(x) f_Y(y),$$

where f is the joint density over X and Y and f_X, f_Y are the individual density functions. We can rewrite this as

$$c(F_X(x), F_Y(y)) = \frac{f(x, y)}{f_X(x) f_Y(y)}. \tag{2.3}$$

The formula here shows how the copula density is obtained from the usual density function f by squeezing it up at places where X and Y have low probabilities and spreading it out at places where X and Y have high probabilities.

Using copulas rather than copula densities makes it easier to handle some of the equations, but we need to check three properties we require of the copula in order to match the properties of the copula density:

- *Increasing in each variable:* C is non-decreasing in each variable, so if $0 \leq a_1 \leq b_1 \leq 1$ and $0 \leq a_2 \leq b_2 \leq 1$ then

$$C(a_1, a_2) \leq C(b_1, b_2).$$

- *The rectangle inequality:* If $0 \leq a_1 \leq b_1 \leq 1$ and $0 \leq a_2 \leq b_2 \leq 1$ then the probability of being in the rectangle $[a_1, b_1] \times [a_2, b_2]$ can be obtained by looking at the right combination of the four possible rectangles with a corner at 0. For c to be a density this must be non-negative and we derive the inequality

$$C(a_1, a_2) + C(b_1, b_2) - C(a_1, b_2) - C(b_1, a_2) \geq 0.$$

- *Uniform marginal distribution:* Notice that $C(v, 1)$ is simply the probability that u_1 is less than v, and similarly $C(1, v)$ is simply the probability that u_2 is less than v. So the condition that the distributions of both u_1 and u_2 are uniform becomes: $C(1, v) = C(v, 1) = v$ for all $0 \leq v \leq 1$.

At first sight it might seem as though the rectangle inequality might follow from the fact that the copula is increasing in each variable, but this is not the case (see Exercise 2.10).

An important copula is that which represents the dependence behavior occurring in a multivariate normal, and this is called the *Gaussian copula*. The copula density is related to the density for the multivariate normal but each component is scaled to ensure that the marginals are uniform. We begin by doing the calculations working with the (cumulative) copula functions. From the fundamental copula equation and the density function for the multivariate normal with correlation ρ and with each variable having zero mean and standard deviation of 1, we have

$$C(F_X(x), F_Y(y))$$

$$= C(\Phi(x), \Phi(y)) = \Pr(X \leq x \text{ and } Y \leq y)$$

$$= \frac{1}{2\pi\sqrt{1 - \rho^2}} \int_{-\infty}^{x} \left(\int_{-\infty}^{y} \exp\left(-\frac{1}{2(1 - \rho^2)} \left(s_1^2 + s_2^2 - 2\rho s_1 s_2 \right) \right) ds_1 \right) ds_2.$$

Here we have used the usual notation in which $\Phi(x)$ is written for the cumulative normal distribution function with mean zero and standard deviation 1. The integrand comes from the multivariate normal with $\mu_X = \mu_Y = 0$ and $\sigma_X = \sigma_Y = 1$. To obtain a formula for $C(v_1, v_2)$ we want to substitute values of x and y for which $F_X(x) = v_1$ and $F_Y(y) = v_2$. So we set $x = \Phi^{-1}(v_1)$ and $y = \Phi^{-1}(v_2)$ to obtain

$$C(v_1, v_2) = \frac{1}{2\pi\sqrt{1-\rho^2}} \int_{-\infty}^{\Phi^{-1}(v_1)}$$

$$\times \left(\int_{-\infty}^{\Phi^{-1}(v_2)} \exp\left(-\frac{1}{2(1-\rho^2)}\left(s_1^2 + s_2^2 - 2\rho s_1 s_2\right)\right) ds_1 \right) ds_2.$$

The equivalent formula for the copula density from Equation (2.3) is

$$c(F_X(x), F_Y(y)) = \frac{\frac{1}{2\pi\sqrt{1-\rho^2}}\exp\left(-\frac{1}{2(1-\rho^2)}\left(x^2 + y^2 - 2\rho xy\right)\right)}{\frac{1}{\sqrt{2\pi}}\exp\left(-\frac{1}{2}x^2\right)\frac{1}{\sqrt{2\pi}}\exp\left(-\frac{1}{2}y^2\right)}$$

$$= \frac{1}{\sqrt{1-\rho^2}}\exp\left(\frac{1}{2}x^2 + \frac{1}{2}y^2 - \frac{1}{2(1-\rho^2)}\left(x^2 + y^2 - 2\rho xy\right)\right)$$

$$= \frac{1}{\sqrt{1-\rho^2}}\exp\left(-\frac{1}{2(1-\rho^2)}\left(\rho^2 x^2 + \rho^2 y^2 - 2\rho xy\right)\right).$$

And hence

$$c(v_1, v_2) = \frac{1}{\sqrt{1-\rho^2}}$$

$$\times \exp\left(-\frac{1}{2(1-\rho^2)}\left(\rho^2\Phi^{-1}(v_1)^2 + \rho^2\Phi^{-1}(v_2)^2 - 2\rho\Phi^{-1}(v_1)\Phi^{-1}(v_2)\right)\right).$$

There are versions of this formula for a higher number of variables. It has become somewhat infamous since the use of Gaussian copulas as a standard approach in measuring risk was blamed in some quarters for the failures of Wall Street 'quants' in predicting the high systemic risks in CDOs that in turn kicked off the financial crisis of 2008. In 2009 the *Financial Times* carried an article discussing the use of Gaussian copula models titled 'The formula that felled Wall Street' (*The Financial Times*, S. Jones, April 24, 2009.)

To help in understanding the Gaussian copula formula we really need to plot it, and it is most useful to plot the copula density function. This is simply a function defined over the unit square and is shown in Figure 2.15.

In this figure the corners $(0, 0)$ and $(1, 1)$ represent what happens to the relationship when both variables are very large and positive or very large and negative. The existence of correlation pushes these corners up and the opposite corners $(0, 1)$ and $(1, 0)$ are pushed down. Remember there is a restriction on the densities integrating to 1 along a line in which one or other variable is held constant, so the lifting of one corner can be seen as balancing the pushing down of an adjacent corner. When $\rho = 0$ the multivariate normal has circular contours, and the variables are independent – as we said earlier, this means a flat copula density at value 1 (which serves as a reminder that the copula abstracts away from the distributions of the underlying variables).

There are many different copula formulas that have been proposed. One example is the Clayton copula given, for two variables, by the formula:

$$C_\theta(v_1, v_2) = \left(\frac{1}{v_1^\theta} + \frac{1}{v_2^\theta} - 1 \right)^{-\frac{1}{\theta}},$$

where θ is a parameter that can take any value strictly greater than zero (we can also make this formula work when θ is strictly less than zero and greater than -1). Figure 2.16 shows what this copula looks like when $\theta = 2$. The copula density is then

$$c_\theta(v_1, v_2) = \frac{3}{v_1^3 v_2^3} \left(\frac{1}{v_1^2} + \frac{1}{v_2^2} - 1 \right)^{-\frac{5}{2}}.$$

In the figure, the peak at $(0, 0)$ shows a very tight correlation between the values that occur in the lower tails of the two underlying distributions. In fact, this peak has been cut off at a value of 8 for the purposes of drawing it.

There is another approach we can take to the way that random variables become highly correlated at extreme values and this is to measure *tail dependence* directly. We know that independence means that information on one variable does not convey anything about the other. So we could say that the probability of X being in the upper decile (i.e. X taking a value where $F_X(X) > 0.9$) is unchanged (at exactly 0.1) even if we also know that Y is also in its upper decile. Thus, with independence

$$\Pr(X > F_X^{-1}(0.9) | Y > F_Y^{-1}(0.9)) = \Pr(X > F_X^{-1}(0.9)) = 0.1.$$

Letting the 0.9 in this expression tend to 1 would make the limit of the conditional probability zero. However, if the variables are not independent, and there is a strong relationship between the variables at the extremes, then this limit will be greater than zero. Notice also that we can rewrite this type of conditional probability as follows (using Bayes' formula) as

$$\Pr(X > F_X^{-1}(\alpha) | Y > F_X^{-1}(\alpha)) = \frac{\Pr(X > F_X^{-1}(\alpha) \text{ and } Y > F_X^{-1}(\alpha))}{\Pr(Y > F_Y^{-1}(\alpha))}$$

$$= \frac{\Pr(X > F_X^{-1}(\alpha) \text{ and } Y > F_X^{-1}(\alpha))}{(1 - \alpha)}.$$

This motivates us to define a coefficient of upper tail dependence through

$$\lambda_u = \lim_{\alpha \to 1} \Pr(X > F_X^{-1}(\alpha) | Y > F_X^{-1}(\alpha))$$

$$= \lim_{\alpha \to 1} \frac{\Pr(X > F_X^{-1}(\alpha) \text{ and } Y > F_X^{-1}(\alpha))}{(1 - \alpha)}$$

and a corresponding coefficient of lower tail dependence as

$$\lambda_\ell = \lim_{\alpha \to 1} \Pr(X \le F_X^{-1}(\alpha) | Y \le F_X^{-1}(\alpha))$$

$$= \lim_{\alpha \to 0} \frac{\Pr(X \le F_X^{-1}(\alpha) \text{ and } Y \le F_X^{-1}(\alpha))}{\alpha}.$$

We say that X and Y have *upper tail dependence* if $\lambda_u > 0$, and that there is no upper tail dependence if $\lambda_u = 0$. The same definitions work for *lower tail dependence* using λ_ℓ instead of λ_u. We can convert this into a statement about the copula functions, since $\lambda_\ell = \lim_{\alpha \to 0} (C(\alpha, \alpha)/\alpha)$.

The existence of tail dependence is quite a strong property, and certainly is much stronger than saying that at high (or low) values the variables fail an independence test. If there is tail dependence then the copula density will go to infinity at the appropriate corner. We will not prove this in a formal way, but we can observe that with lower tail dependence, the definition of λ_ℓ implies that in the limit of small α then $C(\alpha, \alpha) \simeq \alpha \lambda_\ell$. But at this limit $C(\alpha, \alpha) \simeq \alpha^2 c(\alpha, \alpha)$, so we have $c(\alpha, \alpha) \simeq \lambda_\ell / \alpha$. As α goes to zero, the left-hand side approaches $c(0, 0)$ and the right-hand side goes to infinity unless $\lambda_\ell = 0$. Hence, we can deduce that $c(x, y) \to \infty$ when $x, y \to 0$ and $\lambda_\ell > 0$. A similar argument can be made in the case of upper tail dependence.

So, in two-dimensional cases we can look at a plot of the copula density and get a good idea of whether there is tail dependence. Figure 2.15 suggests that there is neither upper nor lower tail dependence for the Gaussian copula for this value of ρ, and from Figure 2.16 we see there is no upper tail dependence for the Clayton copula. However, there seems to be a lower tail dependence for this

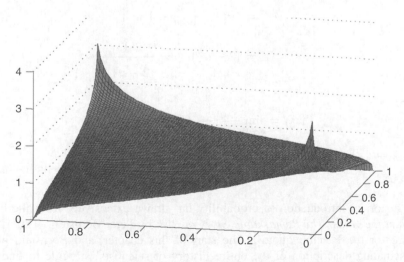

Figure 2.15   Gaussian copula density with $\rho = 0.25$.

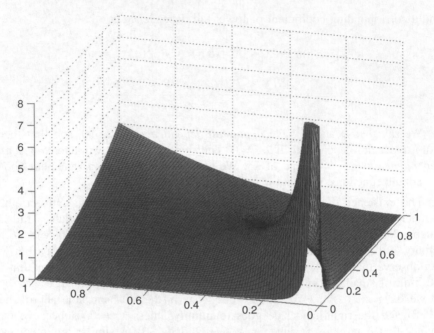

Figure 2.16   Clayton copula density for $\theta = 2$.

copula, and we can confirm this since, for the Clayton copula,

$$C(\alpha, \alpha)/\alpha = (1/\alpha)\left(\frac{1}{\alpha^\theta} + \frac{1}{\alpha^\theta} - 1\right)^{-\frac{1}{\theta}}$$

$$= (\alpha^\theta)^{-\frac{1}{\theta}}\left(\frac{2}{\alpha^\theta} - 1\right)^{-\frac{1}{\theta}}$$

$$= (2 - \alpha^\theta)^{-\frac{1}{\theta}} = \frac{1}{(2 - \alpha^\theta)^{\frac{1}{\theta}}}.$$

Thus

$$\lambda_\ell = \lim_{\alpha \to 0} \frac{1}{(2 - \alpha^\theta)^{\frac{1}{\theta}}} = \frac{1}{2^{\frac{1}{\theta}}} > 0.$$

Notes

The historical introduction to probability has drawn extensively from the book *Against the Gods: The Remarkable Story of Risk* by Peter Bernstein. This is the source for the Paccioli quote at the start of this chapter, and Bernstein gives a fascinating description of the entire history of scientists' struggle to find the right framework to deal with risk. The quote from Jakob Bernoulli is taken from

Oscar Sheynin's translation of his work: *The Art of Conjecturing; Part Four, Showing the Use and Application of the Previous Doctrine to Civil, Moral and Economic Affairs.*

Our discussion of optimal portfolio theory in this chapter is very brief. An excellent and much more comprehensive treatment is given by Luenberger (1998). The theory was originally developed by Harry Markowitz and first published in a 1952 paper; in 1990 Markowitz received a Nobel prize in Economics for his work.

A straightforward introduction to copulas is given by Schmidt (2006). Also the book by McNeil *et al.* (2005) gives a thorough discussion of the topic. The observations we make about the increase in correlation when markets move downwards are well known, for example see Ang and Chen (2002) or Chua *et al.* (2009).

References

Ang, A. and Chen, J. (2002) Asymmetric correlations of equity portfolios. *Journal of Financial Economics,* **63**, 443–494.

Bernoulli, J. *Ars Conjectandi.* Translated by Oscar Sheynin, NG Verlag, 2005.

Bernstein, P. (1996) *Against the Gods: The Remarkable Story of Risk,* John Wiley & Sons.

Chua, D., Kritzman, M. and Page, S. (2009) The myth of diversification. *Journal of Portfolio Management,* **36**, 26–35.

Luenberger, D. (1998) *Investment Science.* Oxford University Press.

McNeil, A., Frey, R. and Embrechts, P. (2005) *Quantitative Risk Management.* Princeton University Press.

Schmidt, T. (2006) Coping with copulas. In *Copulas: From theory to application in finance,* Jorn Rank (ed.), Bloomberg Financial.

Exercises

2.1 Problem of the points

Calculate a fair division of the stake in the 'Problem of the points' described in Section 2.1 if we assume that the current record of success of A against B is an accurate forecast of the probability of winning future games.

2.2 Changing a daily failure rate into a yearly one

A taxi is in use seven days a week. Suppose that the probability that the taxi suffers a mechanical fault on any given day is 0.001 and this stays constant over time.

(a) Find the expected number of failures in a year (of 365 days).

(b) Assuming the failures on different days are independent, find the probability of a failure at least once during the course of 365 days (and show that this probability is less than 0.365).

2.3 Late for the class

James needs to get to his class on time, which means arriving at the university by 10 am. The options are to take the number 12 bus, which takes 40 minutes, or the number 15, which takes 30 minutes, or the express bus, which takes 20 minutes. Arriving at the bus stop at 9.15 am, what is the probability that he will be at his class on time if the number 12 is equally likely to arrive at any time between 9.10 and 9.30, if the number 15 bus is equally likely to arrive at any time between 9.20 and 9.40 and there are two possible express services James may catch: the first is equally likely to arrive at any time between 9.05 and 9.20 and the second is equally likely to arrive at any time between 9.35 and 9.50? Assume that all the buses have arrival times that are independent.

2.4 Combining union and intersection risk

A building project is running late, and if it is more than four weeks late an alternative venue will need to be found for an event planned in the new building. There is a 20% chance of poor weather causing a delay by three weeks, and there is a 10% chance of late delivery of a critical component that would lead to a delay of between two weeks and four weeks, on top of any weather-related delay. The construction involves some excavation of a drain line in an area of some archaeological interest. Archaeologists are at work and there is a small (5%) chance that significant finds will be made that force a delay of around two months.

(a) Use a Venn diagram to show the events that will lead to a delay of more than four weeks.

(b) Assuming that all three events are independent, calculate the probability of more than four weeks' delay.

2.5 Probability for maximum from empirical distribution

You have observed the rainfall amounts on all days in April over the last three years. There are 90 data points and the largest 10 are as follows (in mm): 305, 320, 325, 333, 340, 342, 351, 370, 397, 420. Assuming that rainfall on successive days is independent, use the empirical data to estimate the probability that the maximum daily rainfall during a five-day period next year is less than 350 mm (i.e. rainfall is less than 350 mm on each of the five days).

2.6 Portfolios with large numbers and different variances

There are N independent stock returns X_i, and σ_i is the standard deviation of X_i. If σ_{max} is an upper bound on the size of any σ_i show that by investing an amount W in a way that gives stock i a weight proportional to $1/\sigma_i$, then the overall standard deviation of the portfolio is less than $W\sigma_{max}/\sqrt{N}$, and this expression approaches zero as N gets larger and larger.

2.7 Optimal portfolio

There are three stocks to invest in: A, B and C. In one year the expected increases in price are: 10% for stock A, 15% for stock B, and 5% for stock C. The standard deviations of these numbers are 2% for A and B and 1% for C. In other words, if X is the random variable giving the increase in value for A measured in %, then the standard deviation of X is 2. If the returns are all independent, what is the minimum variance portfolio that achieves a 10% return? (Use a spreadsheet and 'Solver' for this calculation).

2.8 Optimal portfolio with a risk-free asset

Using the same arrangement as for Exercise 2.7, suppose that there is a risk-free investment D that always increases in value by 4% over the course of a year.

(a) Recalculate the minimum variance portfolio for a 10% return, a 7% return, and a 5% return. (Use a spreadsheet and 'Solver' for this calculation).

(b) Show that these three portfolios lie on the same straight line in the standard deviation versus expected profit diagram.

(c) Show that each of the three portfolios is a combination involving some proportion of D and some proportion of a fixed portfolio of the A, B and C stocks (this is an example of the 'one-fund theorem' in portfolio theory).

2.9 Multivariate normal

A company wishes to estimate the probability that a storm and high tide combined will lead to flooding at a critical coastal installation. There are 10 high tide events each year when the installation is at risk. The two critical components here are wind velocity in the shore direction and wave height. The wind velocity has mean 10 km/hr and standard deviation 8 km/hr and the wave height has mean 2 meters and standard deviation 1 meter. The estimated covariance between these two variables is 4. Suppose that the behavior is modeled as a multivariate normal distribution (ignoring issues of negative wave height). Estimate the probability that there will be flooding next year assuming that flooding occurs when the wave height + $0.05 \times$ (wind velocity) is greater than 6.

2.10 Copula properties

Suppose that we define the copula density $c(u_1, u_2) = 2/3$ when both $u_1 < 3/4$ and $u_2 < 3/4$; $c(u_1, u_2) = 2$ when one of $u_1 < 3/4$ and $u_2 < 3/4$; and $c(u_1, u_2) = -2$ otherwise. Show that if $C(u_1, u_2)$ is defined in the usual way from c that it will be increasing in both arguments and have uniform marginals, but will not satisfy the rectangle inequality (since its density is negative on part of the unit square).

2.11 Gumbel copula

The Gumbel copula with parameter 2 is given by

$$C(u_1, u_2) = \exp\left(-\sqrt{(\log(u_1))^2 + (\log(u_2))^2}\right).$$

If two variables each have an exponential distribution with parameter 1 (so they have CDF $F(x) = 1 - e^{-x}$ for $x \geq 0$) and their joint behavior is determined by a Gumbel copula with parameter 2, calculate the probability that the maximum of the two variables has a value greater than 3 and compare this with the case where the two random variables are independent.

2.12 Upper tail dependence

Show that the formula for λ_u given in the text can be converted to

$$\lambda_u = 2 + \lim_{\delta \to 0} \frac{C(1 - \delta, 1 - \delta) - 1}{\delta},$$

and use this formula to check that the Clayton copula with $\theta = 2$ has no upper tail dependence. You will find it helpful to set $g(x) = C(x, x) - 1$ and use the fact that

$$\lim_{\delta \to 0} \frac{g(1 - \delta) - g(1)}{\delta} = -\frac{dg(x)}{dx} \text{ evaluated at } x = 1.$$

3

Measuring risk

The genesis of VaR

Dennis Weatherstone was chairman and later chief executive of J.P. Morgan. In many ways he was an unlikely figure to become one of the world's most respected bankers. As the son of a London Transport clerk who was born in Islington and left school at 16, he was a far cry from the expensively-educated people who typically run major Wall Street firms. He moved into the chairman's role from a position running the foreign-exchange trading desk. He was perceived as an expert on risk, but when he looked at the firm as a whole he found that he had little idea of the overall level of risk at J.P. Morgan. Over a period of several years the concept of 'value at risk' was developed by the analysts and 'quants' working at J.P. Morgan as a way to answer this question. The need was to measure the risk inherent in any kind of portfolio. The value at risk, or VaR, was recalculated every day in response to changes in the portfolio as traders bought and sold individual securities.

This turned out to bring huge benefits when looking across the many activities going on at J.P. Morgan. It became possible to look at profits from different traders and compare them with the level of risk measured by value at risk. In the early 1990s, Weatherstone began to ask for daily reports from every trading desk. This became known as the 415 report: they were created at 4.15 pm every day just after the market closed. These reports enabled Weatherstone not only to compare every desk's estimated profit in comparison to a common measure of risk, but also to form a view for the firm as a whole.

In 1993 the theme of the J.P. Morgan annual client conference was risk, and these clients were given an insight into the value at risk methodology. When clients came to ask if they could purchase the same kind of system for their own companies, J.P. Morgan set up a small group called RiskMetrics to help them. At that stage this was a proprietary methodology that was being given

Business Risk Management: Models and Analysis, First Edition. Edward J. Anderson.
© 2014 John Wiley & Sons, Ltd. Published 2014 by John Wiley & Sons, Ltd.
Companion website: www.wiley.com/go/business_risk_management

away for free, with the aim of helping clients and establishing the reputation of J.P. Morgan in the risk area. VaR became a more and more popular tool and in 1998 RiskMetrics was spun off as a separate company.

This was a time when the regulatory authorities were beginning to pay more attention to risk. For example, the Securities and Exchange Commission was concerned about the amount of risk arising from trading in derivatives, and created new rules forcing financial firms to disclose that risk to their investors. Inevitably the measure that was used was value at risk. All this was part of a slow but inexorable change that took VaR from being a specific set of tools developed within J.P. Morgan and sold by RiskMetrics into a risk management standard applied throughout the financial world.

3.1   How can we measure risk?

In this chapter we will look in more depth at how to measure risk. Does it make sense to talk of one course of action being more risky than another? And if so, what does this mean? In the simplest case we have a range of outcomes all with different probabilities and with different consequences. When the consequences can be accurately turned into dollar amounts, we obtain a distribution over dollar outcomes.

Our discussion in Chapter 2 has essentially assumed that the distribution of outcomes is given by a known probability distribution, but in practice there are great difficulties in knowing the distribution that we are dealing with. It is always hard to estimate the probabilities associated with different outcomes and the monetary consequences of these events. In a financial calculation we may have some chance of estimating the relevant numbers. For example, we might ask how likely it is that the price of gold gets above a certain level and (assuming we are betting against gold prices rising) what we lose if this happens. But in most management roles it is much harder than this. How can I calculate the probability that sales of my new product are less than 1000 units in the first year? How can I know how much it will cost me if the government introduces a revised safety code in my industry? For the moment we set these problems aside and assume that we have access to the numbers we need.

Often the best starting point for the estimation of risk is to consider what has happened in the past. Even if we think that the world has changed, it would still be foolish not to pay any attention to the pattern of results we have observed so far.

To make our discussion more concrete, let us look at some weather-related data, specifically the daily maximum temperature for different cities in Australia. We will look at data from 2010. This information would be of interest if we were trying to sell air-conditioning units, or trying to decide whether to spend money on air conditioning our premises. The demand for air conditioning spikes upwards when temperatures are high. Let's compare Perth and Adelaide weather.

A starting point might be to look at the average of the maximum temperatures. The average daily maximum temperature in Perth in 2010 was 25.27°

(all temperatures are in centigrade), while the average daily maximum temperature for Adelaide was 22.44°, more than 2 degrees cooler. But we could also
measure variability. The usual way to do this is to use the standard deviation σ.
A spreadsheet can be used to calculate this for the two cities. The standard deviation for Perth is 6.41 and the standard deviation for Adelaide is 7.00, which is
significantly larger. But if we are interested in how many very hot days occur,
neither of these figures is very informative. It is better to draw a frequency
histogram of the data. This has been done in Figure 3.1.

The graphs show that there is not much to choose between the two cities as
regards the probability of really hot days. In 2010 the five hottest days in Perth
were 42.9°, 42.7°, 41.5°, 41.1° and 40.0° and the five hottest days in Adelaide
were 42.8°, 42.0°, 41.3°, 41.0° and 40.2°.

The critical point here is that if we are interested in the extreme results
then the mean and standard deviation, which are mainly determined by the mass
of more or less average results, will not give us the information we need. We
must either look at the record of actual historical data or have some idea of the
probability distribution that generates these data.

Now we turn more directly to a risk example. Suppose that we have agreed
to sell 1000 tons of wheat in three years' time at a price of US$300 per metric
ton. If wheat prices are high, we will make a loss, but if wheat prices are low, we
will make a profit. In order to estimate the probability of a loss we consider the

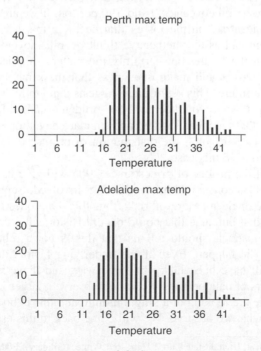

Figure 3.1   Frequency histograms of daily maximum temperature during 2010.

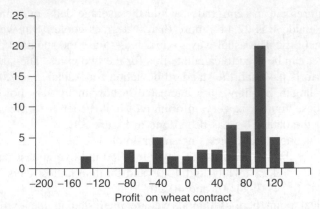

Figure 3.2   Frequency histogram of profits per ton made on a $300 per ton wheat contract.

historical record of wheat prices over the five-year period from January 2005 on a monthly basis.[1] Factoring in the US$300 price, we get the frequency histogram of profits per ton shown in Figure 3.2.

Wheat prices spiked in February and March 2008 to a value of around $450, but by April they had retreated to $387. So for two months (out of the 60) losses would have been $146 and $154 per ton, and then for a further three months losses would have been between $80 and $90 per ton. If we are interested in risk then we must concentrate on the losses that occur in this left-hand tail. It will require a judgment call as to whether we think overall movements are likely to be up or down in the future, but certainly the pattern of price spikes that has been seen in the past would make one guess that the same sort of price spike might occur in the future. This set of data suggests that a price spike may happen about 1 month in 30, so it would certainly be prudent to allow for this happening again! But notice that the mean ($41.8) and standard deviation (69.5) tell us very little about what is going on – the distribution of profits is very far from being a normal distribution in this case.

So, looking at the history of prices since 2005 will give us some idea about the distribution of outcomes and hence the risks involved. Suppose now that we want to extract a single measure of risk from this. One question we might ask is: What is the worst outcome that could occur? Historically, the answer is $154 per ton, but our instincts should tell us that this is not a reliable estimate for the worst that might happen. Even if the underlying factors don't change over time, we may still have been lucky in some way, and if we were to look at a different five years of data perhaps we would observe a larger loss. In looking at the largest loss we are looking at a single month, and this is bound to mean a lot of fluctuation in what we observe. We will come back to this estimation problem

---

[1] Data are for Wheat, United States, no 2 Hard Red Winter (ordinary), FOB Gulf and have been downloaded from www.unctad.org.

in our discussion of extreme value theory in the next chapter. In a sense, the problem of measuring risk, which is a problem of describing what happens in the tails of the distribution, inevitably leads to difficulties when we want to use historical data (since, by definition, it all comes down to the values that occur at only a handful of points).

3.2   Value at risk

When dealing with a distribution of profits, risk is all about the size of the left-hand tail of the distribution, and so it becomes clear that there is no single right way to measure the risk. But by far the most common way is to measure *value at risk*, most often shortened to *VaR*. This is measured at a particular percentage. For example, we might say that the 99% value at risk figure is $300 000. This is equivalent to the statement that 99% of outcomes will lose less than $300 000, or we can be 99% sure that losses will not exceed $300 000. So, giving a VaR number corresponds to picking a single point in the distribution.

Now we set about giving an exact definition for value at risk. Since we are concerned with potential losses, it is easier to describe everything in terms of losses rather than in terms of profits. So the horizontal axis is reversed in order to have higher losses to the right-hand side and the right-hand tail becomes the area of interest. To make all this clearer, Figure 3.3 shows the 95% and 99% VaR numbers for a distribution of losses over the range $(-1.5, 0.5)$ (all values are assumed to be denominated in units of $100 000 dollars). The density function shown is given by the equation

$$f(x) = \frac{15}{16} \left[(x + 0.5)^4 - 2(x + 0.5)^2 + 1 \right]. \tag{3.1}$$

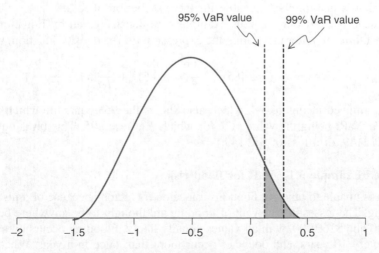

Figure 3.3   95% and 99% value at risk points.

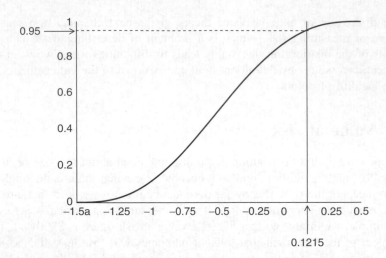

Figure 3.4   A quantile interpretation of value at risk.

Occasionally people will talk about a 1% VaR or a 5% VaR, but this just means the 99% and 95% values. Also, sometimes people will talk about 'VaR with a confidence level of 99%'. This is natural, since a 99% VaR of $100 000 means that we can be 99% confident that losses will not exceed $100 000.

The VaR approach can be seen as an example of using *quantiles* to describe the tails of a distribution. We will use the terminology of the $\alpha\%$ quantile to mean the x value such that $F(x) = \alpha/100$ where F is the cumulative distribution function, so $F(x) = \Pr(X < x)$ where X is the random variable in question. Thus, the 50% quantile is the x value with $F(x) = 0.5$, i.e. the x value for which half the distribution is below it and half above – this is just the median. The 99% VaR value is just the 99% quantile for the distribution of losses.

We can convert our previous example with density given by Equation (3.1) into a CDF form. After integrating the expression for the density function, we get

$$F(x) = \frac{3}{16}(x + 0.5)^5 - \frac{5}{8}(x + 0.5)^3 + \frac{15}{16}(x) + \frac{31}{32}.$$

This is graphed in Figure 3.4, which also shows the 95% quantile which is also the 95% VaR, being the value of x for which $F(x) = 0.95$. This turns out to be $x = 0.12149$, or a loss of $12 149.

Worked Example 3.1   VaR for flood risk

A firm is unable to take out flood insurance on its factory because of a history of flooding. The losses in the event of flooding are thought to be anywhere between $10 000 and $160 000, with all values equally likely. Flooding occurs on average once every 10 years, and does not occur more than once in a year. What is the annual 98% VaR due to flooding?

Solution

We calculate the CDF for the losses (in \$1000s) due to flooding in one year, a random variable we write as L. Then $F(x) = \Pr(L \leq x)$. We have no chance of negative losses, so $F(x) = 0$ for $x < 0$. There is a probability of 0.9 that $L = 0$, so at 0 the CDF jumps up and $F(0) = 0.9$. The probability that losses are less than x for $x \leq 10$ is also 0.9, since flood losses are never less than \$10 000. The probability that losses are less than x when $x > 10$ is given by

$$0.9 + 0.1(x - 10)/150.$$

Observe that with this expression the probability increases linearly from 0.9 at $x = 10$ to 1 at $x = 16$, which is the defining property for the uniform distribution of losses, since when the density function f is constant, its integral, F, is linear. To find the 98% value at risk we solve

$$0.9 + 0.1(x - 10)/150 = 0.98,$$

giving

$$x = 0.8 \times 150 + 10 = 130.$$

Thus, we have shown that $\text{VaR}_{0.98} = \$130\,000$.   □

Writing down a definition for the 95% VaR needs care. If L is the (uncertain) value of the losses, we would naturally set $\text{VaR}_{0.95}$ to be the x value such that $\Pr(L \leq x) = 0.95$, so

$$F(\text{VaR}_{0.95}) = 0.95$$

where F is the CDF of the loss function. We can write this as

$$\text{VaR}_{0.95} = F^{-1}(0.95)$$

where the notation F^{-1} is used for the inverse of F (i.e. $F^{-1}(y)$ is the value of x such that $F(x) = y$).

Unfortunately this definition will not quite work. The problem is that when the losses do not have a continuous distribution there may be no value at which $\Pr(L \leq x) = F(x) = 0.95$. For example, suppose the following losses occur:

loss of \$11 000	probability 0.02
loss of \$10 500	probability 0.02
loss of \$10 000	probability 0.02
loss of \$9000	probability 0.04
loss of \$0	probability 0.9

Then, setting $x = 10000$ gives $\Pr(L \leq 10000) = 0.96$ but for x even slightly less than 10000 the probability is smaller than 0.95, e.g. $\Pr(L \leq 9999) = 0.94$.

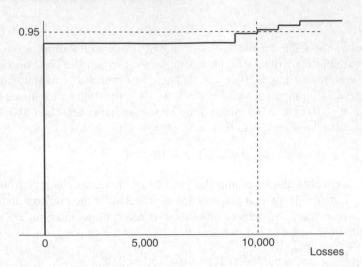

Figure 3.5   95% cumulative distribution function intersection at $10 000.

The best way to resolve this is to draw a graph. With this kind of discrete distribution (i.e. not continuous), the CDF is a step function. Figure 3.5 shows the graph of $F(x) = \Pr(L \leq x)$ for this example. It is easy to see that the graph of $F(x)$ goes through the value 0.95 at $x = 10\,000$ and so this is the 95% VaR.

But what is easy with a graph is a little more complex with a formula. VaR is usually defined as follows

$$\mathrm{VaR}_\alpha = \inf(x : \Pr(L > x) \leq 1 - \alpha),$$

so, for a 95% VaR we have

$$\mathrm{VaR}_{0.95} = \inf(x : \Pr(L > x) \leq 0.05),$$

which we can read as 'The lowest value of x such that the probability of L being larger than x is less than 0.05'.

The idea here is that we think of a very large value of x for which this probability is definitely less than 0.05, then we slowly reduce x until the probability increases to 0.05 or more and that is where we stop. This is exactly the same as taking a large value of x where $F(x)$ is definitely greater than 0.95 and slowly reducing it till the value of $F(x)$ drops to 0.95 or less.

The use of inf here and the choices of inequalities that are strict $(L > x)$ or not $(\leq 1 - \alpha)$ is something that you do not need to worry about: it covers the definition for discrete probability distributions when there might be a range of values for which $F(x) = 0.95$ (i.e. a horizontal section in the graph of F). A simple way to describe the formula for VaR is as follows:

VaR is the loss value at which the graph of $F(x)$ first reaches the correct percentile.

In financial environments the average return is often built into calculations on expected returns, and the thing which is of most interest is the risk that the final outcome will be much worse than the expected returns. In these cases VaR is calculated relative to the mean result. Thus, in the example given in Figure 3.3 the mean profit is 0.5 or $50 000. So it would be normal to quote the 95% relative VaR as a loss of $62 149 rather than the 'absolute VaR' which we calculated previously as $12 149.

Sometimes the relative VaR is called the *mean-VaR* and is written VaR_α^{mean}. In market risk management the period of time involved is very short – for example, one day – and VaR_α^{mean} is called the 'daily earnings at risk'. But in this case the expected market movement is bound to be close to zero and so the two definitions of VaR will, in any case, be essentially the same. The distinction between relative and absolute VaR is more important when dealing with longer time horizons.

As an example of how VaR is reported by companies, we give the following excerpt from the Microsoft annual report for 2012:

(MICROSOFT) QUANTITATIVE AND QUALITATIVE DISCLOSURES ABOUT MARKET RISK

We are exposed to economic risk from foreign currency exchange rates, interest rates, credit risk, equity prices, and commodity prices. A portion of these risks is hedged, but they may impact our financial statements.

Foreign currency: Certain forecasted transactions, assets, and liabilities are exposed to foreign currency risk. We monitor our foreign currency exposures daily and use hedges where practicable to offset the risks and maximize the economic effectiveness of our foreign currency positions. Principal currencies hedged include the euro, Japanese yen, British pound, and Canadian dollar.

Interest rate: Our fixed-income portfolio is diversified across credit sectors and maturities, consisting primarily of investment-grade securities. The credit risk and average maturity of the fixed-income portfolio is managed to achieve economic returns that correlate to certain global and domestic fixed-income indices. In addition, we use 'To Be Announced' forward purchase commitments of mortgage-backed assets to gain exposure to agency and mortgage-backed securities.

Equity: Our equity portfolio consists of global, developed, and emerging market securities that are subject to market price risk. We manage the securities relative to certain global and domestic indices and expect their economic risk and return to correlate with these indices.

Commodity: We use broad-based commodity exposures to enhance portfolio returns and facilitate portfolio diversification. Our investment portfolio has exposure to a variety of commodities, including precious metals, energy, and grain. We manage these exposures relative to global commodity indices and expect their economic risk and return to correlate with these indices.

VALUE-AT-RISK

We use a value-at-risk ('VaR') model to estimate and quantify our market risks. VaR is the expected loss, for a given confidence level, in the fair value of our portfolio due to adverse market movements over a defined time horizon. The VaR model is not intended to represent actual losses in fair value, including determinations of other-than-temporary losses in fair value in accordance with accounting principles generally accepted in the United States ('U.S. GAAP'), but is used as a risk estimation and management tool. The distribution of the potential changes in total market value of all holdings is computed based on the historical volatilities and correlations among foreign currency exchange rates, interest rates, equity prices, and commodity prices, assuming normal market conditions.

The VaR is calculated as the total loss that will not be exceeded at the 97.5 percentile confidence level or, alternatively stated, the losses could exceed the VaR in 25 out of 1000 cases. Several risk factors are not captured in the model, including liquidity risk, operational risk, and legal risk.

The following table sets forth the one-day VaR for substantially all of our positions as of June 30, 2012 and June 30, 2011 and for the year ended June 30, 2012 (in millions):

Risk Categories	June 30, 2012	June 30, 2011	2011–2012		
			Average	High	Low
Foreign currency	$ 98	$ 86	$173	$229	$ 84
Interest rate	$ 71	$ 58	$ 64	$ 73	$ 57
Equity	$205	$212	$194	$248	$165
Commodity	$ 18	$ 28	$ 20	$ 29	$ 15

Total one-day VaR for the combined risk categories was $292 million at June 30, 2012 and $290 million at June 30, 2011. The total VaR is 26% less at June 30, 2012, and 25% less at June 30, 2011, than the sum of the separate risk categories in the above table due to the diversification benefit of the combination of risks.

3.3   Combining and comparing risks

One great advantage of VaR as a way of measuring risk is that it takes the complexity inherent in a probability distribution of possible outcomes and turns it into a single number. In general terms we want a single measure of risk because we want to compare different situations. Is the current environment more risky for our firm than it was a year ago? Is this potential business opportunity more risky than that one? Does our direct report, Tom, have a more risky management approach than Dick, another direct report?

We suppose that X is a random variable giving the *losses*, and we write $\psi(X)$ for the risk measure for a random variable X. We can think of the risk measure ψ as a way of judging the riskiness of the situation described by X. This way of thinking leads to some natural properties that a risk measure should have.

1. *Monotonicity: If losses in every situation get larger then the risk measure increases.* Often we write $X \leq Y$ to mean that under any scenario, the random variable X takes a value that is less than or equal to the value of the random variable Y. So this condition can be expressed succinctly as:

$$\text{If } X \leq Y \text{ then } \psi(X) \leq \psi(Y).$$

2. *Positive homogeneity: Multiplying risks by a positive constant also multiplies the risk measure by the same constant.* Another way to think about this is to say that a change in the unit of currency leads to the risk measure changing in the appropriate way. In symbols:

$$\psi(bX) = b\psi(X) \text{ for any positive constant } b.$$

3. *Translation invariance: If every outcome is changed by a certain fixed amount, this is also the change that occurs in the risk measure.* In financial terms we can see this as a statement that adding a certain amount of cash to a portfolio decreases the risk by the same amount. (This is the property that ties the risk measure back to actual dollar amounts.) We can write this condition as:

$$\psi(c + X) = c + \psi(X) \text{ for any constant } c.$$

Example 3.2   Mean plus three standard deviations

A firm could assess risk by looking at the worst value that might occur, basing this on a calculation of the average loss plus three standard deviations. So, if X is the random variable of losses and this has mean μ, and standard deviation σ, then the risk measure is

$$\psi(X) = \mu + 3\sigma.$$

It is commonplace to take three standard deviations as the largest deviation we are likely to observe in normal operation, and if X has a normal distribution then we can look up the probability: we have $\Pr(X > \mu + 3\sigma) = 0.0013$. Two investment opportunities are compared. Over the last two years investment A has earned a mean return of \$6000 a day with standard deviation \$2500 and investment B has earned a mean return of \$8000 a day with standard deviation \$3000. To compare these we need to remember to convert returns to losses. We get a risk measure for A of $-6000 + 7500 = 1500$ and a risk measure for B of $-8000 + 9000 = 1000$, so on this basis investment A is the most risky. It is surprising that this risk measure is not necessarily monotonic (see Exercise 3.3), though it satisfies the other two conditions.   □

It is not hard to show that VaR satisfies each of these three conditions. Since VaR is a quantile, changing outcomes has no effect unless a scenario moves across the quantile value, and then an increase in loss can only increase the quantile value – this is the monotonicity property. We can also show this property algebraically, if we wish. For random variables X and Y, if $X \leq Y$ then, for any x,

$$\Pr(X > x) \geq \Pr(Y > x)$$

and hence if $\Pr(X > x) \leq 1 - \alpha$ for some α, then $\Pr(Y > x) \leq 1 - \alpha$. From this we can deduce that

$$\inf(x : \Pr(Y > x) \leq 1 - \alpha) \leq \inf(x : \Pr(X > x) \leq 1 - \alpha)$$

which is the inequality we need to show $\mathrm{VaR}_\alpha(X) \geq \mathrm{VaR}_\alpha(Y)$.

The other two properties are also easy to establish.

$$\mathrm{VaR}_\alpha(bX) = \inf(y : \Pr(bX > y) \leq 1 - \alpha)$$
$$= \inf(v : \Pr(X > v/b) \leq 1 - \alpha)$$
$$= b\inf(w : \Pr(X > w) \leq 1 - \alpha) = b\mathrm{VaR}_\alpha(X),$$

so VaR satisfies the positive homogeneity condition. Also

$$\mathrm{VaR}_\alpha(X + c) = \inf(y : \Pr(X + c > y) \leq 1 - \alpha)$$
$$= \inf(y : \Pr(X > y - c) \leq 1 - \alpha)$$
$$= \inf(v : \Pr(X > v) \leq 1 - \alpha) + c$$
$$= \mathrm{VaR}_\alpha(X) + c,$$

and so there is translation invariance as well.

Thus, VaR has the three properties we have discussed as being appropriate for a measure of risk, but the area in which VaR is much less satisfactory relates to the combination of different risks. There is a natural fourth property of risk measures to add to the three properties introduced above.

4. *Subadditivity. Combining two risks together does not increase the overall amount of risk.* This is based on the principle that diversification should lead to a decrease in overall risk. Mathematically we write this as

$$\psi(X + Y) \leq \psi(X) + \psi(Y).$$

A risk measure that satisfies all four properties (monotonicity, positive homogeneity, translation invariance and subadditivity) is called *coherent*. It is important to note that VaR does not satisfy subadditivity and so is not a coherent risk measure. In order to show that VaR fails to be subadditive, we consider a specific example.

Example 3.3   Combining two risks might make the VaR value worse

Suppose that we can invest \$10 000 in a bond A which will normally pay back \$11 000 in a year's time, but there is some credit risk. Specifically, there is a small chance (which we estimate as 4%) that the bond issuer goes bankrupt and then we will get only a fraction of our money back (an amount we estimate as 30% of our investment, i.e. \$3000). Assuming we are right in all our estimates, then the 95% absolute VaR is actually a negative amount −\$1000 (equivalent to a profit of \$1000). This is because the credit risk is too small to appear in the VaR calculation.

Now, consider making a second investment in a bond B with exactly the same characteristics as A and suppose that bond B fails in a way that is quite independent of what happens to A. Then we get the following outcomes

Neither bond fails	Probability $0.96 \times 0.96 = 0.9216$	profit \$2000
A fails, B does not fail	Probability $0.96 \times 0.04 = 0.0384$	loss \$6000
B fails, A does not fail	Probability $0.96 \times 0.04 = 0.0384$	loss \$6000
Both bonds fail	Probability $0.04 \times 0.04 = 0.0016$	loss \$14 000

We can see that the combined portfolio makes a loss with probability 0.0784 and the 95% absolute VaR value is a loss of \$6000. The credit risk is too small to influence the VaR on a single bond, but with a portfolio of bonds it can no longer be ignored. Notice, however, that the diversification benefit does not disappear (see Exercise 3.4), it is just canceled out by the effect of a big loss crossing this particular 95% quantile boundary.   □

The problems with subadditivity highlight one of the limitations of VaR: there is something arbitrary about the confidence level $1 - \alpha$. VaR does not give a full picture of what happens in the tail of the distribution, and it says nothing at all about the maximum losses that may occur. Usually the worst that can happen is that a portfolio becomes worthless; so if we want to know how much we can lose, the answer may well be 'everything'! In a business environment there will usually be some events that lead to losses that simply cannot be estimated in advance. In one sense VaR is helpful, because at least it does not assume any estimates for extreme losses: its treatment of these extreme events means we need to estimate their probability, but allows us to pass over any estimation of the exact consequences.

3.4   VaR in practice

It is odd that VaR is both very widely used and at the same time very controversial. Much of the controversy arises because the basic technique can be used in different ways – and some approaches can be misleading, perhaps even dangerous. However, there is no getting away from VaR – for banks it is part of the Basel II framework which links capital requirements to market risk, and in the US some quantitative measures of risk are mandated by the SEC for company annual reports.

A sense of what is required under Basel II can be seen from the following excerpt from Clause 718 (section 76) (taken from http://www.basel-ii-accord .com).

Banks will have flexibility in devising the precise nature of their models, but the following minimum standards will apply for the purpose of calculating their capital charge.

. . .

(a) 'Value-at-risk' must be computed on a daily basis.

(b) In calculating the value-at-risk, a 99th percentile, one-tailed confidence interval is to be used.

(c) In calculating value-at-risk, an instantaneous price shock equivalent to a 10 day movement in prices is to be used, i.e. the minimum 'holding period' will be ten trading days. Banks may use value-at-risk numbers calculated according to shorter holding periods scaled up to ten days by the square root of time.

(d) The choice of historical observation period (sample period) for calculating value-at-risk will be constrained to a minimum length of one year.

. . .

(f) No particular type of model is prescribed. So long as each model used captures all the material risks run by the bank, banks will be free to use models based, for example, on variance-covariance matrices, historical simulations, or Monte Carlo simulations.

(g) Banks will have discretion to recognise empirical correlations within broad risk categories (e.g. interest rates, exchange rates, equity prices

---

and commodity prices, including related options volatilities in each risk factor category).

...

(i) Each bank must meet, on a daily basis, a capital requirement expressed as the higher of (a) its previous day's value-at-risk number measured according to the parameters specified in this section and (b) an average of the daily value-at-risk measures on each of the preceding sixty business days, multiplied by a multiplication factor.

(j) The multiplication factor will be set by individual supervisory authorities on the basis of their assessment of the quality of the bank's risk management system, subject to an absolute minimum of 3.

In practice there are three different approaches to calculating VaR figures. We may use historical price distributions (non-parametric VaR), we may use mathematical models of prices (perhaps including normal distributions for some risk factors), or we may use Monte Carlo simulation.

The historical approach is simple: we look at our current market portfolio and then use historical information to see how this portfolio would have performed over a period (of at least a year). Assuming that we use a year as the period, we will have about 250 trading days. If, like Microsoft, we want to calculate a 97.5% VaR then this would mean between six and seven occasions during the year when VaR is exceeded. So we could take the seventh smallest daily loss recorded during the year on our portfolio as the VaR estimate. One great advantage of this approach is that by dealing with historical data, we already capture the relationships between the different stocks in our portfolio. So we do not have to start making estimates, for example, of how correlated the stock price movements of Microsoft and Google are.

One disadvantage of this approach is that it is hard to know how long a data series to include. Is there something about current market conditions that is different to the long bull run up to 2008? If so, then we should not include too much of that earlier period in our analysis. But, on the other hand, if we give our historical analysis too short a period to work with, then we may well be overly influenced by particular events during that period. In general, a historical approach is less likely to be appropriate when there has been a significant change in the marketplace.

The parametric approach is flexible and can take account of different correlation structures. A weakness in practical terms is that once the problem becomes of a reasonable size, we will need to make some strong assumptions on the distribution of returns (often that the log returns have a normal distribution). However, once this is done the calculations can be completed very quickly. Typically a parametric approach looks at the response of instruments like options to variations in the underlying securities (a very popular method in this class is provided by RiskMetrics). The danger of a parametric approach is that we are unlikely to model the actual behavior correctly at the tails of the distribution. If there are 'fat tails' then this approach may be very misleading (we will discuss these problems further in the next chapter).

A third option is to use a Monte Carlo simulation. This uses the parametric technique of modeling the individual components that generate risk, but rather than look for analytical solutions, it instead simulates what might happen. A long enough simulation can capture the entire distribution of behavior without the need for very specific choices of distribution and at the same time can represent any degree of complexity in the correlation structure. The weakness of this approach is that it still requires assumptions to be made on distributional forms; it can also be computationally demanding.

Having decided the method that will be used to compute VaR numbers, there are two further decisions that need to be taken. First a risk manager must decide on a *risk horizon*. This is the period of time over which losses may occur. Using a one-day VaR is about looking at price movements during a single day. But Basel specifies a 10-day period, and for many firms an even longer period will be appropriate. However, the longer the period chosen, the longer the time series data that will be needed in order to estimate it. In any event, even with a longer period it is important to ensure that VaR calculations are done regularly and at least as often as the risk horizon (Basel II requires VaR to be calculated daily).

A second decision is the confidence level or quantile that will be used. Basel requires a 99% VaR to be calculated, but we have already seen how Microsoft uses a 97.5% VaR.

It is also important to check how the VaR estimates match actual risk performance, which is called *back-testing*. The simplest way to do this is to apply the method currently in use to the company's past performance, to get an estimate of the VaR that would have been calculated on each day over the last year. The theory of VaR then tells us how many times we would expect to see losses greater than VaR (2.5 times if a 99% VaR level is used for 250 trading days). If we find that VaR limits have been breached more often than this, we need to investigate further and should consider changing the method of calculation.

Whichever approach is used, the generation of VaR numbers can be immensely helpful to managers, and it is worth reviewing why this is so.

- VaR is easily understood and is now familiar to many senior managers.

- VaR provides a single consistent measure of risk that can be used throughout the firm and can form the focus of discussion about risk.

- VaR can be calculated at the level of individual operational entities (or trading desks in a bank). This gives good visibility down to the lower levels of the company. It provides a tool that can be used to impose a consistent risk strategy through the organization, at the same time as enabling senior managers to understand more of where and how risk arises within their organization.

- VaR provides a good tool for assessing capital adequacy (and is required for that purpose by banking regulators).

- VaR has become the standard way of reporting risk externally.

3.5   Criticisms of VaR

As we mentioned earlier, the use of VaR is still controversial, and it is important to understand the criticisms that have been made. The primary problem with VaR is that it does not deal with events within the tail – it gives no guidance on how large extreme losses may turn out to be. There is a lot of difference between saying that on 99% of days I will lose no more than $100 000 and saying that on one day in each year (on average) I will lose $20 million. Yet these two statements are quite consistent with each other. Almost all the time everything is well-controlled and my losses remain modest, but once in a while things will go very badly wrong and I will lose a lot.

David Einhorn, a well-known hedge fund manager, made a speech in 2008 (prophetically warning about Lehman Brothers' potential problems) in which he said that VaR is 'relatively useless as a risk-management tool and potentially catastrophic when its use creates a false sense of security among senior managers and watchdogs. This is like an air bag that works all the time, except when you have a car accident.'

In an influential book called *The Black Swan*, Nassim Nicholas Taleb has argued that we habitually take insufficient account of very rare but very important events – these are, in his terminology, *black swans*. They are sufficiently rare that we have not observed them before and so it makes little sense to talk about predicting their probability. At the same time they have very large effects. They are the unknowns that turn out to be more important than the things we do know about. Who could have predicted the changes that came about after the terrorist attack on the twin towers in 2001? Who could have anticipated the rise of social media on the internet? And these large-scale phenomena are mirrored at the level of the firm by much that comes 'out of left field'.

When we use VaR as a risk measure we deliberately exclude these events and their consequences. Even using a 99% one-day VaR (which sounds quite conservative) we deliberately exclude any events that happen less often than once every six months. For Taleb something that happens twice a year should be regarded as an 'everyday' occurrence. He argues that across many fields the exclusion of the 1% tail involves excluding events and data that turn out to have a very significant effect on the overall picture. For example, if we were to look at iTunes downloads and exclude the 1% most downloaded tunes, our estimate of how much money iTunes makes would probably be very inaccurate.

Another problem with VaR is that it may encourage inappropriate behavior by managers. In a *NY Times* article, Joe Nocera describes how VaR can be gamed.

'To motivate managers, the banks began to compensate them not just for making big profits but also for making profits with low risks. The result was an incentive to take on what might be called *asymmetric risk positions* where there are usually small profits and only infrequent losses, but losses when they do occur can be enormous. These positions made a manager's VaR look good because VaR ignored the slim likelihood of giant losses, which could only come about in the event of a true catastrophe. A good example was a credit-default swap, which is essentially insurance that a company won't default. The gains made from selling credit-default swaps are small and steady – and the chance of ever having to pay off that insurance was assumed to be minuscule. It was outside the 99% probability, so it didn't show up in the VaR number.'

In fact, the incentives to take actions which produce a skewed, or asymmetric, risk position are quite widespread. This will often happen when there is a reward based on relative ranking. Suppose, for example, that we are a fund manager. We may have choices available to us which will make our returns look like the average return for the type of stocks we are investing in. This can be achieved simply by spreading our portfolio widely, and corresponds to a low-risk option if we are being compared with this average performance (or the performance of other fund managers). On the other hand, we could concentrate our portfolio on a few stocks. This would be riskier but may pay off handsomely if these stocks are good performers. Since fund managers are paid partly on the basis of funds under management, and flows into a fund are often determined by its relative ranking, there is a big incentive to do better than other funds. This could lead to behavior which gambles (by stock picking) if things are going badly (perhaps the fund manager can catch up) and plays safe if things are going well ('quit when we are ahead'). Now gambling spreads out the returns, with the probability of catching up being balanced by a chance of doing badly. The end result is that the distribution of returns is spread out on the negative side and compressed on the positive side: it ends up looking like Figure 3.6, with a long negative tail. This will mean a relatively small chance of very poor returns and quite a good chance of reasonably good returns.

The right approach here is to recognize what VaR measures and what it does not measure. It picks a single quantile and estimates where this is: it makes no attempt to say how far the tail stretches (how large the losses may be).

The most common approach to the shortcomings of VaR around extreme events is to use *stress testing*. As with several other risk management approaches, this is a technique that originated within the banking industry. The idea here is that the firm, in addition to a VaR analysis, should try to understand more about what is going on in the tails of the distributions. It does this by looking at a set

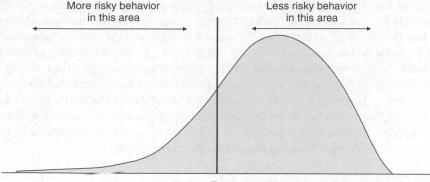

Figure 3.6   Asymmetric return distribution.

of different 'bad' scenarios and working out for each one what the implications might be. As an example, HSBC,[2] after discussing some of the limitations of using 99% daily VaR as a risk measure, states:

> In recognition of the limitations of VaR, HSBC augments it with stress testing to evaluate the potential impact on portfolio values of more extreme, although plausible, events or movements in a set of financial variables. The scenarios to be applied at portfolio and consolidated levels, are as follows:
>
> - sensitivity scenarios, which consider the impact of any single risk factor or set of factors that are unlikely to be captured within the VaR models, such as the break of a currency peg;
>
> - technical scenarios, which consider the largest move in each risk factor, without consideration of any underlying market correlation;
>
> - hypothetical scenarios, which consider potential macro economic events, for example, a global flu pandemic; and
>
> - historical scenarios, which incorporate historical observations of market movements during previous periods of stress which would not be captured within VaR.

The use of stress testing to explore the consequences of risks that occur in the tail is an approach that is complementary to VaR (which ignores the size of these risks) and may be useful for a wide range of firms.

Another approach to stress testing against a range of bad scenarios is to search systematically for the worst scenario amongst a given set; this is called *reverse*

---

[2] HSBC 2010 Annual report, 21 April 2011 http://www.hsbc.com/investor-relations/financial-results

stress testing. For example, we may impose limits on what certain prices or volatilities may be, and then use an optimization procedure (like Solver in a spreadsheet) to find the combination that gives the worst overall result. Particularly when there are financial derivatives involved it can be hard to tell in advance exactly which set of values will be the worst possible. Reverse stress testing identifies, from amongst a defined set of possibilities for various variables, the exact choice that gives the worst result. This is likely to be revealing, but may be only the starting point for further analysis. Once we know the most damaging scenario we can look again to see whether we believe it is possible; if not, it may suggest an adjustment of the constraints on the variables and the procedure can be run again.

3.6   Beyond value at risk

The value at risk measure is concerned with the tail of a distribution in a way that the variance is not. But even though VaR focuses on the tail, it is uninformative about what happens within the tail. The two loss distributions (density functions) drawn in Figure 3.7 have exactly the same 95% VaR value of 25 but the potential losses for the distribution drawn with a solid line are significantly higher. It is natural to talk of a distribution having fatter tails if the probability of getting values at the extremes is higher. The comparator here is the normal distribution (which is the dashed line in the figure) for which the probabilities go towards zero in the same way as $e^{-x^2} = 1/e^{x^2}$ which is a very fast decrease.

One way around this problem is to look at VaR values at different probabilities. The solid line distribution with the fatter tails has a 99% VaR of 41, while the 99% VaR for the dashed line, which is a normal distribution, is 36. So by moving farther out in the tail, the difference between the two distributions becomes more obvious from VaR alone.

An alternative approach is to use what is often called the *expected shortfall*, though other terminology is sometimes used (*tail value at risk* or *conditional value*

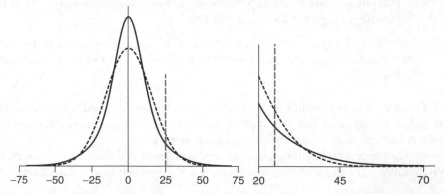

Figure 3.7   Solid line has a fatter tail (shown enlarged in right-hand panel). Both distributions have 95% VaR of 25.

at risk (CVaR)). The expected shortfall at a level α for a random variable X of losses is written $\text{ES}_\alpha(X)$ and is the expected loss conditional on the VaR_α level being exceeded. It is the average value over that part of the distribution which is greater than $\text{VaR}_\alpha(X)$, i.e. over loss values which occur with only a $1 - \alpha$ probability. The expected shortfall is very closely related to value at risk, but captures more about what may happen in the worst cases. In comparison with value at risk the expected shortfall is a more natural measure of risk. The 95% value at risk, is obtained by asking 'What is the minimum loss amongst the 5% of worst outcomes?', whereas the 95% expected shortfall value is obtained by asking 'What is the average loss amongst the 5% of worst outcomes?'

We can more formally write the expected shortfall as

$$\text{ES}_\alpha(X) = E(X \mid X > \text{VaR}_\alpha(X)). \tag{3.2}$$

We should stop and unpack what we mean by an expectation conditional on another event. For a discrete random variable X taking values $x_1, x_2, \ldots, x_N$ with respective probabilities $p_1, p_2, \ldots, p_N$ the expectation has the form $E(X) = \sum_{i=1}^N p_i x_i$. The expected shortfall is obtained by changing the probabilities so that instead of p_i we have

$$p_i' = \Pr(X = x_i \mid X > \text{VaR}_\alpha(X))$$
$$= \frac{\Pr(X = x_i \text{ and } X > \text{VaR}_\alpha(X))}{\Pr(X > \text{VaR}_\alpha(X))}.$$

Now suppose that the x_i are ordered so that they are increasing with the first m being less than the VaR_α level and then the other values all being greater than the VaR_α level (i.e. $x_i \leq \text{VaR}_\alpha(X)$ for $i = 1, 2, \ldots, m$ and $x_i > \text{VaR}_\alpha(X)$ for $i = m+1, m+2, \ldots, N$.) Then we see that $p_i' = 0$ for $i = 1, 2, \ldots, m$ and

$$p_i' = \frac{p_i}{p_{m+1} + p_{m+2} + \cdots + p_N} \text{ for } i = m+1, m+2, \ldots, N.$$

Hence, we have the following formula for the expected shortfall:

$$\text{ES}_\alpha(X) = \sum_{i=1}^N p_i' x_i = \frac{\sum_{i=m+1}^N p_i x_i}{\sum_{i=m+1}^N p_i}.$$

We can rewrite the formula for the expected shortfall when X has a continuous density function $f(x)$. In this case we can write the expectation in terms of the integral of f:

$$\text{ES}_\alpha(X) = \frac{\int_{\text{VaR}_\alpha(X)}^\infty x f(x) dx}{\Pr(X > \text{VaR}_\alpha(X))}$$
$$= \frac{1}{1-\alpha} \int_{\text{VaR}_\alpha(X)}^\infty x f(x) dx.$$

Worked Example 3.4   Calculating VaR and ES with insurance against extreme weather

Suppose that we buy insurance against extreme weather events that occur randomly, with an average of one event every 10 years. We pay $10 000 a year as a premium and receive a payout total of $95 000 in the event of the claim being made. Once a claim is made, the insurance contract ceases. Premium payments are made monthly in advance, and in the event of a claim are refunded for any period after the claim event. Ignoring discounting and any inflationary increases in premiums or a payout, what are the $\text{VaR}_{0.95}$ and $\text{ES}_{0.95}$ values for our losses on this contract?

Solution

With random occurrences the time to the first weather event is a random variable with an exponential distribution. if we take years as units of time then we have an exponential with parameter 0.1. The loss (in $1000s) is given by $L = 10X - 95$, where X is the time till we make a claim and has a density function $f(x) = 0.1e^{-0.1x}$. To calculate $\text{VaR}_{0.95}(L)$ note that

$$\text{VaR}_{0.95}(L) = 10\text{VaR}_{0.95}(X) - 95.$$

Now the probability in the tail of the exponential is

$$\int_u^\infty f(x)dx = \int_u^\infty 0.1e^{-0.1x}dx$$
$$= \left[-e^{-0.1x}\right]_u^\infty = e^{-0.1u}.$$

To find $\text{VaR}_{0.95}(X)$ we want to find a u value that makes this probability 0.05, so we should set u so that $e^{-0.1u} = 0.05$, i.e. $u = -10 \log_e(0.05) = 29.957$. Thus, there is a one in 20 chance that the weather event doesn't happen for about 30 years and the downside risk from the point of view of the person buying insurance is $\text{VaR}_{0.95}(L) = 10 \times 29.957 - 95 = 204.57$ or $204 570.

But looking at expected shortfall gives an even larger figure: we have $\text{ES}_{0.95}(L) = 10\text{ES}_{0.95}(X) - 95$ and

$$\text{ES}_{0.95}(X) = \frac{1}{1 - \alpha} \int_{\text{VaR}_\alpha(X)}^\infty xf(x)dx$$
$$= \frac{1}{0.05} \int_{29.957}^\infty 0.1xe^{-0.1x}dx$$
$$= \frac{1}{0.05} \left[-10e^{-0.1x} - xe^{-0.1x}\right]_{29.957}^\infty$$
$$= \frac{1}{0.05} \left(10e^{-2.996} + 29.957e^{-2.996}\right)$$
$$= 39.946.$$

(You should check that the expression we quoted as the integral here really does differentiate back to $0.1xe^{-0.1x}$). Thus, we have an expected shortfall of

$$\text{ES}_{0.95}(L) = 10 \times 39.946 - 75 = 304.46$$

or \$304 460, which is about \$100 000 more than the value at risk. This is a greater difference than will occur for many distributions and is due to the particular shape of the exponential distribution with a long tail to the right. ☐

Expected shortfall has many advantages over VaR; not only does it give greater visibility into the tails of the loss distribution but it also has better theoretical properties. As we show in the next section, it satisfies the subadditivity property and is coherent. For these reasons it has become more and more popular as a way of keeping track of risk. It has now been proposed as a preferred alternative to VaR by the Basel Committee on Banking Supervision.

Since expected shortfall is much better at giving an indication of tail risk, it may be appropriate to lower the α-level. In other words, the use of 99% VaR rather than a lower level is, in part, because a high alpha level gives a better indication of what is happening relatively far out in the tail. But using expected shortfall with a lower level like 98% or 97.5% can work equally well (and automatically allows for what is happening in the tail). The big advantage of a lower α level is that more data points get considered when testing the expected shortfall estimates.

We need to say more about how a company can back-test for expected shortfall (this has sometimes been raised as a concern in moving from VaR to expected shortfall). The important thing to realize here is that expected shortfall is not generated on its own: it needs an estimate of the VaR value as well. Thus, if we work on a daily basis at a 99% level, we will end up with two daily numbers: the 99% VaR and the 99% expected shortfall. The back-testing process then looks at the occasions when the VaR level is breached and considers the losses on those occasions. These are the losses whose mean is given by the expected shortfall estimates. If the expected shortfall was the same from one day to the next, then the back-testing would be simple: it would simply be a matter of checking the average losses on the days when the VaR level was breached and comparing these with the expected shortfall estimate.

When the expected shortfall estimates vary from day to day, we can still compare these two. If the loss on a certain day with a VaR breach is X then this has mean value of ES_α, and hence $E(X - \text{ES}_\alpha) = 0$. We write X_i for the loss on day i (restricting this to days when there is a VaR breach) and $\text{ES}_\alpha^{(i)}$ for the value of ES_α on day i. Since the difference, $X_i - \text{ES}_\alpha^{(i)}$, has a mean value of zero, we can consider the observed value of the average

$$\frac{1}{N} \sum_{i=1}^{N} (X_i - \text{ES}_\alpha^{(i)})$$

over N days when VaR breaches occur. This will also have mean zero if the model is correct.

There is an argument for scaling these differences to allow for the scaling up of all values that occur on days with a high expected shortfall. This would mean considering the ratio $(X - \mathrm{ES}_\alpha)/\mathrm{ES}_\alpha$. Obviously, since $X - \mathrm{ES}_\alpha$ has a zero mean, the ratio will also have a mean of zero. So we can evaluate this ratio for different VaR breach days and again check the average, which should be close to zero.

Notice that losses on the VaR breach days will have a highly asymmetric form; we expect many data points that are just above the VaR level and progressively fewer at higher and higher levels. So $(X - \mathrm{ES}_\alpha)$ on VaR breach days is much more likely to be negative than positive, with a small probability of a very large positive value making the mean zero. More often than not, a back-test procedure will have limited data on VaR breach days and it will be a matter of chance as to whether an appropriate number of these large positive values have been observed to compensate for the many negative ones. This makes it hard to be confident of the expected shortfall estimates and is behind the concern over the effectiveness of back-testing for expected shortfall.

3.6.1   *More details on expected shortfall

Another way to define expected shortfall is to average the values of VaR_u for all $u \geq \alpha$. To see why this works, we start from the expression

$$\mathrm{ES}_\alpha(X) = \frac{1}{1 - \alpha} \int_{\mathrm{VaR}_\alpha(X)}^{\infty} xf(x)dx. \tag{3.3}$$

Now we are going to make a change of variable from x to a new variable u, which is defined as $u = F(x)$. To do this we use the normal procedure of taking derivatives to see that $du = f(x)dx$. Also we note that

$$x = F^{-1}(u) = \mathrm{VaR}_u(X).$$

Moreover, at the lower limit of the integral $x = \mathrm{VaR}_\alpha(X)$, so $u = \alpha$, and at the upper limit $u = F(\infty) = 1$. Thus

$$\int_{\mathrm{VaR}_\alpha(X)}^{\infty} xf(x)dx = \int_\alpha^1 \mathrm{VaR}_u(X)du$$

and

$$\mathrm{ES}_\alpha(X) = \frac{1}{1 - \alpha} \int_\alpha^1 \mathrm{VaR}_u(X)du. \tag{3.4}$$

Thus, Equations (3.3) and (3.4) provide two alternative expressions for expected shortfall at level α, when the random variable of losses, X, is continuous.

There is an important warning here for discrete distributions, where there is no density function f, then the original definition Equation (3.2) and Equation (3.4) are no longer equivalent and we should use the second definition of Equation (3.4) to avoid problems (or use a more complicated definition instead of Equation (3.2)).

As we discussed earlier, one of the problems with VaR is that it is not a coherent risk measure. Specifically, we do not have the subadditivity property that $\text{VaR}_\alpha(X) + \text{VaR}_\alpha(Y) \geq \text{VaR}_\alpha(X+Y)$. It turns out that expected shortfall *is* a coherent risk measure. There are four properties to check: monotonicity; positive homogeneity; translation invariance and subadditivity. The fact that value at risk has the first three properties means that we can use Equation (3.4) to show that expected shortfall also has these properties, so

$$\text{If } X \leq Y \text{ then } \text{ES}_\alpha(X) \leq \text{ES}_\alpha(Y);$$

$$\text{ES}_\alpha(bX) = b\,\text{ES}_\alpha(X);$$

$$\text{ES}_\alpha(X+c) = \text{ES}_\alpha(X) + c.$$

Now we will explain why expected shortfall is also subadditive. The key observation is that expected shortfall for X is an average of the *highest* values of X that can occur, where the events generating these values have a given probability $1 - \alpha$. If we choose a different set of events which has the same probability $1 - \alpha$ and look at the average of the values of X that occur under these events, then the value must be lower than the expected shortfall. We can represent this in a diagram, as shown in Figure 3.8.

Here we take both X and Y as having finite ranges so that the diagram is easier to draw. We define a new random variable $Z = X + Y$. The regions B and C involve the highest possible values of the loss variable X. Suppose that they in total have a probability of $1 - \alpha$, so that the point shown as x_0 will be at $\text{VaR}_\alpha(X)$. Now suppose that the combined A and B regions are where the highest values of Z occur and A and C have the same probability. Then A and B will also have a total probability of $1 - \alpha$ and the value $Z = z_0$, which marks the lower boundary of this region, will be at $\text{VaR}_\alpha(Z)$. Now notice that taking the expected value of X over the region B and C and comparing it with the expected

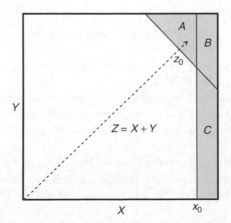

Figure 3.8   Diagram to show subadditivity of expected shortfall.

value of X over the region A and B involves changing a set of events where $X > x_0$ to another set of events (with the same probability) where $X < x_0$. Hence

$$E(X \mid X > \text{VaR}_\alpha(X)) \geq E(X \mid Z > \text{VaR}_\alpha(Z)).$$

Exactly the same argument can be used to show that

$$E(Y \mid Y > \text{VaR}_\alpha(Y)) \geq E(Y \mid Z > \text{VaR}_\alpha(Z)).$$

Then we add these two inequalities together to establish subadditivity:

$$\text{ES}_\alpha(X) + \text{ES}_\alpha(Y) = E(X \mid X > \text{VaR}_\alpha(X)) + E(Y \mid Y > \text{VaR}_\alpha(Y))$$

$$\geq E(X \mid Z > \text{VaR}_\alpha(Z)) + E(Y \mid Z > \text{VaR}_\alpha(Z))$$

$$= E(Z \mid Z > \text{VaR}_\alpha(Z)) = \text{ES}_\alpha(Z).$$

Notes

An excellent introduction for this area is the newspaper article by Joe Nocera and I have drawn on this at various points in this chapter. The book by Crouhy, Galai and Mark is very helpful in understanding what VaR measures really mean in practice and the book by Culp (2001) also treats the more practical aspect of VaR calculation. Nicholas Taleb's famous book called *The Black Swan* is well worth reading for its trenchant views on what is wrong with many of the quantitative approaches to risk measurement. For more on the way that competition for ranking can lead to asymmetric distributions with long negative tails, see Anderson (2012).

The Basel recommendations on a move to expected shortfall are in the Consultative Document issued in May 2012. A more formal proof of the result on the coherence of expected shortfall can be found in Acerbi and Tasche (2002).

References

Acerbi, C. and Tasche, D. (2002) On the coherence of expected shortfall. *Journal of Banking & Finance*, **26**, 1487–1503.

Anderson, E. (2012) Ranking games and gambling: When to quit when you're ahead. *Operations Research*, **60**, 1229–1244.

Basel Committee on Banking Supervision (2012) Fundamental review of the trading book. Consultative Document, May 2012. Available at: http://www.bis.org/publ/bcbs219.pdf

Crouhy, M., Galai, D. and Mark, R. (2006) *The Essentials of Risk Management*. McGraw Hill,

Culp, C. (2001) *The Risk Management Process*. John Wiley & Sons.

Einhorn, D. (2008) Private profits and socialized risk. Speech at Grant's Spring Investment Conference, 8 April 2008 (and Global Association of Risk Professionals Risk Review, 2008).

Nocera, J. (2009) Risk Mismanagement. *New York Times*, 2 January.

Taleb, N. (2010) *The Black Swan*, 2nd edition. Random House.

Exercises

3.1 VaR for normal distributions

(a) If the profits made each day by a trading desk are, on average, $100 000 and have a normal distribution with standard deviation $60 000, calculate a 99% and 95% absolute VaR.

(b) A second trading desk has exactly the same properties as the first (normal distribution with average profit of $100 000 and standard deviation of $60 000). If the second desk makes returns that are completely independent of the first, what are the 99% and 95% absolute VaR values for the combination of the two trading desks?

(c) If the results of the second trading desk are not independent of the first, what is the highest value (i.e. greatest losses) for 99% absolute VaR that might be achieved for the combination of the two trading desks?

3.2 VaR for a triangle distribution

Consider a distribution of losses over the range $-$100 000$ to $100 000 where the density follows a triangle distribution $f(x) = -x/X^2$ for $-X \le x \le 0$ and $f(x) = (X - x)/X^2$ for $0 < x \le X$ where $X = $100 000$. Calculate 99% and 95% absolute VaR figures. (In this case you can calculate the CDF by drawing the graph of f and directly calculating the area under the graph.)

3.3 A non-monotonic measure of risk

Example 3.2 gives a way of calculating a risk measure from the mean and standard deviation. Give an example where increasing the loss on some outcomes would lead to a reduction of the value of $\mu + 3\sigma$. (Hint: Consider a distribution for losses that is z with probability 0.1, 0 with probability 0.8, and 1 with probability 0.1. Start with $z = -1$ and then try increasing z.)

3.4 Diversification reduces VaR

In Example 3.3, use a 98% absolute VaR to show that there is some diversification benefit in investing $10 000 in each of A and B rather than putting $20 000 in two bonds from A.

3.5 From one day to ten days

The Basel II framework asks for a 10-day VaR and then states that 'Banks may use value-at-risk numbers calculated according to shorter holding periods scaled up to 10 days by the square root of time.' By this is meant that if the 1-day VaR is x then the 10-day VaR can be estimated as $x\sqrt{10} = 3.1623x$.

(a) Explain why this formula could only be appropriate for (relative) VaR and not for absolute VaR.

(b) Show that if daily returns are independent and normally distributed then the proposed formula will give the correct result.

3.6 VaR estimates are a process

You are a manager with a VaR system in place to calculate 99% (relative) VaR values on a daily basis. Over the last 500 trading days (two years) there have been five occasions when the VaR values have been breached. A subordinate comes to you with some serious concerns in relation to the current VaR calculations, arguing that they wrongly represent correlations in behavior occurring at times when the markets make large movements. He has carried out a set of alternative calculations of daily (relative) VaR values over the last two years, which also has five occasions when the VaR values have been breached.

(a) Explain why the alternative daily VaR values may differ markedly from the values from the current system, but have the same number of VaR breaches.

(b) Suppose two systems have the same performance on back-test, can we deduce that they are equally good? And what would it mean for one to be better than the other?

3.7 Expected shortfall for a normal distribution

Show that the expected shortfall at the 99% level if losses follow a normal distribution with mean 0 and standard deviation 1 is given by $ES_{0.99} = 2.667$ using the fact that

$$\int_v^\infty x \exp(-x^2/2)dx = \exp(-v^2/2).$$

3.8 Expected shortfall is subadditive

(a) Use the fact that $ES_{0.99} = 2.667$ for a normal distribution with mean 0 and standard deviation 1 (given in Exercise 3.7) to find the 99% expected shortfall if the losses in project A have a normal distribution with mean $-\$5000$ and standard deviation \$3000 (thus, on average, we make a profit).

(b) Now suppose that the losses in project B have a normal distribution with mean $-\$3000$ and standard deviation \$1500. Assuming that the projects are independent, calculate the 99% expected shortfall for the sum of the two projects and hence check subadditivity in this case.

3.9 Bound on expected shortfall

Suppose that the density function f of the distribution for the loss random variable X is decreasing above the 0.95 percentile. Show, by considering the shape of the CDF function F, that $\text{VaR}_u(X)$ is a convex function of u for $u > 0.95$ (i.e. has a slope increasing with u). Use a sketch to convince yourself that the average value of a convex function over an interval is greater than the value half way along the interval (this can also be proved formally). Finally use Equation (3.4) to show that

$$\text{ES}_{0.95}(X) > \text{VaR}_{0.975}(X).$$

4

Understanding the tails

Extrapolating beyond the data

Maria works in a large teaching hospital and has responsibility for ordering supplies. Today she faces what seems like an impossible situation: as she explained to her friend Anna 'I need three years of data but I only have two years'. Anna asks her to explain, and she launches into the problems that she has been mulling over for the last few days.

'For the last two years we have had a diagnostic test for a certain type of infant bronchial condition and have been recording patients with this condition, and now at last there is a drug that is effective. But it is in short supply and expensive. Worse still, it contains some unstable components so can only be used in the first six weeks after manufacture. An urgent order still takes a week to arrive from the manufacturing facility in Switzerland, so we need to keep a minimum of a week's worth of stock and I now have to determine how much that is. In this kind of situation our usual rules suggest that we should run out no more often than once every three years on average. I can see how much we would have needed in each of the last 104 weeks, but that's not a long enough period.'

Anna has just finished a risk management course as part of her MBA and wonders if there is some way to use some of what she has learnt to help her friend. 'So you have 104 weekly data points and your problem is to estimate what the highest value would be if the series were extended to three years, i.e. $3 \times 52 = 156$ data points.'

'That's it exactly,' said Maria. 'Come to think of it, even three years of data would not really be sufficient – just because something didn't happen over the last three years does not mean it will not happen over the next three years.'

'Have you thought of modeling the weekly usage with a normal distribution, or maybe some other sort of distribution?'

Business Risk Management: Models and Analysis, First Edition. Edward J. Anderson.
© 2014 John Wiley & Sons, Ltd. Published 2014 by John Wiley & Sons, Ltd.
Companion website: www.wiley.com/go/business_risk_management

'That was the same idea that I had,' Maria replied 'but I have looked at the numbers and there are more weeks with high demand than seems possible if it really was a normal distribution. What seems to happen is that the incidence of this condition varies according to a whole lot of factors that I don't know about; things like the weather, and the number of chest infections in the population. I don't think that I can use a model to make predictions here, there is just too much uncertainty, so we are left with the inadequate data.'

'So, in essence you need to use the data you have, but extrapolate to higher values that have not yet occurred and at the same time you do not want to make any assumptions about the kind of distribution,' said Anna.

'That's right: without enough data and without a specific model of the distribution it seems impossible to estimate the number I need.'

'I am not so sure,' said Anna, 'even without making an assumption that the weekly numbers match a given distribution, they surely cannot be too badly behaved. Any kind of regularity should give you a handle on the extrapolation problem. It may be hard to extract information about this "upper tail" of the distribution from the data you have, but perhaps not impossible. I heard my professor talk about extreme value theory and perhaps that could help in some way.'

4.1   Heavy-tailed distributions

4.1.1   Defining the tail index

The challenge of using quantitative techniques to measure risk is that it forces us to pay attention to the tails of the distribution. This is exactly the area of greatest difficulty in estimation: we rarely see enough tail events to make firm deductions about the underlying distributions. In this chapter we will look in more detail at some tools for handling the tails of distributions.

When dealing with risk we are most often interested in random variables that have the possibility (at least theoretically) of infinitely large values. Like the normal distribution, the random variables do not have a finite range. Obviously in practice there will most often be a finite limit on the distribution: for example, if dealing with a loss distribution there will be a maximum loss determined by our company's ability to avoid bankruptcy. But it is usually more revealing to model the losses as though there were no maximum limit. In this context it makes sense to look at the shape of the tail of the distribution as it goes to infinity.

The normal distribution gives a natural point of comparison for other distributions. A distribution is called 'heavy-tailed' if it has more weight in the tails than a normal distribution. But we have to stop and think about what we might mean by such a statement. Because the standard deviation of a normal can be set to whatever we like, it is always possible to find a normal distribution which has a zero mean but a large probability of being greater than some given value. We just need to make the standard deviation large enough. So, to make sense of a heavy-tailed distribution we need to think about the behavior of the distribution across a whole range of values and not just at a single point.

One way to understand the behavior of the tail of a distribution is to ask how quickly the CDF of the distribution approaches 1. A good way to do this is to consider the product $(1 - F(x))x^k$ for some power k. The first term $1 - F(x)$ will get closer and closer to zero as x increases, while the second term x^k will get larger and larger, so we can ask which of these two will win? Does the product go to zero or go to infinity? We can guess that as we make the power k larger, there will be some point k_0 at which the x^k term starts to dominate. For $k < k_0$ the product will approach zero and for $k > k_0$ the product will go to infinity. If the tail is heavy then there are high chances of seeing a large value and that means that $F(x)$ approaches 1 only slowly. So we will not be able to multiply by such a large value of x^k and still get the product going to zero. Hence, a heavy tail is associated with a low value of k_0.

When we can define a value of k_0 in this way, with $(1 - F(x))x^k$ going to either zero or infinity according to whether k is either below or above k_0, then we say that the distribution has a *tail index* of k_0. So, for example, a tail index of 2 is roughly equivalent to the statement that $1 - F(x)$ goes to zero in the same way as $1/x^2 = x^{-2}$. But the definition we have given, in terms of a dividing point between two regimes, is more precise.

Example 4.1   A heavy-tailed distribution

A random variable has distribution with density function

$$f(x) = \frac{1}{(2 + |x|)^2}$$

This is symmetric around 0 because the f values do not depend on the sign of x. The density function is graphed in Figure 4.1. We can check that this is a distribution by showing that its integral approaches 1 as $x \to \infty$. To do this we note that for $z > 0$

$$\int_0^z f(x)dx = \int_0^z \frac{1}{(2 + x)^2}dx = \left[\frac{-1}{2 + x}\right]_0^z = \frac{1}{2} - \frac{1}{2 + z},$$

which clearly approaches $1/2$ as $z \to \infty$, and hence the area under the whole curve will be 1 as we require. To determine the CDF we can use symmetry to see that

$$F(0) = \int_{-\infty}^0 f(x)dx = \frac{1}{2}$$

and so for $z > 0$,

$$F(z) = \int_{-\infty}^0 f(x)dx + \int_0^z f(x)dx = 1 - \frac{1}{2 + z}.$$

Hence, $1 - F(z) = (2 + z)^{-1}$ and this distribution has a tail index of 1. This is an extreme case of a distribution with heavy tails.   □

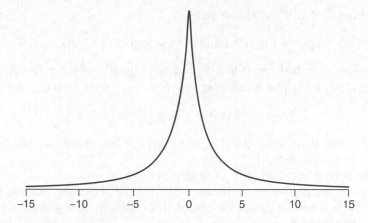

Figure 4.1   An example of a density function with tail index of 1.

It is important to recognize that not all distributions have a well-defined variance and standard deviation. The variance of a random variable X is defined as $E\left[(X - E(X))^2\right]$ and if the distribution has a heavy tail then this expectation may be infinite. In this case it also makes no sense to talk about the standard deviation of the random variable. The smaller the tail index, the fatter the tails of the distribution will be and the more likely it is that the variance will not exist. The key result here is that if the tail index is less than k then $E(X^k)$ will be infinite (we show why this is true in the next section).

There is a whole set of distributions for which the tail index is infinite and all moments exist. These are distributions, like the normal, where the probability in the tail, $1 - F(x)$, approaches 0 faster than any power of x. At the other extreme is a distribution like that of Example 4.1 for which there is a problem even defining the mean. Obviously the distribution is symmetric about zero, and so the mean should be zero, but if we look only at the tail of the distribution then we are interested in integrals of the form $\int_u^\infty xf(x)dx$ for some threshold value u, and it turns out that this integral has an infinite value.

4.1.2   Estimating the tail index

Given a set of data points we may want to estimate how quickly the tail of the distribution goes to zero to find out whether or not we are dealing with a heavy-tailed distribution. For example, this is an important question to answer if we need to estimate an expected shortfall measure from the data. Later in this chapter we will give a more detailed discussion of the way that the estimation of a risk measure can be carried out using extreme value theory, but here we want to give a more elementary approach to estimating tail indexes.

Suppose that a distribution has a tail index of α, then, for large values of x, we expect the CDF to be given approximately by

$$F(x) = 1 - kx^{-\alpha},$$

for some constant k. If this is true then

$$\log(1 - F(x)) = \log(kx^{-\alpha}) = \log(k) - \alpha \log(x).$$

Since $1 - F(x)$ is between 0 and 1, and the log of a number less than 1 is negative, it can be helpful to multiply this equation by -1 to get

$$-\log(1 - F(x)) = -\log(k) + \alpha \log(x).$$

This shows that if we plot $-\log(1 - F(x))$ against $\log(x)$ then we should get a straight line with a slope of α.

Given a set of data points it is simple to form an estimate of $1 - F(x)$ by looking at the proportion of the points in the sample above the value x. Hence, the procedure is to plot minus the log of this estimate against the log of x to estimate the tail index.

Example 4.2   Tail behavior for Forex rates

To illustrate this, Figure 4.2 shows a plot of this form for the exchange rate between the British pound and US dollar for a 500-day period starting in May 2010. The data are for the daily percentage change in the closing price. We are interested in losses from one day to the next if pounds are purchased (pound moving down relative to the dollar), so this gives 499 different data points. During this period the greatest loss was 1.74% on 13 May 2010. This is plotted at an 'x' value of $\log(1.74) = 0.55$ (using natural logs). This point has 498 points below it in terms of losses and so we estimate the $F(x)$ value to be 498/499 and so the y value in the plot is given by

$$-\log(1 - F(x)) = -\log(1/499) = 6.21.$$

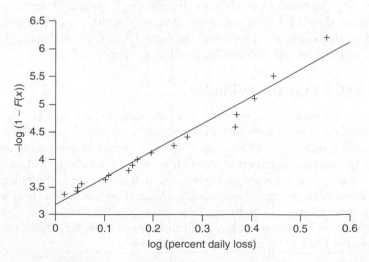

Figure 4.2   Tail data for percentage daily loss in exchange rate GBP/USD.

The figure shows the behavior for the 17 days on which the losses were more than 1%. The straight line fitted to these points has a slope of 4.9. So, from these data alone, we would estimate the tail index as being around 5, and there is some evidence of heavy-tailed behavior.   □

When the tails are not heavy then the tail index will be infinite. Of course, in practice this translates simply into a high slope, and drawing this type of log-log plot for data drawn from a normal distribution we would expect a higher slope than we see in Example 4.2. However, since we only ever have a finite sample to look at, from a practical perspective it is impossible to know what the exact tail behavior is. The larger the sample, the better our estimate will be, but the tail behavior is, by definition, something that happens at the extreme, and we will never have enough data to be sure about this.

Example 4.3   Looking at tail behavior for a normal distribution

We can easily simulate data drawn from a normal distribution. Figure 4.3 shows the result of two different samples from a normal distribution with mean 0 and standard deviation 1. On the left is a plot based on a sample of 500 points. The figure shows $-\log(1 - F(x))$ plotted against $\log(x)$ for the top 20 points; in this sample the highest value is 2.859. The fitted straight line to these points has a slope of 5.38. Note that a different sample could well produce a different estimate.

On the right of the figure, a new sample of 4000 points has been made. Again, the plot shows the top 20 points. This time the highest point is 3.48 and corresponds to an estimated value for $F(x)$ of 3999/4000, leading to a y value in the plot of

$$-\log(1 - F(x)) = -\log(1/4000) = 8.29.$$

The straight line fitted to the points on the right-hand plot has slope 9.57, substantially higher than the number obtained from the smaller sample. We can see

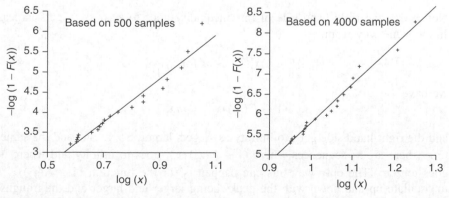

Figure 4.3   Two different estimates for the tail index using samples from a normal distribution.

that, by taking a much larger sample, the top 20 points correspond to part of the distribution that is farther out in the tail. This gives the method more of a chance to pick up the thin tail behavior and the tail index estimates become larger. But this example also shows the difficulty of being sure about the tail index on the basis of a sample of a few hundred points. It demonstrates, in particular, that the evidence for heavy tails in the foreign exchange data of Example 4.2 is quite weak. □

4.1.3   *More details on the tail index

We have assumed in our definition of a tail index that there is a single value of k_0 that lies between the region where $(1 - F(x))x^k$ goes to infinity and the region where it goes to zero. But what if there is a range of values where $(1 - F(x))x^k$ neither approaches zero nor approaches infinity? In this case there will not be a well-defined tail index. In fact, this is a very unlikely occurrence; sufficiently unlikely that we can really ignore the possibility. Let's try to see why this is so.

We start by being more precise about what it means for a function $G(x)$ not to go to infinity and not to go to zero. Not going to infinity means that there is some value M such that $G(x) < M$ for an infinite sequence of x values: $x_1, x_2, \ldots$. Thus, no matter how large we take x we can always find a larger point where $G(x) < M$. This is the logical negative of a statement that $G(x)$ tends to infinity, which is equivalent to saying that for every M, $G(x)$ will end up above M and stay there. In the other direction, $G(x)$ not going to zero means there is some value of m such that $G(x) > m$ for an infinite sequence of x values: $x_1, x_2, \ldots$. Now suppose that this happens for $G(x) = (1 - F(x))x^k$ for all k in a range $[k_0 - \delta, k_0 + \delta]$. Now if

$$(1 - F(x_i))x_i^{k_0+\delta} < M \text{ for } x_i \to \infty,$$

then

$$(1 - F(x_i))x_i^{k_0} < Mx_i^{-\delta}.$$

Notice that the right-hand side of this inequality goes to zero as x_i gets larger. In the same way from

$$(1 - F(x_i))x_i^{k_0-\delta} > m \text{ for } x_i \to \infty,$$

we have

$$(1 - F(x_i))x_i^{k_0} > mx_i^{\delta}$$

and the right-hand side goes to infinity as x_i gets larger. So, at one and the same time, there are x_i sequences where $(1 - F(x_i))x_i^{k_0}$ goes to infinity and where it goes to zero. The only way this can happen is for the function $(1 - F(x_i))x_i^{k_0}$ to oscillate up and down with the peaks being larger and larger and the troughs getting closer and closer to zero. We could construct such a function if we tried, but it is definitely not something that would occur in practice.

When we talk of a tail index of k then we do not have to specify whether the critical expression $L(x) = (1 - F(x))x^k$ itself goes to zero or infinity. In fact, either of these options is possible, but the function $L(x)$ must not go to infinity or to zero too quickly. There is a specific condition required for this: $L(x)$ must be *slowly varying*, meaning that

$$\lim_{x \to \infty} \frac{L(tx)}{L(x)} = 1$$

for any value of $t > 0$. So, for example, taking $t = 2$, doubling the value of x cannot (in the limit of large x) look like applying any particular multiplier other than 1. Notice that if $L(x) = kx^\beta$ then $\lim_{x \to \infty} L(tx)/L(x) = t^\beta$ and so this can only equal 1 (and L be slowly varying) if $\beta = 0$.

There are some complications here that we don't want to get sucked into. The condition that $(1 - F(x))x^\alpha$ is slowly varying is actually a stronger condition than saying that the exponent α marks the dividing point between functions approaching zero and functions approaching infinity. (We can see this by observing that a periodic function like $2 + \sin x$ is not slowly varying but will be dominated by x^ε for even tiny values of ε.) The 'slowly varying' condition is the one that is required to prove the extreme value results that we give later on, even though the way we have defined a tail index is a bit simpler.

As we have mentioned already, there is a very close connection between the existence of moments for a distribution and the tail indices involved. We start by looking at the condition that the second moment $E(X^2)$ exists. When X has a density function f and a CDF F, we can write

$$E(X^2) = \lim_{R \to \infty} \int_{-R}^{R} f(x)x^2 dx.$$

Here we have written the upper and lower limits of the integral as $-R$ and R (rather than infinity) because the integral from $-\infty$ to ∞ is only defined when the limit as $R \to \infty$ exists, and the question of existence or not is precisely what we are interested in. Now, choosing an arbitrary point u and integrating in the range above u (noting that in this range $x \geq u$) we get the inequality

$$\int_{u}^{R} f(x)x^2 dx \geq \int_{u}^{R} f(x)u^2 dx = (F(R) - F(u))u^2.$$

Letting R go to infinity shows that $\lim_{R \to \infty} \int_{u}^{R} f(x)x^2 dx \geq (1 - F(u))u^2$.

Now consider a distribution with a tail index α strictly less than 2. Then $(1 - F(u))u^2$ approaches infinity. Hence, for any large number M we can choose a u with $(1 - F(u))u^2 > M$ and so, for this u, $\lim_{R \to \infty} \int_{u}^{R} f(x)x^2 dx > M$. The integral over the whole range $-R$ to R has to be larger than this, i.e.

$$\lim_{R \to \infty} \int_{-R}^{R} f(x)x^2 dx > M.$$

But, since M can be chosen to be any number we like, this limit as $R \to \infty$ cannot exist, i.e. $E(X^2) = \infty$.

More generally, we can use this argument to show that if a distribution has a tail index of α then the moments $E(X^k)$ will not exist for any $k > \alpha$.

4.2   Limiting distributions for the maximum

In this section we look at problems where the maximum of a number of values is of interest. Suppose that we are concerned with predicting the maximum of N different random variables $X_1, X_2, \ldots, X_N$ all with the same distribution. For example, we might be interested in the maximum wealth from 100 individuals that we sample at random, so that X_i is the wealth of the ith individual. We know that if all the X_i are independent, there are N of them, and each has the same CDF, $F(x)$, then the distribution of their maximum, which we call $F_{\max}$, has CDF $(F(x))^N$. When N is large, it is only the tail of the original distribution that matters, since any value of x where $F(x)$ is substantially below 1 will automatically have $(F(x))^N$ very small. This makes sense: when we take the maximum over a large number of draws from a random variable we are pretty much bound to see a value in the right-hand tail of the distribution, and so the behavior of the distribution in the tail is the only thing that counts.

We are going to show that, in a way that we will make clear in a moment, the distribution of the maximum of N values, for N large, will only depend on the tail index of the original distribution. We consider a distribution with tail index α, and in particular we suppose that F is a distribution where, in the tail of the distribution, $1 - F(x) = kx^{-\alpha}$, i.e. $F(x) = 1 - kx^{-\alpha}$. We need to define a specific multiplier b_N which will act as a scaling parameter. Set

$$b_N = (kN)^{1/\alpha}.$$

Then
$$F_{\max}(b_N x) = (F(b_N x))^N = (1 - kb_N^{-\alpha} x^{-\alpha})^N.$$

But
$$b_N^{-\alpha} = ((kN)^{1/\alpha})^{-\alpha} = (kN)^{-1},$$

so
$$F_{\max}(b_N x) = \left(1 - \frac{x^{-\alpha}}{N}\right)^N. \tag{4.1}$$

We will need to make a lot of use of powers of e in this section and it is sometimes clearer to use the notation $\exp(x)$ to mean e^x. We will swap between these two (and also combine them) with the aim of avoiding superscripts of superscripts so that the expressions are easier to read. One fact about the exponential function is that it can be given as the limit:

$$\exp(x) = \lim_{N \to \infty} (1 + x/N)^N, \tag{4.2}$$

so, from Equation (4.1) we see that when N is large, we have $F_{max}(b_N x)$ is approximately $\exp(-x^{-\alpha})$.

We have shown that if we take the maximum of, say, 50 independent identically distributed random variables where $1 - F(x)$ goes to zero, like $x^{-\alpha}$, then the distribution is approximately a scaled version of one with distribution function $\exp(-x^{-\alpha})$, where the shape stays the same but the horizontal axis is multiplied by $b_{50} = (50k)^{1/\alpha}$.

Thus, the tail index, α, is the only thing that determines the shape of the distribution, and the distribution of maximum values has a CDF which is of the form $F(x) = \exp(-x^{-\alpha})$, but this gets scaled by an amount that depends on N and the particular distribution that is involved.

The distribution with a CDF given by $F(x) = \exp(-x^{-\alpha})$ is called the *Fréchet distribution*, after the very eminent French mathematician Maurice Fréchet. The density function for a Fréchet is obtained using the rules of calculus: we have

$$(d/dx)\exp(g(x)) = g'(x)\exp(g(x))$$

where we write $g'(x)$ for the derivative of a function $g(x)$. Thus, the density is

$$f(x) = \alpha x^{-\alpha-1}\exp(-x^{-\alpha}).$$

Figure 4.4 shows the density of the Fréchet distribution for different values of α. The three curves on the left of the figure show the density function when $\alpha = 1$, $\alpha = 2$ and $\alpha = 3$ (with higher values of α giving higher peaks).

Also shown in Figure 4.4 is the distribution of the maximum of 16 draws from a distribution with $F(x) = 1 - x^{-2}$, for $x > 1$. This should approximately match the Fréchet with $\alpha = 2$ scaled horizontally. Since $F(x) = 1 - kx^{-\alpha}$ with $k = 1$ and $\alpha = 2$, the scaling parameter is $a_N = (kN)^{1/\alpha} = 16^{1/2} = 4$. This comparison is also shown (the two curves with peaks at around 3): the appropriately scaled Fréchet density is the dashed line and the distribution of the maximum is the solid line. The horizontal scaling applied to the CDF will require a matching vertical scaling of the density function by a factor of 4 (so that the density function still integrates to 1) and this is why the peak here is lower than the standard Fréchet with $\alpha = 2$. The match between the solid and dashed lines is extremely good, especially given the relatively small size of $N = 16$.

We now have a clear picture of what happens when the original distribution has a fixed tail index, but what will happen if we try the same approach with tails that approach zero faster than any polynomial (i.e. they have infinite tail index)? The best way to understand this situation is to work through an example, so

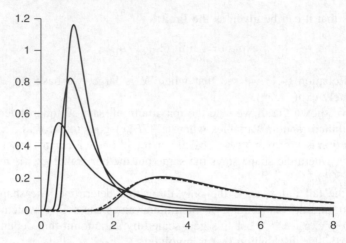

Figure 4.4   Fréchet densities for different values of α.

we consider the exponential distribution, which has CDF $F(x) = 1 - e^{-x}$ (and density function $f(x) = e^{-x}$) for $x > 0$. This does indeed have the property of an infinite tail index. With a finite tail index we used a horizontal scaling of $b_N = (kN)^{1/\alpha}$, but when α is effectively infinite this becomes a scaling by 1, i.e. no scaling at all. In fact, rather than scale the horizontal axis we will shift it by a certain amount that depends on N. For this example we want to shift by an amount $\log(N)$. Note that

$$F(\log(N) + x) = 1 - e^{-\log(N)}e^{-x} = 1 - \frac{\exp(-x)}{N}.$$

Thus

$$F_{\max}(\log(N) + x) = (F(\log(N) + x))^N$$
$$= \left(1 - \frac{\exp(-x)}{N}\right)^N.$$

From our previous observation, Equation (4.2), about the exponential as a limit we can see that as N gets large,

$$F_{\max}(\log(kN) + x) \to \exp(-e^{-x}).$$

Again we have a distribution that is reached in the limit. In fact this distribution where $F(x) = \exp(-e^{-x})$ is called the *Gumbel distribution*. We have shown that the maximum of, say, 50 draws from the exponential distribution where $F(x) = 1 - e^{-x}$ has a distribution that is approximately Gumbel shifted by an amount $\log(50) = 3.91$.

At first sight the Gumbel distribution looks forbidding: it has a CDF where e is raised to a power which itself involves e raised to a power. The density

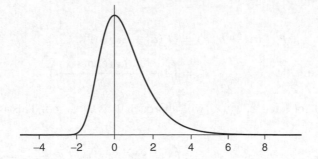

Figure 4.5   The density function for the Gumbel distribution.

function for the Gumbel is

$$f(x) = (d/dx)\exp(-e^{-x}) = e^{-x}\exp(-e^{-x}).$$

The density function for the Gumbel is shown in Figure 4.5 and we can see that it looks pretty 'ordinary' (given all the exponentials in its definition).

Though the exponential distribution has an infinite tail index, it still has tails that are thicker than the normal distribution which involves an $\exp(-x^2)$ behavior for the density function. The cumulative for a normal distribution is hard to deal with, so we will consider a simpler distribution but which also has an $\exp(-x^2)$ characteristic. So we take as our second example the distribution with CDF given by $F(x) = 1 - \exp(-x^2)$ (this has density $f(x) = 2x\exp(-x^2)$). To find the limiting distribution for a maximum of N draws from this we will need to use both scaling (by a factor b_N) and shifting (by an amount a_N). We set

$$a_N = \sqrt{\log N} \text{ and } b_N = 1/(2\sqrt{\log N}).$$

We seem to have pulled these values for a_N and b_N out of the air, but later we will give an indication of how they are chosen. Now note that

$$F(a_N + xb_N) = 1 - \exp\left(-(a_N + xb_N)^2\right)$$

$$= 1 - \exp\left(-\left(\sqrt{\log N} + \frac{x}{2\sqrt{\log N}}\right)^2\right)$$

$$= 1 - \exp\left(-\left(\log N + x + \frac{x^2}{4\log N}\right)\right)$$

$$= 1 - \frac{1}{N}\exp(-x)z_N(x),$$

where

$$z_N(x) = \exp(-x^2/(4\log N)).$$

So

$$F_{\max}(a_N + b_N x) = (F(a_N + xb_N))^N$$
$$= \left(1 - \frac{z_N(x)\exp(-x)}{N}\right)^N.$$

Observe that for fixed x, $z_N(x)$ will approach 1 as $N \to \infty$. This is enough to show that

$$\lim_{N\to\infty} F_{\max}(a_N + xb_N) = \lim_{N\to\infty}\left(1 - \frac{\exp(-x)}{N}\right)^N = \exp(-e^{-x}). \qquad (4.3)$$

We end up with a Gumbel distribution: exactly the same limit distribution as for the exponential case. This might be surprising, since we would not expect to reach the same limit distribution for two such different tail behaviors: there is an enormous difference between $\exp(-x)$ and $\exp(-x^2)$ (for example, when $x = 5$ we get $e^{-5} = 6.7 \times 10^{-3}$ and $e^{-25} = 1.4 \times 10^{-11}$). An important observation is that the multiplier b_N goes to zero as $N \to \infty$. This corresponds to a bunching up of the distribution that does not occur with the exponential. So here there is a qualitative difference between the two cases: with the thinner tail associated with $\exp(-x^2)$ we find that we can be more and more accurate with our prediction of what the maximum value of N draws from the distribution will be.

The two examples we have given are particular instances of a more general result called the *Fisher–Tippett theorem*. This theorem shows that, for most distributions with an infinite tail index (essentially exponential decay), the Gumbel distribution occurs as the limit when considering maxima of repeated draws. In fact, the conditions for this to occur are sufficiently general that we can assume it will happen with any infinite tail index distribution. To make this more precise we define a_N as the $1 - 1/N$ quantile for the distribution, i.e.

$$a_N = F^{-1}(1 - 1/N)$$

and

$$b_N = 1/(Nf(a_N)),$$

where f is the density function for F. Then the theorem shows that, when the tail index is infinite,

$$\lim_{N\to\infty} F_{\max}(a_N + xb_N) = \exp(-e^{-x}).$$

We will give a formal statement of the Fisher–Tippett theorem with some further discussion in the next subsection.

Example 4.4   Application of Fisher–Tippett theorem to a normal

Now we will apply the Fisher–Tippett theorem to the case of finding the distribution of the maximum of 50 independent samples from a standard normal distribution. This has density function

$$\varphi(x) = \frac{1}{\sqrt{2\pi}} e^{-x^2/2}$$

and we write $\Phi(x)$ for the standard normal CDF, $\int_{-\infty}^{x} \varphi(s)ds$. The first step is to find the right values of a_N and b_N for $N = 50$. We have

$$a_N = \Phi^{-1}(1 - 1/N) = \Phi^{-1}(0.98) = 2.054,$$

$$b_N = 1/(N\varphi(a_N)) = 1/(50\varphi(2.054)) = 0.413.$$

With these values we can say that $F_{\max}(a_N + b_N x)$ is approximately Gumbel, i.e.

$$F_{\max}(2.054 + 0.413x) \approx \exp(-e^{-x}).$$

We can also reverse this and say that $F_{\max}(y)$, the chance that the maximum of 50 samples is less than y, is approximately

$$\hat{F}(y) = \exp\left(-\exp\left(-\frac{(y - 2.054)}{0.413}\right)\right).$$

This has a corresponding density

$$\hat{f}(y) = \frac{1}{0.413} \exp\left(-\frac{(y - 2.054)}{0.413}\right) \exp\left(-\exp\left(-\frac{(y - 2.054)}{0.413}\right)\right).$$

In Figure 4.6 we show the density of the maximum of 50 draws from a normal distribution (the solid line) compared with the Gumbel distribution given by $\hat{f}$ (the dashed line). This is not a particularly good match, which shows that convergence in the case of the normal distribution is rather slow. Using different scaling constants a_N and b_N can improve this slightly, but no matter what values are used, the approximation will not be exact. (Note that the values of a_N and b_N are not unique – different choices of these sequences can give the same result in the limit and may do a slightly better job for lower values of N.)   □

Up to this point we have assumed that we are dealing with distributions having an infinite range. Whether the tail index is finite or infinite is a matter of how quickly the density in the tail drops to zero, but we have assumed that the density remains positive. Now we turn to the case that there is a maximum value for the distribution. So there is a fixed right-hand end point, $x_{\max}$ say, and

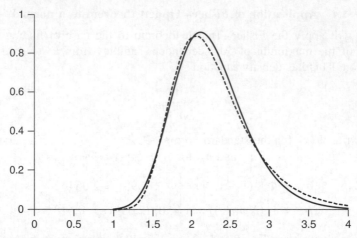

Figure 4.6   Gumbel compared with maximum from 50 samples from the standard normal distribution.

obviously the distribution of the maximum of N samples from this distribution will also have this as its maximum value. Now it makes sense to look at how quickly the CDF of the original distribution, $F(x)$, approaches 1 as x approaches the point $x_{\max}$, or (equivalently) how quickly the density function approaches zero there. It turns out that the distribution of the maximum also approaches a particular type of distribution in this case. In fact, the limit becomes a *Weibull distribution*, which has a CDF $H(x) = \exp(-(-x)^{\alpha})$ for $x < 0$ for some choice of $\alpha > 0$. The density function for the Weibull is $h(x) = \alpha(-x)^{\alpha-1}\exp(-(-x)^{\alpha})$ for $x < 0$.

In fact this case is also described by the Fisher–Tippett theorem. Just as with the other cases we have shifting and scaling parameters a_N and b_N to apply. Since the Weibull finishes at 0, the shift must be by an amount $x_{\max}$. So we end up with a set of numbers b_N where

$$\lim_{N \to \infty} F_{\max}(x_{\max} + xb_N) = \exp(-(-x)^{\alpha}) \text{ for } x < 0.$$

The parameter α for the Weibull will be determined by the behavior of F near $x_{\max}$. If, for example, $F(x) = 1 - k(x_{\max} - x)^2$ for some k, then the limiting distribution is Weibull with $\alpha = 2$.

4.2.1   *More details on maximum distributions and Fisher–Tippett

In this section we want to fill in some of the missing mathematical details. We start with the question of why a distribution having a certain tail index has a maximum (of N independent samples) which approaches a Fréchet distribution.

The discussion we gave in the previous subsection assumes a particular form for the CDF, but the definition of a tail index is more general than this.

We need to work with a slowly varying function. Suppose that $L(t) = (1 - F(t))t^\alpha$ is a slowly varying function of t (this implies that α is the tail index of F but is a slightly stronger claim). Thus, $\lim_{z \to \infty} L(tz)/L(z) = 1$ for any value of $t > 0$. In this expression we set $z = b_N$ and $t = x$ and we see that when we define

$$z_N(x) = \frac{(1 - F(b_N x))b_N^\alpha x^\alpha}{(1 - F(b_N))b_N^\alpha} = \frac{(1 - F(b_N x))x^\alpha}{(1 - F(b_N))}, \qquad (4.4)$$

then the slowly varying condition implies that, for any fixed x, $z_N(x) \to 1$ provided that $b_N \to \infty$.

Now we set

$$b_N = F^{-1}\left(1 - \frac{1}{N}\right),$$

which is the $1 - (1/N)$ quantile for the distribution, and the right choice for the scaling factor. This will indeed approach ∞ as N gets large. Now we notice that we can rearrange Equation (4.4) to get

$$1 - F(b_N x) = z_N(x)(1 - F(b_N))x^{-\alpha}$$

$$= \frac{z_N(x)x^{-\alpha}}{N}.$$

Thus

$$F_{\max}(b_N x) = (F(b_N x))^N = \left(1 - \frac{z_N(x)x^{-\alpha}}{N}\right)^N.$$

Because of the limit behavior of z_n, we know that any $\varepsilon > 0$ will have $1 - \varepsilon < z_N < 1 + \varepsilon$ for N large enough. Hence

$$\lim_{N \to \infty}\left(1 - \frac{(1 - \varepsilon)x^{-\alpha}}{N}\right)^N > \lim_{N \to \infty} F_{\max}(b_N x) > \lim_{N \to \infty}\left(1 - \frac{(1 + \varepsilon)x^{-\alpha}}{N}\right)^N.$$

Since the left- and right-hand sides can both be made as close as we like to $\exp(-x^{-\alpha})$ by choosing ε small enough, we have established that

$$\lim_{N \to \infty} F_{\max}(b_N x) = \exp(-x^{-\alpha}),$$

and this is the Fréchet distribution we have been working towards.

This argument, which sandwiches the $F_{\max}$ limit between two expressions that can be chosen as close as we like to the exp function, is also exactly what we need to show that the limit in Equation (4.3) is correct.

Now we want to move towards a more formal description of the Fisher–Tippett theorem. Notice that when we say that the distribution of maxima approaches a limiting distribution, what we mean is that the distribution is obtained by a scaling and shifting procedure (with this scaling and shifting depending on the number N of draws from the original distribution). So anything that is a scaled and shifted version of the limiting distribution could be used instead. We use the terminology of types: two distributions F_1 and F_2 are of the same *type* if they can be obtained from each other by scaling and shifting, i.e. if $F_1(x) = F_2(a + bx)$ for the right choice of a and b. For example, any normal distribution, no matter what its mean and standard deviation, is of the same type.

We say that the distribution F is in the *maximum domain of attraction* of a distribution H if there are sequences a_N and b_N with $(F(a_N + b_N x))^N \to H(x)$ for every x. We write this as $F \in MDA(H)$. The terminology here can be misleading – the maximum domain of attraction sounds like it is the largest domain of attraction in some sense, but instead what is meant is that it is the domain of attraction under a maximum operator. Saying $F \in MDA(H)$ is exactly the same as saying that the maximum of N samples from F has a distribution which is the same shape as H (after scaling by b_N and shifting by a_N).

Because the choice of a_N and b_N is arbitrary, this is really a statement about types of distribution: if $\tilde{F}$ is a distribution of the same type as F, and $\tilde{H}$ is of the same type as H, then $F \in MDA(H)$ and $\tilde{F} \in MDA(\tilde{H})$ are exactly equivalent statements.

Now we are ready to state the Fisher–Tippett theorem in the form it is usually given.

Theorem 4.1 If F is in $MDA(H)$ for some distribution H (and H is not concentrated on a single point) then there is a parameter ξ for which H is the same type as

$$H_\xi(x) = \begin{cases} \exp(-(1 + \xi x)^{-1/\xi}) & \text{for } \xi \neq 0, \\ \exp(-e^{-x}) & \text{for } \xi = 0. \end{cases}$$

The formula for H_ξ defines a set of distributions usually called *generalized extreme value* (or GEV) distributions. The formula also implies the range of x values that apply, since we need $H_\xi(x)$ increasing from 0 to 1. When $\xi < 0$, we need $x \leq -1/\xi$ and when $\xi > 0$, we need $x \geq 1/\xi$.

Essentially there are three cases to consider according to whether the ξ value is positive, zero, or negative. We have already met two examples of the generalized extreme value distributions. The $\xi = 0$ case is the Gumbel distribution. When $\xi > 0$ we can set $\alpha = 1/\xi$ and get a distribution function $H(x) = \exp(-(1 + x/\alpha)^{-\alpha})$, which is of the same type as the Fréchet distribution $\exp(-x^{-\alpha})$ (we just have to scale by $1/\alpha$ and shift by 1). The reason for giving this result in the more complicated form, with ξ rather than α, is that it makes it clearer that the $\xi = 0$ case can be reached in the limit as $\xi \to 0$ from either above or below. In both cases we are just using our standard observation, Equation (4.2), that $(1 + x/n)^n$ approaches e^x as $n \to \infty$.

We need to say more about the case with $\xi < 0$. In this case the GEV distribution has a fixed right-hand end point, since $\exp(-(1 + \xi x)^{-1/\xi})$ reaches a maximum value of 1 when $x = -1/\xi$, marking the right-hand end of the distribution. It is not hard to see that this GEV is of the same type as a Weibull distribution. To link the parameter ξ to the behavior of the original distribution function, we need to go back to a slowly varying function. However, the idea of a slowly varying function is all about how the function behaves as it goes to infinity, so we need to do some manipulation and work with the inverse of the difference between x and x_{max}. If

$$(1 - F(x_{\text{max}} - (1/z)))z^\alpha$$

is a slowly varying function then the maximum of N draws from the distribution is approximately (after scaling and shifting) a Weibull with parameter α (or GEV with parameter $-1/\alpha$).

The Fisher–Tippett theorem as it stands doesn't say how likely it is that there will be a limiting distribution for F_{max}, it just specifies the form of that distribution, if it occurs. But in fact it is extremely difficult to find any example without some limiting distribution. Remember that we allow scaling and shifting, so it is only the shape of the distribution that is important. We could, for example, use scaling and shifting to fix the fifth percentile and ninety fifth percentile points. Then we can watch what happens to the other quantile points as N increases. If they all converge then we will have a limiting distribution. If we imagine that there is no limiting distribution then it is hard to see how we could ever cycle between different shapes, but if there is no cycling then the quantile points which are constrained to be between the 5th and 95th percentile are bound to converge. So it is only at the extremes that things could go wrong, e.g. by having tail behavior like x^{-k} with k getting larger and larger as N increases, but doing this in a way that does not lead to some other limiting distribution. It turns out that this will not happen for distributions that occur in practice, and this can be demonstrated by looking in more detail at conditions sufficient to guarantee that F is in $MDA(H_\xi)$. We have already shown how a slowly varying condition is enough when there is a tail index of α (and a similar condition applies with a fixed right-hand limit), but we need something more complicated to deal with the $\xi = 0$ case and we will not give any details here.

4.3   Excess distributions

Now we turn to a different, but related, topic. Instead of looking at the maximum of N random variables, in this section we will look more directly at the behavior of the tail of a distribution. The idea is similar to the discussion of the previous section, where we were concerned with the shape of the distribution of maximum values. Now we are interested in the shape of the distribution above a threshold u. As this threshold increases, we ask whether there is a limiting distribution for the excess (i.e. the part of the distribution above the threshold u) always assuming that we scale appropriately. The idea is illustrated in Figure 4.7 where two different threshold values u have excess distributions that look similar.

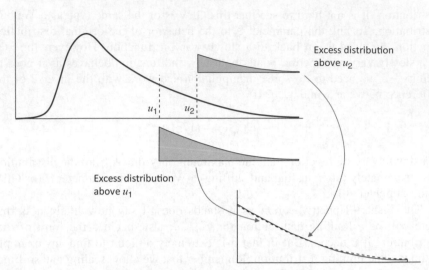

Figure 4.7   The distribution above threshold u_1 and the distribution above threshold u_2 can be scaled to become very similar.

The excess distribution is obtained by considering the distribution of $X - u$ for some threshold u, conditional on X being greater than u. We write this as $F_u(x)$, thus

$$F_u(x) = \Pr(X - u \leq x \,|\, X > u), \text{ for } x > 0.$$

This is simply the probability that the random variable X is in the range u to $x + u$ given that it is greater than u, i.e. using Bayes' rule

$$F_u(x) = \frac{F(x + u) - F(u)}{1 - F(u)}, \text{ for } x > 0.$$

This expression for F_u can be rewritten as

$$F_u(x) = 1 - \frac{1 - F(x + u)}{1 - F(u)}.$$

We will discover that for many distributions F there will be a limiting distribution (defined up to scaling) for F_u as u gets very large. This theory closely parallels the theory of the distribution of the maximum of N samples given in the previous section.

To illustrate this, suppose that $F(x) = 1 - kx^{-\alpha}$, so that F has a tail index of α. Then

$$F_u(x) = 1 - \frac{k(x + u)^{-\alpha}}{k(u)^{-\alpha}}$$

$$= 1 - \left(1 + \frac{x}{u}\right)^{-\alpha}, \text{ for } x > 0.$$

Thus, if we scale by u we see that

$$F_u(ux) = 1 - (1 + x)^{-\alpha}, \text{ for } x > 0.$$

Here there is no approximation; the excess distribution (after scaling) is exactly given by this formula. The distribution is called a Type 2 Pareto distribution. A more exact treatment (which we give in the next subsection) shows that this form of limiting distribution is the only one that can arise with a tail index of α.

We can carry out the same kind of calculation with distributions having an infinite tail index. So we suppose that

$$F(x) = 1 - \exp(-x^2).$$

Then

$$\begin{aligned}
F_u(x) &= 1 - \frac{\exp(-(x + u)^2)}{\exp(-u^2)} \\
&= 1 - \frac{\exp(-x^2 - 2ux) \exp(-u^2)}{\exp(-u^2)} \\
&= 1 - \exp(-x^2 - 2ux), \text{ for } x > 0.
\end{aligned}$$

Now consider scaling by $1/(2u)$. We have

$$F_u\left(\frac{x}{2u}\right) = 1 - \exp\left(-\frac{x^2}{4u^2}\right) \exp(-x), \text{ for } x > 0.$$

Since $\exp(-x^2/(4u^2))$ approaches 1 as $u \to \infty$ we see that

$$F_u\left(\frac{x}{2u}\right) \to 1 - \exp(-x), \text{ for } x > 0.$$

as $u \to \infty$. Thus, in this case scaling by $1/(2u)$ gives approximately an exponential distribution. It turns out that the right choice of scaling constant will achieve an exponential distribution in the limit for any reasonable distribution with an infinite tail index.

Thus, when the tail index is finite a Pareto distribution emerges in the limit, and when the tail index is infinite an exponential distribution occurs. This is reminiscent of the kind of thing that happens with the Fisher–Tippett theorem, for the distribution of the maximum of N samples from the same distribution. In fact, as we discuss in the next subsection, the conditions for convergence are also exactly the same.

Again we can summarize the final position in a single theorem, but this time we use a generalized Pareto distribution to capture the different distributional forms that may occur. We state the theorem more carefully in the discussion of the next subsection. But for the moment we just want to see how to use the result. The idea is that we will fix on a large value of u and then approximate

$F_u(x)$ with the *generalized Pareto distribution* (GPD) which has the following distribution function:

$$G_{\xi,\beta}(x) = \begin{cases} 1 - (1 + \xi x/\beta)^{-1/\xi} & \text{for } \xi \neq 0 \\ 1 - \exp(-x/\beta) & \text{for } \xi = 0 \end{cases}, \tag{4.5}$$

where $\beta > 0$. When $\xi \geq 0$ we require $x \geq 0$. The $\xi < 0$ case corresponds to a distribution for F with a maximum value $x_{\max}$, and here we require $0 \leq x \leq -\beta/\xi$.

In general, the parameter ξ captures the shape and β simply acts as a scaling constant. The parameter ξ is really the inverse of the tail index, so $\xi = 1/\alpha$. The reason for working with ξ, rather than α, is that it becomes clearer that the form of the generalized Pareto is consistent for different ξ values, since the $\xi = 0$ case is obtained in the limit as ξ approaches zero.

If F has a tail index of $\alpha < \infty$ (i.e. $\xi > 0$) then we know that

$$F_u(x) = 1 - \left(1 + \frac{x}{u}\right)^{-\alpha}, \text{ for } x > 0.$$

We can compare this expression to the GPD given by Equation (4.5). Converting from ξ to α shows that the scaling parameter β is given by $u\xi$ in this case.

With an infinite α (i.e. $\xi = 0$) the scaling parameter will be determined by the exact shape of the tail. The example that we discussed above with $F(x) = 1 - \exp(-x^2)$ has a β value of $1/(2u)$. Other examples of distributions with infinite tail index will give different values of β.

The density function for the generalized Pareto distribution is given by

$$g_{\xi,\beta}(x) = (1/\beta)(1 + \xi x/\beta)^{-(1/\xi)-1},$$

as is simple to see by differentiating Equation (4.5). The density when $\xi = 0$ is given by $g_{\beta}(x) = (1/\beta)\exp(-x/\beta)$, which is the limit of $g_{\xi,\beta}(x)$ as ξ approaches zero. Figure 4.8 plots this density for different values of ξ, in each case with a β value of 1.

If the excess has a GPD distribution, then it is natural to ask about the mean of the distribution: this will be the mean value of the excess over u, i.e. $E(X - u \mid X > u)$. When X is distributed as $G_{\xi,\beta}$ it turns out that the mean value is given by

$$E(X) = \beta/(1 - \xi).$$

Exercise 4.4 asks you to carry out the integration to show this formula is true. The same formula holds when $\xi = 0$ since the exponential distribution with density $(1/\beta)\exp(-x/\beta)$ has mean β. Notice that in Figure 4.8 the differences between the three distributions do not seem enormous, but the mean of the density shown with a solid line is 5; the mean of the dashed line is 5/3 and the mean of the dotted line is 1.

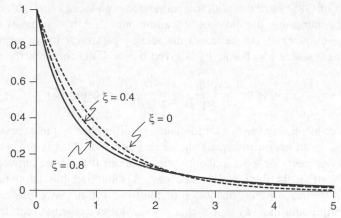

Figure 4.8   Density functions for generalized Pareto distribution with different parameters. All β values are 1.

In the case that $\xi \geq 1$ then the formula no longer makes sense (giving a result which is negative or infinite). This is exactly what we would expect: the tail index is $1/\xi$, and if this index is less than 1 then the mean does not exist.

Given a distribution defined by F, the GPD describes the excess distribution above a threshold u. Now we ask what happens when a different threshold is used. Suppose that F_u is given. Then the formula which defines F_u is

$$1 - F_u(x) = \frac{1 - F(x + u)}{1 - F(u)}$$

and this can be rewritten

$$1 - F(x + u) = (1 - F_u(x))(1 - F(u)). \qquad (4.6)$$

Suppose we have a $v > u$. Then, from Equation (4.6), we deduce that

$$
\begin{aligned}
1 - F_v(x) &= \frac{1 - F(x + v)}{1 - F(v)} \\
&= \frac{(1 - F_u(x + v - u))(1 - F(u))}{(1 - F_u(v - u))(1 - F(u))} \\
&= \frac{(1 - F_u(x + v - u))}{(1 - F_u(v - u))}.
\end{aligned}
$$

In the case where $\xi \neq 0$ and F_u is given by the GPD formula, we get

$$1 - F_v(x) = \frac{(1 + \xi(x + v - u)/\beta)^{-1/\xi}}{(1 + \xi(v - u)/\beta)^{-1/\xi}} = \left(1 + \frac{\xi x}{\beta + \xi(v - u)}\right)^{-1/\xi}.$$

This is also of GPD form but with the parameter β replaced by $\beta + \xi(v - u)$. In other words, increasing the threshold by an amount $\Delta = v - u$ leaves the shape parameter ξ unchanged, but increases the scaling parameter by an amount $\xi\Delta$.

In the case where $\xi = 0$ and F_u is given by the GPD formula, then

$$1 - F_v(x) = \frac{\exp((x + v - u)/\beta)}{\exp((v - u)/\beta)} = \exp(x/\beta).$$

When $\xi = 0$, increasing the threshold makes no difference to the excess distribution; it stays as an exponential with the same parameter. This is the well-known memoryless property of the exponential distribution: if the time between events is an exponential distribution then knowing that nothing has happened so far makes no difference to the distribution of the time to the next event.

This result about the way that β changes (or doesn't change) with the threshold needs to be treated with care, since it assumes that the excess distribution is exactly a GPD, and this is only an approximation. For example, in our discussion earlier of the case where $F(x) = 1 - \exp(-x^2)$ we found that the excess distribution F_u had the property that $F_u(x/(2u))$ approaches the CDF for an exponential distribution. In other words, $F_u(x)$ is given approximately by $G_{\xi,\beta}(x)$ with $\beta(u) = 1/(2u)$ and $\xi = 0$. So, in this case, we do see a change in β as u changes, even though this is the case where β is supposed to be fixed. However, expressed as a multiple of the change in u, the difference is very small. If we take $v > u$ and let $\Delta = v - u$ then

$$\beta(v) = \frac{1}{2v} = \frac{1}{2u} + \frac{u - v}{2uv} = \beta(u) - \frac{1}{2uv}\Delta.$$

4.3.1   *More details on threshold exceedances

First we return to the question of the behavior of the excess distribution when F has a tail index of α. As we have done before, we use the marginally stronger idea of a slowly varying function and suppose $(1 - F(t))t^\alpha$ is a slowly varying function of t (this implies that α is the tail index of F). Thus, if

$$z_u(k) = \frac{(1 - F(ku))u^\alpha k^\alpha}{(1 - F(u))u^\alpha} = \frac{(1 - F(ku))k^\alpha}{(1 - F(u))},$$

then, for any fixed k, $z_u(k) \to 1$ as $u \to \infty$. Thus

$$F_u(ux) = 1 - \frac{1 - F(ux + u)}{1 - F(u)}$$

$$= 1 - (1 + x)^{-\alpha}z_u(1 + x),$$

where we have used the definition of $z_u(k)$ with $k = 1 + x$. So, as $u \to \infty$ we have

$$F_u(ux) \to 1 - (1 + x)^{-\alpha} \text{ for } x > 0. \tag{4.7}$$

In general, we say that X is distributed as a Pareto distribution $\text{Pa}(\alpha, \kappa)$ if the CDF is given by

$$F(x) = 1 - \left(1 + \frac{x}{\kappa}\right)^{-\alpha},$$

so, in this case we have shown that, after scaling by u, the excess distribution is a Pareto with $\kappa = 1$.

Now we will give the equivalent of the Fisher–Tippett theorem in this setting, which demonstrates the role of the generalized Pareto distribution introduced above.

Theorem 4.2 If F is in $MDA(H_\xi)$ then there is a function $\beta(u)$ with

$$\lim_{u \to \infty} F_u(\beta(u)x) = \begin{cases} 1 - (1 + \xi x)^{-1/\xi} & \text{for } \xi \neq 0, \\ 1 - e^{-x} & \text{for } \xi = 0. \end{cases}$$

Notice that there is not only a similarity of form between this result and the Fisher–Tippett theorem; it is exactly the set of distributions whose maximum converges to a distribution of the form H_ξ that turns out to have an excess distribution converging to $G_{\xi,\beta}(x)$ for some choice of function $\beta(u)$.

With this formulation we obtain the exponential distribution in the limit as $\xi \to 0$, so the distribution shape changes in a way that is continuous with ξ. However, since there is an arbitrary $\beta(u)$ involved, we could scale x by a further factor of ξ and get back to the form we gave earlier, i.e. with $1 - (1 + x)^{-\alpha}$ as the limit. For $\xi \geq 0$, the limiting distribution is defined over the range $x \geq 0$ and when $\xi < 0$ we require $0 \leq x \leq -1/\xi$.

It is interesting to compare this result with the Fisher–Tippett theorem. There we work with a maximum over N samples, and look at the limit as $N \to \infty$, while here we consider an excess over u and look at the limit as $u \to \infty$. In the Fisher–Tippett theorem the limiting CDF has the form $\exp(-Z(x))$ where the CDF moves from 0 to 1 as the expression $Z(x)$ moves from ∞ to 0, while here the limiting distribution is simply $1 - Z(x)$ and is defined over the range where Z moves from 1 to 0. Moreover, in the Fisher–Tippett theorem we may require both scaling and shifting (i.e. both a_N and b_N non-zero) whereas here we only need to scale by $\beta(u)$.

4.4 Estimation using extreme value theory

An understanding of the tail of a distribution is particularly valuable if we want to estimate probabilities associated with events that we have not seen, or have only seen rarely. The idea here is to estimate the shape and scale parameters from the data and then use these to estimate the probability of interest. Of course, we will need to assume that there is sufficient consistency in the tail behavior that the theory we have discussed applies, but this is a caveat that needs to be borne in mind whatever estimation we carry out. Estimation is always in some sense a deduction about what happens at a point where we do not have data, and this deduction is bound to make use of some modeling assumptions.

This takes us back to the opening scenario of this chapter, in which the requirement is to estimate a quantile of a distribution of weekly requirements for a particular drug. In that example, Maria wanted to know the x value which would be exceeded on only one week in 156. This is extrapolation well beyond the 104 data points that she has observed, and so any estimate she makes is bound to be very uncertain. Nevertheless the ideas we are exploring here allow us to use quite unrestrictive assumptions to say something about what happens well out in the tail. We don't need to use a particular model of the distribution (whether that be normal, exponential or something else) since the estimates we make will apply to all these distributions and many more. All we need to assume is that there is enough regularity in the behavior in the tail – the existence of a specific tail index would be enough.

The idea of using assumptions on reasonably regular behavior in order to make extrapolations is not too different in kind from what we do when we interpolate (i.e. when we estimate behavior between existing data points). Extrapolation makes an assumption that the behavior of the distribution does not change suddenly for extreme values, while making estimates from within the range of values that we have observed makes an assumption that there is no sudden change of behavior within that range. The difference is that when we extrapolate there is, by definition, nothing in the data which can warn us about changes, whereas a big change in the distribution at a point within the range of data we observe might be detected by looking at the data. This might seem like an important distinction, but in practice we would need quite a lot of data to spot such a change. For the problem that Maria faces, even four or five years of data would give much the same issue. If, for some reason, there is a change in behavior of the distribution of weekly requirements that occurs at the 99% point, then looking at empirical data where only two or three data points are expected to be larger than this value will never reveal what is going on. So, having double the amount of data will certainly help in making a good estimate, but there will still be a need to assume some regularity in the data.

In this section we will introduce a three-step method for estimating value at risk or expected shortfall (or other risk measure). However, it is important to realize that this is just one of several methods that might be employed. Particularly for time series data there are good arguments for using a more sophisticated approach that takes account of changes in volatility over time (as often happens with financial data). Nevertheless when the correlations between data points are not too strong and when there is no reason to expect changes in the underlying distribution over time, the method we will describe can be very effective.

4.4.1   Step 1. Choose a threshold u

The idea is to approximate the distribution of losses above a certain threshold level, u, using a generalized Pareto distribution. We need to make u reasonably large so that the extreme value theory will apply. However, since we will end up estimating the parameters of the GPD from the data that occur above the threshold, we must ensure that there are enough data points to do this. It is not so hard to choose a reasonable value of the threshold given a large data set, but it can be difficult if the data set is small. The approach we will illustrate uses a sample mean excess plot. For any threshold value v, we look at the average amount by which the data points exceed this threshold, averaging just over the data points larger than v. The sample mean excess is an estimate of the true value, and, as we have already seen, this will be $\beta/(1-\xi)$ if the excess distribution is a GPD $G_{\xi,\beta}$. The discussion in the previous section also demonstrates that once we have chosen a threshold large enough for the GPD approximation to be accurate then we expect that increasing the threshold by an amount Δ leaves ξ unchanged but increases β by an amount $\xi\Delta$. Thus, we expect that in the tail of the distribution the mean excess will be linear as a function of the threshold with a slope of $\xi/(1-\xi)$ provided that $\xi < 1$.

Hence, if we plot the mean excess for different thresholds, we should get a rough straight line for values of the threshold large enough for the GPD approximation to apply but small enough for there to be enough points above the threshold for the sample estimate of the mean to be accurate. It is reasonable to carry out the estimation procedure with a value of u that is near the start of this final straight line section if it exists.

Example 4.5   Estimating risk measures for Amazon.com

To illustrate this, we look at some stock price data: Figure 4.9 shows the percentage weekly loss (negative values are gains) for the share price of Amazon.com over the 14-year period from the start of 1998 to the end of 2011. There are 730 weekly data points from which we calculate 729 percentage losses. Obviously there was significantly higher volatility in the early part of this period, but we will not try to use these changes in volatility within our estimation procedure. Figure 4.10 shows the mean excess data as the threshold varies. We can see that for thresholds in the range 0 to 2%, the mean excess stays around 6% loss, but

Figure 4.9   Weekly losses/gains for Amazon.com stock over period 1998 to 2011.

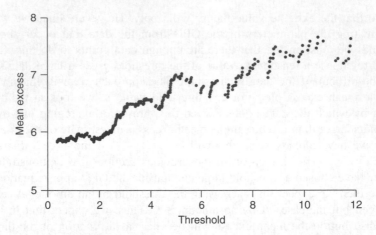

Figure 4.10   Mean excess plot for Amazon stock price data.

for thresholds larger than a 2% loss, the mean excess starts to rise. It does this in an erratic way, but there is a definite upward trend. This suggests that we can use a value of 2% loss for the threshold u. But the best choice of threshold is fairly arbitrary, and for these data we could also take a threshold value around 6%, which would still give 125 exceedances.   □

4.4.2   Step 2. Estimate the parameters ξ and β

Having fixed a value of u, the second step in our procedure is to estimate the parameters ξ and β in the generalized Pareto distribution that applies above u. We can do this using a maximum likelihood approach. This is a fairly standard method from statistics, but we will give a summary of how it works.

Suppose that we have M observed values and we ask what is the probability that we would observe these values if the distribution were really GPD $G_{\xi,\beta}$ and each observation was chosen independently? Clearly the probability of getting any exact set of values is zero but we can consider the *likelihood* defined as the probability density of the overall distribution of M possible values evaluated at the set of observations we observe. Another way to think of the likelihood is as the limit of the probability of getting points in small intervals around those we actually observe, but normalized to allow for the effect of the interval sizes.

For example, suppose there are just three points that are above the level u and the amounts by which they exceed u are Y_1, Y_2 and Y_3. Since the density of the GPD is $(1/\beta)(1 + \xi x/\beta)^{-1/\xi-1}$ the density of the joint distribution assuming independence is the product of this, i.e. the likelihood is

$$(1/\beta^3)(1 + \xi Y_1/\beta)^{-1/\xi-1}(1 + \xi Y_2/\beta)^{-1/\xi-1}(1 + \xi Y_3/\beta)^{-1/\xi-1}.$$

The idea is now to look for the values of the two parameters which make this as large as possible. The product form here is awkward to deal with, particularly when we have many more than three observations. But maximizing likelihood will end up at the same choice of ξ and β as maximizing any increasing function of likelihood, such as the square of the likelihood. The increasing function which works best is to maximize the log likelihood in order to turn the product into a sum. So we maximize

$$\log((1/\beta^3)(1+\xi Y_1/\beta)^{-1/\xi-1}(1+\xi Y_2/\beta)^{-1/\xi-1}(1+\xi Y_3/\beta)^{-1/\xi-1})$$
$$= -3\log\beta + (-1/\xi - 1)(\log(1+\xi Y_1/\beta) + \log(1+\xi Y_2/\beta) + \log(1+\xi Y_3/\beta)).$$

More generally, with M observed excess values $Y_1, Y_2, \ldots Y_M$ we make the estimate of ξ and β by choosing the values $\hat{\xi}$ and $\hat{\beta}$ which maximize

$$-M\log\beta - \left(1+\frac{1}{\xi}\right)\sum_{i=1}^{M}\log\left(1+\frac{\xi Y_i}{\beta}\right). \qquad (4.8)$$

subject to the constraints $\beta > 0$ and $1 + \xi Y_i/\beta > 0$ for all Y_i. This calculation can be carried out using Solver in a spreadsheet (all you need is guesses of β and ξ, a column of Y_i values, a column in which $\log(1+\xi Y_i/\beta)$ is calculated, and then a cell holding the objective given by Formula (4.8)).

Example 4.5   (continued) Estimating risk measures for Amazon.com

Returning to the Amazon.com data, we have 264 weeks in which the loss was greater than 2%. The log likelihood expression is maximized by taking $\hat{\xi} = 0.27$ and $\hat{\beta} = 4.51$. This corresponds to a tail index value of $1/\hat{\xi} = 3.7$. Figure 4.11 shows how well the fitted CDF fits the data for the 264 data points above the threshold. We show the estimated value of $F_u(x)$ as a line and the empirical data as points. For example, the 10th largest loss is 20.73 and this is plotted with a y value of $254/264 = 0.962$.   □

4.4.3   Step 3. Estimate the risk measures of interest

The third and final step is to use the fitted distribution to estimate values of interest. We will consider estimating both value at risk and expected shortfall. First consider value at risk. $\text{VaR}_\alpha(X)$ is the x such that $F(x) = \alpha$. If we are working with a fixed threshold u and $x > u$ then

$$F(x) = 1 - \Pr(X > x)$$
$$= 1 - \Pr(X > u)\Pr(X > x \mid X > u)$$
$$= 1 - (1 - F(u))(1 - F_u(x - u)).$$

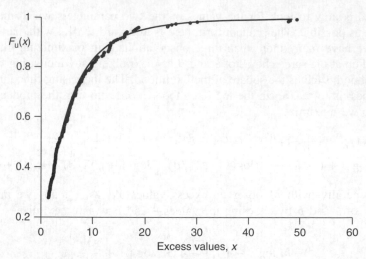

Figure 4.11   The estimated GPD distribution compared with empirical data for the Amazon.com % weekly loss with a threshold of 2.

Since we assume that the F_u distribution is $G_{\xi,\beta}$ with the estimated values of $\hat{\xi}$ and $\hat{\beta}$, then for $F(x) = \alpha$ we require

$$\alpha = 1 - (1 - F(u))(1 + \hat{\xi}(x - u)/\hat{\beta})^{-1/\hat{\xi}}. \tag{4.9}$$

Solving for x gives the value at risk we want, and we can rearrange Equation (4.9) to get x in terms of the other parameters and so obtain the following estimate

$$\text{VaR}_\alpha(X) = u + \frac{\hat{\beta}}{\hat{\xi}}\left(\left(\frac{1-\alpha}{1-F(u)}\right)^{-\hat{\xi}} - 1\right).$$

The expected shortfall at a level α is $\text{ES}_\alpha(X) = E(X|X > \text{VaR}_\alpha(X))$. To calculate this we can set $v = \text{VaR}_\alpha(X)$ and use our previous calculation that in the tail of the distribution for $v > u$ the excess distribution is $F_v(x) = G_{\xi,\beta'}(x)$ where $\beta' = \beta + \xi(v - u)$. Moreover, we know that the mean value of the distribution $G_{\xi,\beta}$ is $\beta/(1 - \xi)$. Hence, the expected shortfall is

$$E(X|X > \text{VaR}_\alpha(X)) = \text{VaR}_\alpha(X) + E(\text{excess over } \text{VaR}_\alpha(X))$$

$$= \text{VaR}_\alpha(X) + \frac{\beta_V}{1 - \hat{\xi}},$$

where β_V is the estimated β value for the tail above $\text{VaR}_\alpha(X)$, so $\beta_V = \hat{\beta} + \hat{\xi}(\text{VaR}_\alpha(X) - u)$. Thus

$$\text{ES}_\alpha(X) = \text{VaR}_\alpha(X) + \frac{\hat{\beta} + \hat{\xi}(\text{VaR}_\alpha(X) - u)}{1 - \hat{\xi}},$$

which can be simplified to give the final formula:

$$\text{ES}_\alpha(X) = \frac{\text{VaR}_\alpha(X) + \hat{\beta} - \hat{\xi} u}{1 - \hat{\xi}}.$$

Example 4.5   (continued) Estimating risk measures for Amazon.com

We can calculate the estimated 99% VaR and expected shortfall values for the Amazon.com data in exactly this way. We use our earlier estimates of $\hat{\xi} = 0.27$ and $\hat{\beta} = 4.51$ with the threshold value of $u = 2$. We also need $1 - F(u)$, which is the probability of a loss greater than u, and we estimate this from the proportion of observations over the threshold of 2, i.e. $1 - F(u) = 264/729$. This gives

$$\text{VaR}_{0.99}(X) = 2 + \frac{4.51}{0.27}\left(\left(\frac{0.01}{(264/729)}\right)^{-0.27} - 1\right) = 29.3.$$

The expected shortfall is therefore estimated by

$$\text{ES}_{0.99}(X) = \frac{29.3 + 4.51 - 0.27 \times 2}{1 - 0.27} = 45.6.$$

We can compare these figures with the observations. Out of 729 observations we would expect to see 7 above the $\text{VaR}_{0.99}$ level: the seventh highest observation is 27.86 and the eighth highest is 26.35. So the GPD method is giving a slightly higher VaR than is suggested by looking just at the handful of highest losses. It would be a matter of argument as to whether the 29.3 estimate is substantially better than an estimate based simply on the seventh highest observation. However, there is no doubt that the more sophisticated approach we have given will be better at predicting the expected shortfall, since this is where the extrapolation element in the estimation becomes more critical. The empirical data give an average loss amongst the top seven data points of 38.4, whereas the GPD estimate suggests a higher figure of 45.6.   □

Notes

This chapter has the most sophisticated analysis of any part of this book, but at the same time it covers ground where much more could be said. The book by McNeil, Frey and Embrechts also gives a comprehensive introduction to many different techniques in this area, and this is a good starting point for anyone who wants to know more.

We have given a simple approach for estimating tail indices at the start of the chapter, and at the end of the chapter, we have a more complex approach using maximum likelihood to estimate ξ (which is the inverse of the tail index). But there are other approaches; the most common technique is called the Hills estimator.

The distribution given in Example 4.1 is quite closely related to the well-known Cauchy distribution which has a density function

$$f(x) = \frac{1}{\pi(1 + x^2)},$$

and which also does not have a properly defined mean.

The Fisher–Tippett theorem goes back to a paper by these authors in 1928, but there was not a complete proof until the work of Gnedenko in 1943 and his name is often added to the name of the result, which becomes the Fisher–Tippett–Gnedenko theorem. But even Gnedenko's research did not answer all the questions in relation to this result, and work on the theorem, with new proofs, continued into the 1970s.

The material we have described on excess distributions is often called 'peaks over threshold' theory (POT). This is now the dominant version of extreme value theory, and is more useful than its cousin dealing with the maximum of N samples.

The book by Embrechts, Kluppelberg and Mikosch covers the main theoretical ideas of extreme value theory and contains proofs of the two theorems we quote. The paper by De Haan also gives an accessible proof of the Generalized Extreme Value result and provides the sequence of normalizing constants $a_N = F^{-1}(1 - 1/N)$ and $b_N = 1/(Nf(a_N))$, which have the advantage of being relatively simple to use in examples. These references also provide more specific conditions for a distribution to be in $MDA(H_0)$ (for example, this will be guaranteed if $\lim_{x \to \infty} f'(x)(1 - F(x))/(f(x)^2) = -1$).

References

De Haan, L. (1976) Sample extremes: an elementary introduction. *Statistica Neerlandica*, **30**, 161–172.

Embrechts, P., Kluppelberg, C. and Mikosch, T. (1997) *Modelling Extremal Events for Insurance and Finance*. Springer.

Fisher, R. and Tippett, L. (1928) Limiting forms of the frequency distribution of the largest or smallest member of a sample. *Mathematical Proceedings of the Cambridge Philosophical Society*, **24**, 180–190.

Gnedenko, B. (1943) Sur La Distribution Limite Du Terme Maximum D'Une Serie Aleatoire. *Annals of Mathematics*, **44**, 423–453.

McNeil, A., Frey, R. and Embrechts, P. (2005) *Quantitative Risk Management*. Princeton University Press.

Exercises

4.1 Comparing upper and lower tails for exchange rate

Figure 4.2 is generated using the data in spreadsheet BRMch4-ExchangeRateData.xlsx. Use these same data to carry out a similar analysis for movements in exchange rate in the opposite direction (a drop in value of the US$ in comparison with the pound). Do you think a tail index of about 5 is also appropriate in this case?

4.2 Fréchet distribution

A company is concerned about the possible bad publicity arising out of a guarantee made on its website ('We will fix your router- related problem in less than 20 minutes or we will give you free internet for a year'). Assume that the manager looks at the data on the 20 working days to assess, for each day, the maximum time that a router- related problem took to fix. On each day there were between 25 and 30 customer enquiries of this sort made. The maximum from amongst these times varies from day to day but the average value for the 20 different maximum times is 12 minutes. Assume that the time required is heavy-tailed with a tail index of $\alpha = 5$, and hence determine the distribution of daily maximum times. Use this distribution to estimate the probability that the guarantee will be broken on a given day and hence the expected number of days before this occurs. [You will need to use a formula for the mean of a Fréchet distribution. If $F(x) = \exp(-x^{-\alpha})$ then the mean is given by a gamma function, $\Gamma(1 - (1/\alpha))$. Here we need $\Gamma(0.8) = 1.164$.]

4.3 Tail behavior in a mixture of normals

Suppose that we model the cost of gold at some fixed time in the future (say 10 January 2020) as given by a normal distribution with mean μ and standard deviation σ. Our idea is that there will be an average value that gold has in 2020 but that the price will fluctuate around that value. We do not know what either of these numbers will be, but we think that μ will be about $1500 per oz and we think that the daily volatility, which is measured by σ, will be about $100. More precisely, we represent our uncertainty about μ by saying that μ is drawn from a normal distribution with mean 1500 and standard deviation 100 and we represent our uncertainty about σ by saying that σ is drawn from a normal distribution with mean 100 and standard deviation 20. This process produces what is called a mixture of normal distributions. Use the spreadsheet model BRMch4-MixtureOfNormals.xlsx to explore the way that mixtures of normal distributions impact on tail behavior.

4.4   Calculating the mean of the GPD

Show that if a distribution has density $f(x) = (1/\beta)(1 + \xi x/\beta)^{-1/\xi-1}$ then its mean is $\beta/(1-\xi)$. [Hint: use the indefinite integral

$$\int (x/\beta)(1 + \xi x/\beta)^{-(1/\xi)-1} dx = \frac{x+\beta}{\xi-1}(1 + \xi x/\beta)^{-(1/\xi)}$$

and consider an integral from 0 to R, and then let the upper limit R go to infinity.]

4.5   Mean excess plot when means are not defined

Generate some data from a distribution with a value of $\xi = 1.2$ by using the cell formula $=1/(1-\text{RAND}())^\wedge 1.2$ in a spreadsheet (with such a low tail index the mean of the distribution will not exist). Check what happens to the mean excess plot in this case. You should find that it seems surprisingly well-behaved with a straight line feel through most of the range. Can you explain what is going on? (This shows the value of checking the fit obtained from the maximum likelihood estimator in the way that is done in Figure 4.11.)

4.6   Estimating parameters from mean excess figures

An analyst is looking at data on fee costs from winding up businesses after firm bankruptcy events. He has data on 900 such events and he calculates the mean excess figures using thresholds of $10 million and $20 million. There are 50 events with total fee costs greater than $10 million, with an average for those 50 of $19 million (giving a mean excess of $9 million), and there are 15 events with total fee costs in excess of $20 million, with an average for those 15 of $32 million (giving a mean excess of $12 million). Estimate the values of β and ξ for the generalized Pareto distribution for the excess above $25 million.

4.7   Estimating VaR and ES using extreme value theory

Use the process described for the Amazon stock data to estimate the 99% VaR and 99% ES for daily losses on the S&P 500 index. The data are given in the spreadsheet model BRMch4-S&P500.xlsx.

5

Making decisions under uncertainty

Do we want stable prices?
Rico Tasker is in charge of fresh fruit and vegetable sales at a large retail chain. An important product for Rico is tomatoes. The price the retailer pays is fixed at wholesale fruit markets and varies according to the weather and the season. The retailer makes a markup of between 60% and 72% on sales, with an average markup of 66%. These high markups are needed to cover the cost of storage, handling and other retail expenses. Any tomatoes not sold after five days are thrown away and, on average, this amounts to 10% of the tomatoes bought. The average wholesale price of tomatoes is $3 per kilo. After discussions with the particular grower who currently provides about three quarters of the tomatoes that the retailer sells, Rico goes to his boss, Suvi Marshall, with a proposal that the grower be offered a fixed price of $3 per kilo throughout the year and that the retailer sell the product with a price promise at $4.99 a kilo. This guarantees the same markup (1.99/3 = 0.663) and by working with a single grower with fixed prices the whole process will be simpler. The grower has agreed to meet all the retailer's requirements up to a limit given by the grower's entire output in any week. Rico argues that, in addition to reducing management costs, making this choice will remove part of the risk faced by the retailer by eliminating volatility in price.

'But what happens when there are a large number of tomatoes and everyone else has low prices?' says Suvi 'Won't that make our sales much lower?'

'Yes, I guess there will be a drop, but many people come to our shop for all their fruit and vegetables,' Rico says, 'so we will still have healthy sales. Besides, we should make more sales at times when there is a shortage when everyone else has higher prices.'

Business Risk Management: Models and Analysis, First Edition. Edward J. Anderson.
© 2014 John Wiley & Sons, Ltd. Published 2014 by John Wiley & Sons, Ltd.
Companion website: www.wiley.com/go/business_risk_management

Suvi is still not entirely convinced. 'What about the minimum level of supply from the grower?' she asks. 'If there is a general shortage, don't we normally sell most of what we have anyway? So will we really sell more in those periods?'

On the other hand, Suvi is quite attracted by the idea of a price promise, seeing the marketing potential of a guarantee that prices will be fixed for a full year. But this is a commitment that could lead to a bad outcome. Sometimes weather patterns persist for a long time. What if there was six months of high availability and low wholesale prices? Or six months of shortage? Moreover, she knows that the actual profit made depends critically on the amount that goes to waste. If the 10% average was to creep up to 12% or 13% it would make a big difference. Overall she faces a difficult decision: paradoxically it is uncertainty about outcomes that makes this hard, even though the proposed change is designed to reduce uncertainty.

5.1   Decisions, states and outcomes

In the previous chapters we were concerned with understanding (and measuring) the characteristics of risk in terms of probabilities, and the consequences in terms of costs. Now we turn to the question of how a manager should behave in a risky environment. In Chapter 6 we will focus on how individuals *actually* behave when there are uncertainties. But we start with a normative, rather than descriptive, view: given the need to make a decision between alternatives, each of which carries risks, how should a manager make this decision?

It is helpful to distinguish carefully between the things that we can control, these are the *decisions* we take, and the things that happen that are outside of our control, these are the *events* that occur. We can think of the complete decision problem as having five elements: decisions; events; different possible outcomes; probabilities of different events; and the value we place on different outcomes. We will deal with these in turn.

5.1.1   Decisions

A decision is actually a choice between possible actions. If only one thing can be done, then there is no decision to be made. In this chapter we will focus on decisions made at a single point in time and that makes things simpler. (In Chapter 7 we will look in more detail at dynamic problems where a succession of decisions needs to be made.) A decision could involve the choice of a variable, for example we might decide how high to build a sea wall, or how much inventory of some product to hold. In these cases there are effectively an infinite number of possible choices, but we will concentrate on the situation in which there is only a finite set of possible actions. This will make our discussion much simpler, and in practice a decision with an entirely free choice for some continuous variable can usually be approximated through giving a large enough menu of possible choices.

5.1.2   States

We treat events as random; we may have knowledge of the likelihood of different events, but we cannot forecast exactly what will happen. We refer to the uncertain events, or the values of any uncertain variables, as the *state*. The state is unknown to the decision maker at the time when the choice of an action is made. The list of all possible states that could possibly occur is called the *state space*. One good way to think about the states is to imagine a game in which one of the players is the decision maker and the other player is 'nature'. Both players get a chance to move: the decision maker chooses his actions and nature chooses hers. In this view the state is simply what nature chooses to do.

The state captures all of the uncertainty involved in the decision problem. For example, suppose that we want to model a situation in which we invest $1000 in a stock on Wednesday if its price is higher at the close on Tuesday than it was at the close on Monday. We will need to decide what to do if the closing prices on the two days are the same: suppose that we toss a coin to decide whether to invest in this case. We might decide to model this by saying that the states are the difference in prices between Monday and Tuesday and there are three actions: invest, not invest and toss a coin. But this leaves some uncertainty out of the state description, and instead we need to include both the change in price and the coin toss result within the description of the state.

5.1.3   Outcomes

The action we take, and the random events that occur, together determine what happens to us: this is the *outcome* or consequence of taking a particular decision. The outcome will often contain several dimensions: for example, an investment decision will lead to a dividend income stream as well as a final portfolio value; a decision to abandon a new product launch will lead to consequences not only for profit but also for reputation; a decision to relocate a manufacturing facility will lead to both direct costs and more indirect costs associated with travel to work times for employees. The outcome needs to take into account everything that can have an impact on the decision we make.

We can summarize this framework through the diagram of Figure 5.1, which shows:

- The decision maker choosing between a set of available actions.

- The possible states that can occur.

- The outcome that is reached as a result of the combination of the choice of action by the decision maker and the state that occurs.

All this is relatively straightforward, but a word of warning is in order. The terminology that we have used is not quite universal. Sometimes the word 'outcome' is used simply to refer to the particular state that occurs rather than the consequence of a decision. Also it is quite common for people to use the word

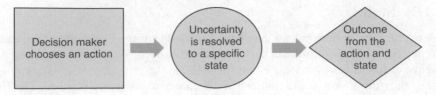

Figure 5.1   Framework for actions, states and outcomes.

'state' to refer to the information available to a decision maker at a certain point in time (this happens often when dealing with problems which evolve over time). So, for example, we might follow the amount of money that a gambler has after a number of different gambles at the roulette wheel and refer to this as the state at time t. This is really the state of the gambler and not a state of nature. When the term 'state' is used in this way it will be determined by the random states of nature as well as the decisions made by the gambler; so, in our terminology, this is more of an outcome than a state.

Once we have set in place the framework of decisions and states together with the outcomes that arise from every combination of action and state, we need two further components to complete our decision model.

5.1.4   Probabilities

We need to know the probability that different states occur. In previous chapters we have been happy to talk about probabilities for events (i.e. probabilities for subsets of the state space). In doing so we have sidestepped what is really a topic of considerable debate. It may be obvious what we mean when we say that the probability of rolling a six with a dice is 1/6, but few real events can have their probabilities calculated so simply. We may say that the probability that the US has a recession (two quarters of negative growth) at some point in the next 10 years is 60%. But this is simply an informed guess: either this event happens or it doesn't, and the chances are greatly influenced by innumerable factors both within and outside the US (for example, policy decisions by governments around the world). Moreover, there are factors that can affect the likelihood of a recession that go beyond a simple political calculation, such as climate change, natural disasters, terrorist action and wars. However, if we have to make a business decision that is impacted by the state of the economy in the future, then there is an implication that we need to take account of the possibility of a recession. This leads us to make some sort of subjective judgment of this probability, and even if we don't write down probability values (or include them in spreadsheets), if a business decision has outcomes that depend on this uncertain event then the probabilities are likely to be taken account of in our decision process in some implicit way – in which case it is probably better to have them out in the open and subject to debate. Here we will proceed on the assumption that the decision maker has an agreed set of probabilities for the events involved in the decision problem. There are some alternative approaches to decision problems that can be used when the uncertainties are large enough that working with a specific subjective probability is dangerous: these techniques of robust optimization are discussed in Chapter 8.

5.1.5 Values

Finally, and critically, we need to know the value that we place on different outcomes. Our decision model has to have a well-defined way of comparing different outcomes. In fact, we need to go beyond the comparison of simple outcomes; one choice of action might be certain to lead to outcome A, while an alternative is equally likely to lead to either outcome B or outcome C. If C is preferable to A, but A is preferable to B then making a decision between the two possible actions will be hard. This is the central problem that we address in this chapter: 'How can a decision maker choose between different actions when each possible choice leads not to a single outcome but to a range of outcomes with different probabilities?'

Another way to represent a situation like this is to draw a decision tree with different paths in the tree indicating different decisions and different states that can occur. This has been done for a simple problem in Figure 5.2. In this case there are just two choices of action for the decision maker and two states that can occur. This gives a combination of four different outcomes. Often we will write probabilities against the different states so that we can see the likelihood of different outcomes.

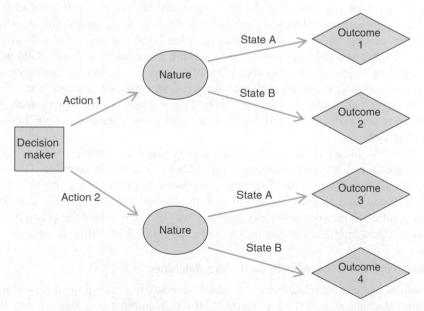

Figure 5.2   A decision tree with two actions and two states.

In the model we have described, the arrows suggest a movement in time as well. It makes sense to start at the point where a choice of action is made, since our whole interest is in the best choice for the decision maker. Any uncertainty in the state of nature that is resolved before the decision point will no longer be relevant. The state space needs to deal with all the uncertainty about the state of nature which will be resolved *after* the decision is made. Often the choice of action determines what happens, and thus the probabilities and possible states evolve differently depending on the action that is taken (we will see this in some of the examples that we look at later). From a conceptual point of view, this makes things more complicated since it introduces a dependence between actions and states (rather than the interaction between the two only occurring with the outcome), but from a practical point of view, there is no difficulty with drawing an appropriate decision tree and making the required calculations.

5.2   Expected Utility Theory

5.2.1   Maximizing expected profit

We want to start with the simplest situation, so let us suppose that the only thing which matters for our business is the overall profit. Hence, we assume that there is a single dollar number associated with each possible outcome and we do not need to allow for any of the less quantifiable aspects of a decision. Sometimes, even if there are other factors to consider, we can price these to give a final result expressed in dollars. For example, if we are considering moving a call center operation overseas then the lower standard of spoken English could lead to our customers being less satisfied and this needs to be taken account of in our decision process. We need to ask ourselves what is the real cost arising from this lower level of service – perhaps we should think about how much we would need to spend on improvements on other parts of our operations in order to make our customers as happy overall with our service under the new arrangements with an overseas call center as they are now. In any event, if we can convert the service issues into a dollar number then we will have a better basis for making this decision.

The first proposal we might make, and perhaps the simplest choice for a manager, is to maximize expected profit. This is entirely reasonable. Given a number of possible courses of action, a manager calculates the expected profit for each and then chooses the action with the highest expected profit value. The decision tree framework gives a convenient way of representing the problem and also of calculating and comparing the expected profit from different actions.

Example 5.1   Investment in tunneling machine

Consider Matchstock Enterprises, which is considering investing in a new tunneling machine at a cost of $520 000. This is required for a specific job that Matchstock has taken on and the machine will be sold in two years' time for

$450 000. A similar second-hand machine is available right now at $450 000, and will be able to do the job satisfactorily. After two years this (now four-year-old) machine could be sold for $400 000. The main difference is in reliability and what happens if there is a breakdown. With a new machine parts will be covered throughout the first two years, leaving only the labor costs and the cost of lost working time. Matchstock has taken expert advice and believes that with a new machine there will be a 0.25 chance of a single breakdown and a 0.05 chance of two breakdowns during the two-year project (with a probability of 0.7 of no breakdowns) and each breakdown will cost $45 000 in total. With a second-hand machine, the company believes that there will be a 0.3 probability of a single breakdown and a 0.1 probability of two breakdowns (with 0.6 probability of no breakdowns). In this case each breakdown is estimated to cost $55 000.

Figure 5.3 shows how this is represented as a decision tree. In comparison with Figure 5.2 we have shrunk the nodes and put information (including probabilities) on the arrows. In this case we can calculate the total costs related to the tunneling machine for each of the possible outcomes, and this is shown in the right-hand column of the figure. For example, buying a new machine and getting two breakdowns leads to a cost of $70 000 from the loss in value of the machine over two years plus a cost of $45 000 incurred twice, giving $160 000 in total. These total costs are taken off the profit to give a final profit, and so maximizing expected profit is equivalent to minimizing expected cost. We have

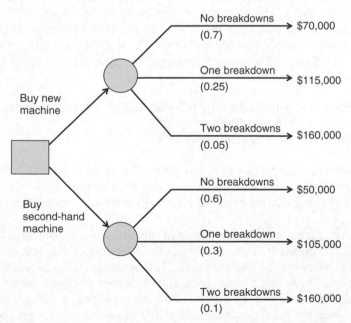

Figure 5.3   The possible outcomes and their probabilities for Matchstock Enterprises.

taken no account in this of the 'time value of money', so there is no discounting of costs.

What should Matchstock do to maximize its expected profit? Buying a new tunneling machine incurs expected costs of $0.7 \times 70 + 0.25 \times 115 + 0.05 \times 160$ (in \$1000s), which works out to \$85 750. Buying a second-hand machine gives expected costs of $0.6 \times 50 + 0.3 \times 105 + 0.1 \times 160 = 77.5$, i.e. \$77 500. So, choosing the second-hand tunneling machine has lower costs and will maximize the expected profit. $\qquad\square$

This methodology could also be applied to the decision facing Suvi Marshall in the tomato-purchasing scenario we gave at the beginning of the chapter. The uncertainty relates to the weather and hence the conditions of shortage or surplus. Looking at the problem in this way will encourage Suvi to make a more detailed investigation of likely waste figures alongside sales estimates for different scenarios. Where there is lack of information, that too can be included as uncertainty in the analysis. In this case it seems unlikely that the decision tree analysis will produce an unequivocal recommendation to go with the fixed-price scheme or not, but it will certainly help in establishing the critical parameters for this decision.

5.2.2   Expected utility

Rather than dealing directly with money, or profit made, it is useful to introduce the idea of a utility derived from the profit made. To see why we might want to do this, we will show that most people, especially when dealing with personal decisions rather than decisions they take as a manager, will not just maximize expected profit. For example, suppose that you wish to choose between the following two options:

Choice A: With probability 0.5 gain \$1000; and with probability 0.5 lose \$800,

Choice B: With certainty gain \$99.

In this case choice A has an expected profit of $0.5(1000) + 0.5(-800) = \$100$ and this is greater than the profit from choice B (a certain profit of \$99). So, a decision maker maximizing expected profit will definitely choose A in preference to B.

However, facing exactly this choice in practice, most people would have no hesitation in choosing B. This is not a misunderstanding of the choices available, or a failure to do the simple arithmetic. Instead it represents a reaction to the unpleasantness of having to hand over \$800 if a coin toss goes the wrong way. Given that there is only a \$1 difference in the expectations, the great majority of people will opt for the certainty of a \$99 payoff. If we reflect on what is happening here then we can see that our choice depends partly on our current financial circumstances and partly on our taste for gambling.

One approach to this question is to define an individual *utility function* that determines how valuable (or how costly) it would be for us to gain (or lose) different amounts of money. In some form this idea can be traced back to the cousins Nicolas and Daniel Bernoulli who considered how players should behave in games of chance. The Swiss mathematician called Gabriel Cramer, in writing to Nicolas Bernoulli in 1728, talked of a utility function in the following way:

> 'The mathematicians estimate money in proportion to its quantity, and men of good sense in proportion to the usage that they may make of it.'

The idea here is that we may value $2000 as being worth to us less than twice as much as $1000 – it all depends on how we are likely to use the money.

One of the problems or paradoxes that had exercised the Bernoulli cousins is a game in which we are offered a prize that depends on tossing coins. If the first toss comes up heads then we win a dollar and the game finishes. If the first toss is a tail and the next toss is a head then we win two dollars and the game finishes. But if the first two tosses are tails and the next is heads then we win $4 and the game finishes. More generally, if we have n tails tossed and the $n + 1$th toss is a head, we will win 2^n. If this is the arrangement, how much would we be prepared to pay to enter this game? Most people might pay $5 or maybe $10, but no more.

If we use an expected profit calculation then we should be happy to pay any amount less than the expected prize value in the game, so we need to calculate this expectation. Suppose that we play just three rounds. It is easy to see that we have a 1/2 probability of getting $1; a 1/4 probability of getting $2 and a 1/8 probability of getting $4. Our expected winnings are $(1/2) + (1/4) \times 2 + (1/8) \times 4 = 3/2$. But if we play on for more tosses we can see that each extra toss adds a term like $(1/2) \times (1/2^n) \times 2^n = 1/2$. After 50 tosses our expected winnings would be $25. As we keep playing the sums of money become exponentially large, but the probabilities of winning these amounts become tiny. Overall it is clear that the expected value is infinite. (If we were using the terminology of Chapter 4 we would say that this is a very heavy-tailed distribution so that expectations don't exist.)

We might well object that the amounts of money here are crazy – after 30 throws without a head we are set to receive 2^{30} which is more than a billion dollars. But the resolution that Daniel Bernoulli gave to this paradox (usually called the Petersburg paradox) rests on the idea of utility. For each extra toss, the amount won doubles but the probability of winning this amount halves. From an expected profit viewpoint the contribution of each term is the same (and equal to 0.5 in the way we have done this calculation above). But Bernoulli argued that even if having $2 was twice as valuable as $1, receiving a prize of $200 million dollars was not twice as valuable as receiving a prize of $100 million. For most people, $100 million is such a large amount (enough to live in luxury without having to work) that it would be foolish to think that the additional benefit they receive from the extra $100 million is of the same value as the first $100 million. But this is the implication of a pure expected profit calculation; we should be indifferent between receiving $100 million for sure and tossing a coin and getting $200 million only if we win.

So these arguments lead us to a theory of decisions based on utilities. Instead of expected profit we need to look at expected utility, and the theory is called *Expected Utility Theory* (EUT). In EUT, in order to compare the two choices A and B introduced above, we need to know the utility function that we have for different possible wealth values. Figure 5.4 shows a possible utility function. We have made this zero at our current wealth, so that the horizontal axis simply gives the profit (or loss) from the choice made. On the positive side, the utility function curves downwards, so that gaining $1000 is less than twice as good as gaining $500. On the negative side, it also curves downwards, so that losing $1000 is more than twice as bad as losing $500.

Looking at Figure 5.4 it is easy to see that, in this case, the negative utility at −$800 is bigger than the positive utility at $1000. Since choice A gives equal chance to these two possibilities, the expected utility for A will be negative. Using EUT with this utility function we would not choose A even if the alternative was to receive nothing. Thus, we certainly prefer choice B with a guaranteed gain.

Notice that adding a constant to all the outcomes amounts to looking at the utility function at a different point on the curve and, because utility functions are nonlinear, this might lead to a different choice being made in a decision problem. For that reason we should think of the utility as a function of the *total wealth* for an individual, rather than being associated with just the changes from the current wealth. However, we can always add or subtract a constant from all the utility values without changing the preferences between different options. This means that we can scale things to make the utility of our current wealth zero: effectively

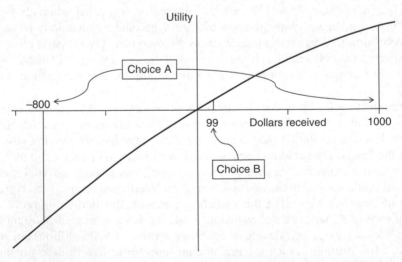

Figure 5.4   Comparing choices A and B using a utility function.

we can work with utilities for changes in wealth provided that we keep in mind that everything is ultimately tied back to a utility for overall wealth.

Worked Example 5.2   Deciding between two funds

Suppose that an investor is deciding between two investment funds: a growth fund and a stable fund. In the past three years there has been one bad year in which the stable fund lost 5% and the growth fund lost 25%; one good year in which the stable fund gained 15% and the growth fund gained 30%; and one medium year in which the stable fund gained 5% and the growth fund gained 10%. The investor has no way of knowing what will happen next year, but by looking at the relative probabilities of different outcomes she estimates that bad years occur three years in ten, good years occur three years in ten, and medium years occur four years in ten. Assume that the investor's utility for x thousand dollars is given by $\log(x)$. Given she has \$100 000 to invest, which option has the highest expected utility?

Solution

Table 5.1 shows, for each fund option, the outcomes, the utilities (calculated as log to base e) and the expected values. For example, the expected utility for the growth fund is calculated from $0.3 \times 4.317 + 0.4 \times 4.701 + 0.3 \times 4.867 = 4.636$.

We can see that, even though the growth fund has a higher expected value than the stable fund, the expected utility is a little lower than for the stable fund. So we can conclude that, with this utility function, the investor should choose the stable fund. □

5.2.3   No alternative to Expected Utility Theory

Expected Utility Theory is a powerful way to think about making decisions in a risky environment: and it represents a rational approach when maximizing expected profits is inappropriate. But it is interesting to ask whether there are other formulations we might use. For example, we might consider some way in which we take account directly of the variance of the dollar outcomes rather than doing this indirectly through the shape of the utility function. Is our risk-taking

Table 5.1   Different outcomes for the investment choice of Example 5.1.

	Bad year	Medium year	Good year	Expected value
Probability	0.3	0.4	0.3	
Value	95	105	115	105
Utility	$\log(95) = 4.554$	$\log(105) = 4.654$	$\log(115) = 4.745$	4.651
Value	75	110	130	105.5
Utility	$\log(75) = 4.317$	$\log(110) = 4.700$	$\log(130) = 4.867$	4.636

propensity something that should be considered over and above the utility function? It is a remarkable fact that Expected Utility Theory can be derived from three (very reasonable) axioms for choices: ordering, continuity and independence. The consequence is that, if we accept the axioms, then we do not need to consider any more complex decision algorithm.

Before embarking on a description of the axioms, we need to introduce some notation and terminology. We will use the term *prospect* to describe a choice with a whole set of possible outcomes, each with a probability (like the choices A and B in the previous example). Each prospect is a single (circle) node in the decision tree, representing a single choice for the decision maker.

More formally, we assume that a prospect has a finite set of outcomes. A prospect q involves an outcome x_1 with probability p_1, an outcome x_2 with probability $p_2, \ldots$ and an outcome x_n with probability p_n. It is helpful shorthand to write a prospect as $(x_1, p_1; x_2, p_2; \ldots; x_n, p_n)$ so that each outcome is followed by its probability. Often we will leave out a zero outcome, so that the prospect ($100, 0.3; -$200, 0.2$) is taken as meaning a 0.3 probability of receiving $100; a 0.2 probability of losing $200; and a 0.5 probability of getting nothing.

If the set of consequences simply contains dollar amounts, then a prospect is just a discrete random variable with a probability distribution on dollar outcomes. There are other terms that are used instead of 'prospect', (for example, some authors use the terminology of lotteries). For most of this chapter we will be discussing distributions that are discrete; we will not need to deal with any kind of continuous prospect.

We use the symbol $\succeq$ to indicate preference between two prospects. Thus, we say $q \succeq r$ when we mean that q is preferred to r. The way this is written ($\succeq$ rather than $\succ$) indicates that it should be taken as a weak inequality. What this means is that q might be chosen if both options are available; it does not mean that q is always chosen when both options are available. Thus, it incorporates the possibility that the decision maker is indifferent between the two prospects. In fact, if we are indifferent between q and r then both $q \succeq r$ and $r \succeq q$.

Now we introduce the three axioms. In each case we might want to consider how reasonable they are – if we accept all of them, then it turns out that we have to accept Expected Utility Theory as giving a complete description of the way that a rational decision maker should behave.

Axiom 1: Ordering

The ordering axiom really has two parts: *completeness* and *transitivity*. Completeness entails that for all choices q, r: either $q \succeq r$ or $r \succeq q$ or both. This just means that the decision maker has a consistent way of making decisions between prospects.

Transitivity requires that for all prospects q, r, and s: if $q \succeq r$ and $r \succeq s$, then $q \succeq s$. This again seems entirely obvious. If a decision maker prefers q to r, but finds r preferable to s, then it is hard to see how it could be wrong to choose q in preference to s.

Axiom 2: Continuity

Continuity requires that for all prospects q, r, and s where $q \succcurlyeq r$ and $r \succcurlyeq s$: we can find a probability p such that we are indifferent between r and the (compound) prospect which has q with probability p, and s with probability $1 - p$. That is, we have $r \succcurlyeq (q, p; s, 1 - p)$ and $(q, p; s, 1 - p) \succcurlyeq r$.

Axiom 3: Independence

Independence requires that for all prospects q, r, and s: if $q \succcurlyeq r$ then $(q, p; s, 1 - p) \succcurlyeq (r, p; s, 1 - p)$, for all p. In other words, knowing that we prefer q to r, we should still prefer an option with q rather than r even if there is some fixed chance of a different prospect, s, occurring.

If all three of the axioms hold, then it can be proved that preferences can be obtained from expected utilities for some function $u(\cdot)$ defined on the set of outcomes. This function u can then be taken as representing the utility of the decision maker. This is a famous and important theorem proved by von Neumann and Morgenstern. Essentially what the theorem shows is that it doesn't matter how a decision maker decides between prospects: if all decisions are consistent in the way implied by the three axioms, then there must be some utility function, $u_0(x)$, for which all the decisions made will exactly match those that would have occurred if the decision maker had been maximizing expected utility (using u_0 to calculate these expectations). This will be true even if the decision maker never consciously formulates the utility function u_0.

We will give a more formal statement of the result, as well as an idea of how it is proved, in the subsection below. But now we want to think about why we would expect the axioms to hold. The ordering axiom is fairly straightforward, but we look at the other two in more detail.

5.2.3.1   Continuity

The continuity axiom supposes that there are three options. First there is a very attractive option q: this could be a prize of a 100-day round the world cruise. Then there is a less attractive option s: we can think of no holiday prize at all. Finally there is an intermediate option r: this might be a prize of a 14-day cruise around the Mediterranean. We don't have to worry about the fact that some people who hate spending time at sea might find the 100-day option q worse than the 14-day option r. In this thought experiment we are dealing with an individual who definitely has $q \succcurlyeq r$ and $r \succcurlyeq s$. Continuity, as its name suggests, is really all about watching what happens to the preferences when the chance of the better outcome is slowly increased.

We suppose that there is a lottery in which there is a probability p of winning the prize of a 100-day round the world cruise, but if we don't win that prize we get nothing. This is the prospect $(q, p; s, 1 - p)$. If p starts at zero then we cannot win the prize in the lottery and we would therefore prefer to have the result r than a ticket in the lottery. On the other hand, by the time p reaches 1, the lottery is no longer a lottery; it always produces the prize. So, with $p = 1$ a lottery ticket will be preferred to having the result r. The axiom simply states that there must be an intermediate value of p at which the decision maker is indifferent between having a ticket for the lottery and taking the prize of r, a 14-day cruise. This is just the value of p at which we swap from preferring the outcome r to preferring the lottery ticket.

This axiom could perhaps fail if there was an infinitely bad outcome, so, however small the probability of it occurring, the prospect automatically becomes very bad. Of course, death is the ultimate bad outcome here, and it is sometimes argued that this would make the continuity axiom break down. If q is getting a single dollar; r is getting nothing, but s is dying, then obviously $q \succcurlyeq r$ and $r \succcurlyeq s$. But does it therefore follow that there will be some probability p very close to 1 where we are indifferent between (A) getting nothing and (B) getting a dollar with probability p but losing our life otherwise?

Against this objection we may observe that we often, in practice, operate in exactly this way. A coffee costs \$4 on this side of the street, but just as good a coffee costs only \$3 on the other side of the street. However, if we cross the street for the cheaper coffee aren't we running some tiny risk of being killed by a reckless driver? Gilboa observes that whether or not we cross the street may depend on how the decision is put to us:

'...If you stop me and say "What are you doing? Are you nuts to risk your life this way? Think of what could happen! Think of your family!" I will cave in and give up [the dollar saving].'

This demonstrates the importance of the way that a choice is framed and we will discuss this a little further in the next chapter: it would be foolish to assume that a real decision maker always makes the same decision when faced with the same two options.

5.2.3.2 Independence

One way to think of the axiom of independence is to see it as related to decisions made ahead of time. If we prefer q to r, the 100-day cruise to the 14-day option, then this should still be true even if a third option may end up occurring. In this example, if s is the option of no prize at all, the axiom states that if given a straight choice between prizes q and r we prefer q, then given two lotteries, each with, say, one in a 1000 chance of winning, we would prefer to be holding a ticket for the lottery with q as the prize than holding a ticket for the lottery with r as the prize; and this remains true no matter what the probability of winning, provided it is the same for the two lotteries. If we play the lottery and lose then there is no decision to be made. If we play the lottery and win and are then offered a choice between the prizes q and r, we have already said that we would choose q. There would thus be something odd about choosing the r lottery, if the q lottery were available. It would amount to a kind of dynamic inconsistency where we make a different decision ahead of time to the decision we make when it comes to the final choice. Of course, this sort of inconsistency does sometimes occur in our choices, but we might prefer to be more consistent and the independence axiom is simply a method to enforce that. In the next chapter we will explore in much more detail the way that the axiom of independence may fail for real decision makers facing real choices.

5.2.4   *A sketch proof of the theorem

The von Neumann–Morgenstern theorem can be stated as follows:

Theorem 5.1   Suppose that a preference relation $\succeq$ on the set of all prospects satisfies the axioms of ordering, continuity and independence. Then there is a utility function, u, defined on all the possible outcomes, such that the utility function U on prospects, derived from u by the formula

$$U(x_1, p_1; \; x_2, p_2; \ldots; \; x_n, p_n) = \sum_{i=1}^{n} p_i u(x_i),$$

has the property that for any prospects q and r, $U(q) \geq U(r)$ if and only if $q \succeq r$. Moreover, any other utility function v on outcomes which has this property must be a positive linear transformation of u (i.e. $v(x) = a + bu(x)$ for some a and b with $b > 0$).

We will not give a complete proof of Theorem 5.1, but (if you have the mathematical interest) it is worthwhile looking at the most important steps in such a proof. To do this we need to manipulate prospects and the first thing to note is that when prospects occur inside other prospects, we can expand them in order to get to a single list of outcomes and their probabilities. For example, if q_1 is the prospect which has $100 with probability 0.5 and $50 with probability 0.5 and $q_2 = (\$100, 0.5; q_1, 0.5)$ then we can expand the prospect q_1 to get

$$q_2 = (\$100, 0.5; \; 0.5(\$100, 0.5; \; \$50, 0.5)) = (\$100, 0.75; \$50, 0.25).$$

We suppose that we have a finite set of prospects, which implies that there is a finite set of outcomes amongst which different prospects distribute their probabilities. We will work in stages.

Step 1   The ordering axiom can be used to establish a best and a worst outcome amongst the set of outcomes. We do this by taking the outcomes one at a time and comparing them with the best from amongst the outcomes already considered. The overall best must have been compared directly with every outcome which came after it in the order and, by transitivity, will also be better than everything that came before (we will not try to give any details). Here we are dealing with outcomes rather than prospects, but an outcome is essentially the same as the prospect that assigns a probability 1 to that outcome. The same procedure works to find the worst outcome as well. We call the best outcome x^* and the worst outcome x_*.

Step 2 Now we assign the utility value of 0 to x_* and a utility value of 1 to x^*. For any other outcome x in the list we assign a utility $u(x)$ equal to the probability p such that we are indifferent between the prospect $(x, 1)$ and the prospect $(x^*, p; x_*, 1 - p)$ (using the continuity axiom).

Step 3 Next we will use the independence axiom, which allows a free choice of the probability p and the exact alternative chosen, s. Suppose that we are given numbers α and β with $1 > \alpha \geq \beta \geq 0$. We choose $p = \alpha - \beta$ and $s = (x^*, \beta/(1 - p); x_*, (1 - \beta/(1 - p)))$ and substitute these values into the independence axiom Thus, since $x^* \succcurlyeq x_*$ we get

$$(x^*, \alpha - \beta; \ x^*, \beta; \ x_*, 1 - \alpha) \succcurlyeq (x_*, \alpha - \beta; \ x^*, \beta; \ x_*, 1 - \alpha),$$

which simplifies to

$$(x^*, \alpha; \ x_*, 1 - \alpha) \succcurlyeq (x^*, \beta; \ x_*, 1 - \beta). \tag{5.1}$$

This is a useful intermediate result which will apply even when x^* and x_* are not the best and worst outcomes. Essentially we have shown that if we form a prospect from two outcomes, then increasing the probability of the better of the two makes the prospect more attractive.

Step 4 Now we show that the utility of the outcomes matches the preference ordering between them. Suppose that $u(x) \geq u(y)$, then from Formula (5.1)

$$(x^*, u(x); \ x_*, 1 - u(x)) \succcurlyeq (x^*, u(y); \ x_*, 1 - u(y)).$$

But if we look back at how $u(x)$ and $u(y)$ are defined, we see that this is equivalent to

$$(x, 1) \succcurlyeq (y, 1).$$

Step 5 Now suppose that we have a prospect $(x, \alpha; \ y, 1 - \alpha)$: we want to show that this has utility $\alpha u(x) + (1 - \alpha)u(y)$; in other words, we want to show indifference between $(x, \alpha; \ y, 1 - \alpha)$ and $(x^*, \alpha u(x) + (1 - \alpha)u(y); \ x_*, 1 - \alpha u(x) - (1 - \alpha)u(y))$. We do this by observing that

$$(x, 1) \succcurlyeq (x^*, u(x); \ x_*, 1 - u(x)),$$

and hence

$$(x, \alpha; \ y, (1 - \alpha)) \succcurlyeq (x^*, \alpha u(x); \ x_*, \alpha - \alpha u(x); \ y, (1 - \alpha)). \tag{5.2}$$

But because we also have

$$(y, 1) \succcurlyeq (x^*, u(y); \; x_*, 1 - u(y)),$$

we can take the prospect on the left-hand side and obtain the following from the independence axiom

$$(y, (1 - a); \; x^*, au(x); \; x_*, a - au(x))$$
$$\succcurlyeq (x^*, (1 - a)u(y); \; x_*, (1 - a)(1 - u(y)); \; x^*, au(x);$$
$$x_*, a - au(x)).$$

The prospect on the right-hand side can be simplified to

$$(x^*, \alpha u(x) + (1 - \alpha)u(y); \; x_*, 1 - \alpha u(x) - (1 - \alpha)u(y)).$$

Thus, we can combine Formulae (5.2) and (5.1) by transitivity to obtain the relationship:

$$(x, \alpha; \; y, 1 - \alpha)$$
$$\succcurlyeq (x^*, \alpha u(x) + (1 - \alpha)u(y); \; x_*, 1 - \alpha u(x) - (1 - \alpha)u(y)).$$

We can repeat all of this with the preference orders reversed to show

$$(x^*, \alpha u(x) + (1 - \alpha)u(y); \; x_*, 1 - \alpha u(x) - (1 - \alpha)u(y))$$
$$\succcurlyeq (x, \alpha; \; y, 1 - \alpha),$$

which finally establishes the indifference we require.

There are a number of things we need to do in order to fill in the gaps here. First the result holds without any restriction on there being just a finite set of possible outcomes or prospects. This requires us to start with a finite set of prospects and then to add another set which lies outside this range (say they are all worse than the worst outcome in the first set) and stitch together the two utility functions we generate.

Second we have not fully included all the components of the argument we need in step 4, which shows that an inequality in utilities for outcomes translates into a preference order. We need to show that the same thing is true for prospects and we also need an 'if and only if' argument.

Also in step 5 we have demonstrated what we want for a prospect with just two alternatives – we need to extend this to prospects with any number of alternatives.

Finally we have not dealt with the uniqueness of the utility function (up to positive linear transformations) – the theorem will not allow us to, for example, square all the utility values. This would leave the ordering of individual outcomes unchanged, but we would lose the ability to get the utility of a prospect as the probability-weighted combination of the individual outcome utilities.

5.2.5   What shape is the utility function?

Where outcomes are monetary, then the utility function is simply a real-valued function defined on the real line. It is helpful to use the terminology of convex and concave functions. A convex utility function has the property that its slope is increasing (or, to put it another way, it has a second derivative which is non-negative). Another way to characterize a convex function is to say that a straight line joining two points on the graph of the function curve can never go below the function. This can be put into mathematical form by saying that a function $u(\cdot)$ is convex if, for any p between 0 and 1,

$$pu(x_1) + (1 - p)u(x_2) \geq u(px_1 + (1 - p)x_2).$$

The left-hand side is the height of a point a proportion $1 - p$ along the straight line between points $(x_1, u(x_1))$ and $(x_2, u(x_2))$ and the right-hand side is the point on the curve itself at this x value. This is illustrated in Figure 5.5. This property of convex functions can be generalized to any number of points. So, a convex function u has the property that

$$\sum_{j=1}^{n} p_j u(x_j) \geq u\left(\sum_{j=1}^{n} p_j x_j\right), \tag{5.3}$$

if the p_j are nonnegative and $\sum_{j=1}^{n} p_j = 1$.

The connection with expected utility is that a convex utility function implies risk-seeking behavior. If there is a prospect having a probability p_1 of achieving x_1 and a probability $(1 - p_1)$ of achieving x_2, then the expected utility is $p_1 u(x_1) + (1 - p_1)u(x_2)$, which we prefer to (its value is greater than) the utility $u(p_1 x_1 + (1 - p_1)x_2)$ that we obtain from the expected result.

More generally, we can say that when the utility function is convex it will always be preferable to choose a prospect involving uncertainty, say

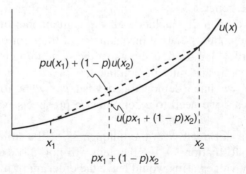

Figure 5.5   If $u(x)$ is convex then a straight line joining two points on the graph of u lies above the u function.

$(x_1, p_1; x_2, p_2; \ldots; x_n, p_n)$ rather than having the expected outcome, $\sum p_j x_j$ with certainty. We can see this from Inequality (5.3), where the expected utility of the prospect, on the left-hand side, is greater than the utility of the expected outcome, on the right-hand side.

The reverse is true for a concave utility function. A concave function is one where a straight line joining two points on the graph of the function curve can never go above the function, and this is equivalent to the function having a decreasing slope. In this case, having the expected outcome $\sum p_j x_j$ with certainty is always preferable to facing the gamble involved in the uncertain prospect. In other words, a concave utility function like the one shown in Figure 5.4 implies risk-averse behavior and this is the usual pattern for utility functions.

Of course, we can also have utility functions which are convex in some areas and concave in others. For example, suppose that the utility for a wealth of x measured in \$100 000 units is

$$u_A(x) = \sqrt{x} - 0.9 \log(x + 1).$$

Though it is not obvious, this utility function turns out to be positive for positive wealth: we draw it in Figure 5.6 for x in the range 0 to 3. We can see that the curve is concave for x below about 1. In fact, though hard to see from the graph, it becomes convex for values above 1. This particular utility function has a derivative

$$u'_A(x) = (1/2)x^{-0.5} - 0.9/(x + 1)$$

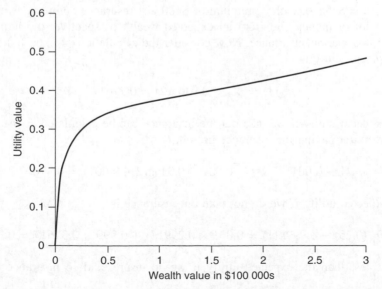

Figure 5.6   A curve showing utilities for different wealth values, using the formula $u(x) = \sqrt{x} - 0.9 \ln(1 + x)$.

which approaches zero as x gets large. If we plot the utility function for much larger values of x, we can see that it is actually concave (the second derivative becomes negative for $x \geq 8.163$). Hence, the function moves from concave to convex and back to concave.

With this kind of utility function we can expect mildly risk-seeking behavior for wealth positions between \$100 000 and \$8 163 000 and risk-averse behavior when wealth levels are below \$100 000. Worked Example 5.3 demonstrates that this will indeed occur through showing that, with this utility function, the same individual could be prepared both to take out insurance (which is risk-averse behavior) and also engage in a lottery for a big prize (which is risk-seeking behavior).

Worked Example 5.3   Insurance and lotteries can coexist

Suppose that an individual with the utility function $u_A(x)$ currently has wealth \$100 000 (corresponding to $x = 1$). There is a small risk of a fire destroying \$75 000 worth of property. This happens next year with probability $1/1000$ and the insurance company charges \$100 to fully insure against this loss. There is also an opportunity to enter a lottery where a single ticket costs \$500 but there is a one in a thousand chance of winning a prize worth \$500 500 (i.e. we get \$500 000 and also the price of our ticket back). Show that the individual will take out insurance and also enter the lottery, though neither choice will increase expected wealth.

Solution

First consider the insurance option. The expected loss is only \$50 000 $\times$ $(1/1000) = \$75$, so, with a premium of \$100, the insurance company is making quite a lot of money and from an expected wealth perspective, the individual should not take out insurance. Now consider the calculation of expected utility. The current utility is

$$u_A(1) = \sqrt{1} - 0.9\log(2) = 0.37617.$$

The expected utility if we take out the insurance can be calculated as the utility we have after paying the insurance premium:

$$u_A(1 - 0.001) = \sqrt{1 - 0.001} - 0.9\log(2 - 0.001) = 0.37612.$$

The expected utility if we do not take out insurance is

$$0.001 u_A(0.25) + 0.999 u(1) = 0.001 \times 0.299171 + 0.999 \times 0.37617 = 0.37609.$$

This is less than the expected utility if we do insure, and so insurance makes sense.

Now we look at the lottery choice. Not buying the lottery ticket leaves us at the current utility of 0.37617; if we do buy the ticket we have a 0.001 probability of winning \$500 000 and a 0.999 chance of losing \$500. This gives a final expected utility of

$$0.001u(6) + 0.999u(0.99500)$$
$$= 0.001 \times (\sqrt{6} - 0.9\log(7)) + 0.999 \times (\sqrt{0.995} - 0.9\log(1.995))$$
$$= 0.37624.$$

So, entering the lottery gives a slightly higher expected utility. The same individual is risk averse on losses (enough to buy a rather expensive insurance product), but risk-seeking enough on gains to enter a lottery. □

If the utility function is linear (giving a straight line graph) then there is neither risk aversion nor risk seeking and we say that the decision maker is *risk neutral*. It is not hard to see that maximizing the expectation of a function of the form $u(x) = a + bx$ has the same solution as maximizing the expectation of x (provided $b > 0$), so a risk neutral decision maker will just maximize expected profit.

When there is risk aversion, the degree to which the decision maker is risk averse depends on the curvature of the utility function; if the utility function is almost linear, the amount of risk aversion will be minimal. To capture the degree of risk aversion we need to look at $u''(x)$, which is the second derivative of $u(x)$, i.e. the rate at which the slope of u changes. With risk aversion, the slope of the utility function is decreasing and therefore the second derivative is negative. It is common to quantify risk aversion using the *Arrow–Pratt measure of absolute risk aversion*, given by

$$A(x) = -\frac{u''(x)}{u'(x)}.$$

We write $u'(x)$ for the derivative of u and dividing by this is a way of normalizing $A(x)$. With this normalization, scaling u by a constant will not affect the absolute risk aversion.

As an example, consider the logarithmic utility function defined by $u(x) = \log(x)$. Then $u'(x) = 1/x$ and $u''(x) = -1/x^2$. Thus, the Arrow–Pratt risk-aversion measure is

$$A(x) = \frac{(1/x^2)}{(1/x)} = \frac{1}{x}.$$

This decreases towards zero as x gets large. The idea that individuals are less risk averse as their overall level of wealth gets larger is captured by this logarithmic utility function.

5.2.6 *Expected utility when probabilities are subjective

The development of von Neumann–Morgenstern style Expected Utility Theory is fundamental to our understanding of decision making under uncertainty, but it is, in a way, a lopsided development. The assumption is that utilities are unknown (they are deduced from the choices made between prospects) but that probabilities are known precisely.

As we pointed out earlier, decisions often involve choices that are impacted by events that we can have no control over and where even the idea of a probability needs to be treated carefully. It is interesting to ask how a decision maker might behave if she was not prepared to specify fixed probabilities for different events. After all, in everyday life we routinely make decisions without stopping to ask about probabilities. So, if a manager makes a decision without consciously thinking about probabilities, but at the same time is entirely rational and thoughtful about those decisions, can we deduce that there is some consistent decision framework that doesn't use probabilities?

The answer to this question is a qualified 'No' In an important piece of work by Leonard 'Jimmie' Savage published in 1954, it is shown that if choices satisfy some reasonable-seeming axioms then the decision maker must act as though he or she were assigning a subjective probability to each of the possible outcomes that can arise from a choice of action, as well as a utility function on those outcomes, and then decide between choices on the basis of maximizing expected utility as determined by the subjective probabilities.

Putting all this into a solid theory requires a great deal of care. In Savage's model a decision maker has a choice between different actions and these actions will determine the outcomes that go along with the states. Formally, we list all possible states as a set S and an action is treated as a function taking states to outcomes, where X is the set of outcomes. The set of states has to resolve all uncertainty, so that the action simply specifies what outcome occurs in each of the possible states. This way of thinking does not fit well with a decision tree framework, where we imagine taking the decision first followed by the uncertainty being resolved to lead to a particular outcome. In some ways it is like reversing this process: we think of the uncertainty being resolved to a single state and then the decisions map each state to an outcome. The two ways of thinking are not really so different; in each approach a decision and a state of nature together produce an outcome. The reason for proceeding with the more complex idea of actions as maps from states to outcomes is that we will need to make comparisons between all possible actions. That is, we need to be able to imagine an action that specifies particular outcomes for each possible state of nature, without restricting in any way which outcomes go with which states. Having once imagined such an action, we then need to be able to compare it with any other imagined action and answer the question: Which would be preferable?

To get a flavor of the axioms, we will describe three of them (there are seven in total).

Axiom P1

This states that there is a weak order on the actions. So, for actions $f : S \to X$ and $g : S \to X$, either $f \succcurlyeq g$ or $g \succcurlyeq f$ or both, and this relationship is transitive.

Axiom P2

This axiom shows that in comparing actions we only care about the states where the two actions produce a different result. So, if $f \succcurlyeq g$ and there is a set $A \subset S$ with $f(s) = g(s)$ for all states s in A, then the preference between f and g is determined by what is happening for $s \notin A$. More precisely, we can say that if $f'(s) = f(s)$ for $s \notin A$ and $g'(s) = g(s)$ for $s \notin A$ and $f(s) = g(s) = h(s)$ for $s \in A$, then $f' \succcurlyeq g'$.

Before giving the third axiom we need two preliminaries. First we need to be able to make a comparison between outcomes rather than actions. This is simple enough, we say that for outcomes $x, y \in X$, $x \succcurlyeq y$ if the action that takes every state to x is preferred to the action that takes every state to y. If we know we are going to end up with x under some action f and we know we are going to end up with y under the action g, then it no longer matters what is happening with the states.

Second we need to define a property of a set of states (i.e. an event) $A \subset S$ which amounts to saying there is a non-negligible possibility that one of the states in A occurs. The most natural way to describe this is to say that the probability of A is greater than zero, but with the Savage theory we do not have probabilities to work with. Instead we say that the event A is *null* if, for any f and g that differ only on A, these two actions are equivalent (both $f \succcurlyeq g$ and $g \succcurlyeq f$). In other words, what happens for null events makes no difference to our preferences.

Axiom P3

This states that if outcome x is preferred to outcome y, and two actions f and g differ only on a set of states A which is not null, and, moreover, on A, f produces x and g produces y, then f must be preferable to g. Actually, the axiom says more than this, since it also says that the reverse implication is true (equivalently we require that if x is strictly preferred to y then f is strictly preferred to g). Formally, we can write this as follows. For an event A that is not null, if $f(s) = x$ for $s \in A$, $g(s) = y$ for $s \in A$, and $f(s) = g(s) = h(s)$ for $s \notin A$, then

$$x \succcurlyeq y \text{ if and only if } f \succcurlyeq g$$

At this point you may well feel that we have stretched our minds enough without the need to go into further details. The remaining four axioms are a mixed bunch. Axiom P4 relates to a situation where outcome x is strictly preferred to y, and outcome z is strictly preferred to w. The axiom states that if the action which delivers x on A and y otherwise is preferred to the action which delivers x on B and y otherwise, then the action which delivers z on A and w otherwise is preferred to the action which delivers z on B and w otherwise. This is more or less the same as saying that A happens more often than B, but of course we cannot use this language since we do not have a notion of probability yet defined. Axiom P5 simply states that there must be two actions which are not equivalent to each other. Axiom P6 is related to continuity and implies that we can always partition the set of states S sufficiently finely that a change on just one component of the partition leaves a strict preference unchanged. This is quite a strong assumption and it can only work if there is an infinite state space S, and each individual state s in S is null. Often this is not true for problems of interest, but we can make it true by expanding our state space to consider some infinite set of (irrelevant) other events occurring in parallel with our original states. The final axiom P7 is only needed when the outcome set X is infinite, and we will not give a description of it.

To state Savage's theorem we have to understand what might be meant by an assignment of probabilities to all the states in S. This requires a measure μ which assigns a probability to every event $A \subset S$. The measure has to be finitely additive (so $\mu(A \cup B) = \mu(A) + \mu(B)$ if A and B are disjoint), and it also has to be non-atomic in the sense that if there is an event A with $\mu(A) > 0$ and we choose any r, a number between 0 and $\mu(A)$, then we can find a subset $B \subset A$ with $\mu(B) = r$. Once we have a measure μ on the states S and a scalar function defined on the states, we can evaluate the expectation of that function by writing its integral with respect to μ. Thus, finally we are ready to state the theorem.

Theorem 5.2 (Savage) When X is finite, the relationship $\succcurlyeq$ satisfies axioms P1 to P6 if and only if there exists a non-atomic finitely additive probability measure μ on S and a non-constant function $u(x)$, for $x \in X$ such that, for any actions f and g

$$f \succcurlyeq g \text{ if and only if } \int_S u(f(s))d\mu(s) \geq \int_S u(g(s))d\mu(s).$$

Furthermore, in this case μ is unique and u is unique up to a positive linear transformation.

5.3 Stochastic dominance and risk profiles

Suppose that a prospect has n possible outcomes and we put these in order from the worst x_1 to the best x_n. We can do this even if the outcomes do not have monetary values. We can then draw a risk profile and use this to compare two different prospects. For example, suppose that a short-term contract employee Raj currently has a contract position for six months. If Raj does nothing, he believes that there is a 10% chance of his job not being renewed at the end of the contract period, a 35% chance of the job being renewed for another three months, a 35% chance of the job being renewed for another six months and a 20% chance of the job being made permanent. What happens will depend both on Raj's performance and the trading performance of the company. Raj still has two months of his existing contract to run, but believes he has some skills that will help in getting a permanent position and that might be overlooked in the normal process. So he is considering going to his boss and making the case for a permanent appointment straight away. He believes that this will increase the

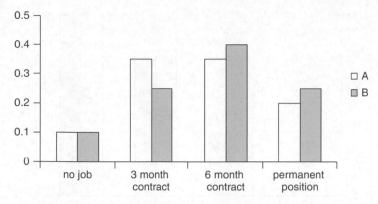

Figure 5.7   Risk profiles for the two options for Raj.

chance of his being made permanent to 25%, and in this case the probabilities of
the other outcomes are 10% job not renewed; 25% job renewed for three months;
40% job renewed for six months. What should Raj do? The risk profiles for these
two choices are shown in Figure 5.7.

It is not obvious how to make a comparison between the actions $A = $ 'do
nothing' and $B = $ 'ask for permanent position' from these risk profiles alone.
The complication here is that we have not specified how much more valuable a
permanent appointment is than a six-month one. And we also have to consider
the relationship of a six-month to a three-month appointment; all that we know is
that there is a preference order: permanent is best, then six months is better than
three months and losing the job is worst. The best approach here is to draw a
cumulative risk profile, as shown in Figure 5.8. Here it becomes clear that action
A has a cumulative risk (of a worse outcome) that is always higher or equal to
the cumulative risk for action B; there is a kind of dominance between the two.
Later we will show that this implies that it is better for Raj to choose option B
and make the case for a permanent appointment, no matter what the exact utility
values he places on the different outcomes. But first we need to more carefully
define what is meant by stochastic dominance.

Given two prospects q and r we can take the combined set of all possible
outcomes and put them in order of increasing value or utility. Notice that there
is no loss of generality in comparing two prospects in assuming they have the
same set of possible outcomes, and if one of the prospects does not include one
of the outcomes, we simply assign zero probability to that outcome.

We say that q stochastically dominates r if, when we take the combined set
of outcomes with probabilities q_i for q and r_i for r (and where q_i and r_i relate
to the same outcome x_i), the following inequalities all hold

$$\sum_{j=m}^{n} q_j \geq \sum_{j=m}^{n} r_j \text{ for } m = 2, 3, \ldots, n \qquad (5.4)$$

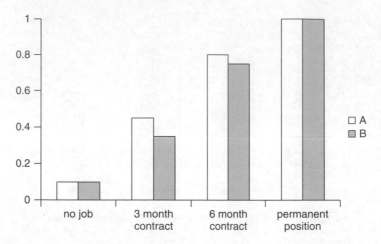

Figure 5.8   Cumulative risk profile for the two options for Raj.

with at least one of these inequalities being strict. Each of these sums goes from an outcome m through all the better outcomes up to the best. Thus, prospect q gives higher probabilities to the higher x_j which are preferable.

There is an alternative definition which uses the fact that the sum of all the q_j values is 1 (since they are probabilities). Using this, we see that Inequality (5.4) can be rewritten

$$1 - \sum_{j=1}^{m-1} q_j \ge 1 - \sum_{j=1}^{m-1} r_j \text{ for } m = 2, 3, \dots, n.$$

This can be rearranged (by swapping the sums to the other side of the inequality and setting $l = m - 1$) to become

$$\sum_{j=1}^{l} r_j \ge \sum_{j=1}^{l} q_j \text{ for } l = 1, 2, \dots, n - 1. \tag{5.5}$$

Thus, we have converted an expression involving probabilities of doing 'better than something' into one involving doing 'worse than something', and this has meant a change in the sign of the inequality. This alternative definition of stochastic dominance is now in terms of the cumulative risk profile. We can see from Figure 5.8 that in the case of Raj's choice between actions A and B, the cumulative probabilities for A are larger than those for B, and this implies that B stochastically dominates A.

This method works when there are no monetary values associated with outcomes, but of course we can also apply the same logic when there are dollar values. When the outcomes are values, we can think of a prospect q as equivalent to a discrete random variable, X say, taking values $x_1, x_2, \dots, x_n$. If we take the prospect r as equivalent to a random variable Y, then Inequality (5.5) can be written as

$$\Pr(Y \le x_l) \ge \Pr(X \le x_l), \quad \text{for } l = 1, 2, \ldots, n - 1,$$

which converts stochastic dominance into a statement about the CDFs of the two random variables. More generally, we say, for two random variables X and Y:

> X *stochastically dominates* Y if the cumulative distribution functions F_X and F_Y, for X and Y, satisfy $F_X(x) \le F_Y(x)$ for every x, and this inequality is strict for at least one x.

Note that sometimes people forget which way round the inequality goes here: it is the random variable (or prospect) with the lower cumulative distribution that stochastically dominates the other. This definition of stochastic dominance does not need the random variables to be defined over just a finite set of outcomes, and we can apply it equally well when X and Y are continuous rather than discrete random variables.

Another point to note is that if $X = Y + k$ for some constant $k > 0$ then

$$F_X(x) = \Pr(Y + k \le x) = F_Y(x - k) \le F_Y(x),$$

since the CDF is an increasing function. So, as we would expect, if X is obtained from Y by adding a positive constant, then X stochastically dominates Y.

Worked Example 5.4   Checking stochastic dominance

Check whether or not there is stochastic dominance between the prospects A and B defined as follows:

$$A = (\$100, 0.2; \$160, 0.2, \$200, 0.6),$$

$$B = (\$100, 0.3; \$120, 0.1; \$180, 0.1; \$200, 0.5).$$

Solution

We start by forming the complete set of payment amounts that occur in either of the two prospects: {\$100, \$120, \$160, \$180, \$200}. These form the rows in a table giving, for each prospect, the probability of receiving that amount or less. This is shown in Table 5.2 which gives, for each prospect, the sums of the form in Inequality (5.5). Each element in the column that shows the sum for A is less than or equal to the corresponding element in the sum for B, and so prospect A stochastically dominates prospect B. (Note that an alternative approach is to look at the sums of probabilities of receiving more than a certain amount and use Inequalities (5.4)).   □

A key result is that, under Expected Utility Theory, if the prospect q stochastically dominates the prospect r, then q will be preferred to r no matter what utility

Table 5.2   Comparison of prospects A and B.

Dollar amount	A probabilities	Sum for A	B probabilities	Sum for B
$100	0.2	0.2	0.3	0.3
$120	0	0.2	0.1	0.4
$160	0.2	0.4	0	0.4
$180	0	0.4	0.1	0.5
$200	0.6	1.0	0.5	1.0

is given to the individual outcomes. The only requirement is that the utilities are strictly increasing, so $u(x_1) < u(x_2) < \cdots < u(x_n)$. In fact, this implication also works in the other direction. If the expected utility for a prospect q is greater than the expected utility for a prospect r for every possible choice of increasing utility function (that has $u(x_1) < u(x_2) < \cdots < u(x_n)$) then q stochastically dominates r. We will show how to prove this result in the next subsection. Sometimes this property (of dominance for any increasing utility function) is used as a definition of stochastic dominance.

Besides proving these results, the next section also introduces a second type of stochastic dominance – called *second order stochastic dominance*. This can be defined by saying that a random variable X second order stochastically dominates Y if $E(u(X)) \geq E(u(y))$ for every increasing concave utility function u (i.e. whenever the decision maker is risk averse). As we will show later, the condition to achieve second order stochastic dominance is weaker than that to achieve ('first-order') stochastic dominance. Rather than requiring $F_X(x) \leq F_Y(x)$ for every x, we need

$$\int_a^x F_X(z)dz \leq \int_a^x F_Y(z)dz, \text{ for every} x,$$

where a is the lower limit for the distribution. So we compare the integrals of the CDFs rather than the CDFs themselves.

5.3.1   *More details on stochastic dominance

We will work with random variables rather than prospects; then we can state the result as follows.

Theorem 5.3   For random variables X and Y, both taking values in the finite range $[a, b]$, the stochastic dominance condition:

$$F_X(x) \leq F_Y(x) \text{ for all } x \in [a, b],$$

holds if and only if

$$E(u(X)) \geq E(u(Y))$$

for all increasing functions u.

Sketch Proof
We will switch to continuous random variables, where the quickest way to prove
this result is to use integration by parts, i.e. we use the fact that

$$\int_a^b g(x)h'(x)dx = [g(x)h(x)]_a^b - \int_a^b g'(x)h(x)dx.$$

Thus, we will assume that the derivative of u exists and there are well-defined
densities f_X and f_Y. The result holds more generally and the sketch proof we
give can be extended to these other cases (e.g. when X and Y are discrete).
    Using integration by parts we have

$$E(u(X)) - E(u(Y)) = \int_a^b u(x)f_X(x)dx - \int_a^b u(x)f_Y(x)dx$$

$$= \int_a^b u(x)(f_X(x) - f_Y(x))dx$$

$$= [u(x)(F_X(x) - F_Y(x))]_a^b - \int_a^b u'(x)(F_X(x) - F_Y(x))dx$$

$$= -\int_a^b u'(x)(F_X(x) - F_Y(x))dx. \qquad (5.6)$$

Here we have used the fact that $F_X(b) = F_Y(b) = 1$ and $F_X(a) = F_Y(a) = 0$.
    Now suppose that $F_X(x) \leq F_Y(x)$ throughout the range $[a, b]$ and $u'(x) \geq 0$.
Then the integral in Equation (5.6) is non-negative and so $E(u(X)) \geq E(u(Y))$.
    To show what we want in the other direction, we suppose that $F_X(c) >
F_Y(c)$. Continuing with our assumption that densities exist, the CDF functions
are continuous, so there will be a small range $c - \delta$ to $c + \delta$ on which F_X is
greater than F_Y, specifically suppose that $F_X - F_Y > \varepsilon > 0$ on this range. Now
consider a utility function defined so that $u'(x) = \varepsilon/(b - a)$ except in the range
$c - \delta$ to $c + \delta$, where $u'(x) = 1/\delta$. In other words, the utility function is almost
flat except in the interval where F_X is greater than F_Y, where the utility function
jumps up sharply. Then

$$E(u(X)) - E(u(Y)) = \int_a^b u'(x)(F_Y(x) - F_X(x))dx$$

$$\leq \int_a^{c-\delta} u'(x)dx + \int_{c-\delta}^{c+\delta} u'(x)(-\varepsilon)dx + \int_{c+\delta}^b u'(x)dx$$

$$\leq \varepsilon - (2\delta)\frac{\varepsilon}{\delta} < 0.$$

Thus, we have shown that whenever there is a point where F_X is greater than
F_Y, we can find a utility function that makes Y preferable to X. □

Now we turn to the second order stochastic dominance result that parallels Theorem 5.3 but applies to concave utility functions. Just as for Theorem 5.3, we will give a continuous version of this, but the result is true also for a discrete random variable.

Theorem 5.4 For random variables X and Y, both taking values in the finite range $[a, b]$, the second order stochastic dominance condition:

$$\int_a^x F_X(z)dz \leq \int_a^x F_Y(z)dz \text{ for all } x \in [a, b], \qquad (5.7)$$

holds if and only if

$$E(u(X)) \geq E(u(Y))$$

for all increasing concave functions u.

Sketch Proof
Again we will use integration by parts to prove this result. We write $\tilde{F}(x)$ for the integral of F. Thus

$$\tilde{F}_X(x) = \int_a^x F_X(z)dz, \text{ and} \tilde{F}_Y(x) = \int_a^x F_Y(z)dz.$$

Then, integration by parts of Equation (5.6) gives:

$$E(u(X)) - E(u(Y)) = \int_a^b u'(x)(F_Y(x) - F_X(x))dx$$

$$= \left[u'(x)(\tilde{F}_Y(x) - \tilde{F}_X(x))\right]_a^b - \int_a^b u''(x)(\tilde{F}_Y(x) - \tilde{F}_X(x))dx.$$

Now note that $\tilde{F}_X(a) = \tilde{F}_Y(a) = 0$, so

$$E(u(X)) - E(u(Y)) = u'(b)(\tilde{F}_Y(b) - \tilde{F}_X(b)) - \int_a^b u''(x)(\tilde{F}_Y(x) - \tilde{F}_X(x))dx.$$
$$(5.8)$$

Hence, if $\tilde{F}_Y(x) \geq \tilde{F}_X(x)$ and $u''(x) < 0$ for all $x \in [a, b]$ then, since $u'(b) \geq 0$, we have $E(u(X)) \geq E(u(Y))$.

To prove the result in the other direction, we suppose that $\tilde{F}_X(c) > \tilde{F}_Y(c)$ at some point c. Then we can find an $\varepsilon, \delta > 0$ with $\tilde{F}_X(x) - \tilde{F}_Y(x) > \varepsilon$ for $x \in (c - \delta, c + \delta)$. Now construct a utility function u with

$$u(x) = x, \text{ for } a \leq x < c - \delta,$$

$$u(x) = x - (x - c + \delta)^2/(4\delta), \text{ for } c - \delta \leq x < c + \delta,$$

$$u(x) = c, \text{ for } c + \delta \leq x \leq b.$$

With this choice of u the derivative is continuous, having $u'(x) = 1$ for $x \leq c - \delta$, $u'(x) = 0$ for $x \geq c + \delta$ and u' linear decreasing from 1 to 0 in the range $(c - \delta, c + \delta)$. Thus, u'' is negative on $(c - \delta, c + \delta)$ and zero outside this range.

Hence, since $u'(b) = 0$, we have, from Equation (5.8)

$$E(u(X)) - E(u(Y)) = \int_{c-\delta}^{c+\delta} (-u''(x))(\tilde{F}_Y(x) - \tilde{F}_X(x))dx$$

$$\leq \int_{c-\delta}^{c+\delta} (-u''(x))(-\varepsilon)dx$$

$$= \varepsilon(u'(c + \delta) - u'(c - \delta))$$

$$= -\varepsilon$$

and the random variable Y has a higher expected utility than X.   □

If X second order stochastically dominates Y, then the condition (5.7) implies that F_X starts by being lower than F_Y. For ordinary stochastic dominance (sometimes called first order stochastic dominance), F_X remains lower than F_Y. But with second order stochastic dominance, F_X can become larger than F_Y provided its integral stays below the integral of F_Y. In other words, the area under the graph of F_X up to a point x is less than the area under the graph of F_Y up to the same point x. This can be seen as a statement about what happens to the area of the region *between* the two graphs of F_X and F_Y. The situation is illustrated in Figure 5.9. In this diagram, the area marked A is larger than the area marked B. It is not hard to see that $\tilde{F}_Y(x) - \tilde{F}_X(x)$ is equal to $A - B$ when $x = x_0$ (the second crossing point) and greater than this for x on either side of x_0. Thus, with the CDFs shown in the diagram, the random variable X second order stochastically dominates Y.

Finally we will show that for a distribution that is symmetric about its mean (like the normal distribution), the process of shrinking the distribution closer to the mean will produce a new random variable that second order stochastically dominates the first one. To do this we make use of the diagram shown in Figure 5.10. In this figure, the CDF F_X is for a random variable X with normal

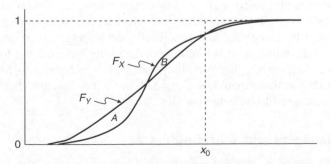

Figure 5.9   The CDF for X second order stochastically dominates the CDF for Y.

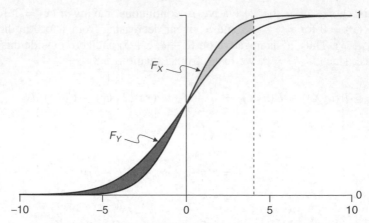

Figure 5.10   The normal distribution with smaller variance is second order stochastically dominant.

distribution (mean 0 and standard deviation 2), while F_Y is for a random variable Y having normal distribution with larger standard deviation (3, in this case), but the same argument holds no matter what symmetric distribution is used. It is clear from symmetry that the light-shaded region to the right of the axis has a smaller area than the dark-shaded region to the left of the axis. Moreover, this remains true no matter where the vertical dashed line is drawn (far enough to the right and the two areas become the same). This is enough to show, using the approach of Figure 5.9, that X second order dominates Y.

The process of shrinking towards the mean will produce, in the limit, a single point mass at the mean, and we already know that this will be preferred to the original distribution by a risk-averse decision maker (i.e. the certainty of getting the mean result is a prospect which second order stochastically dominates the original prospect).

5.4   Risk decisions for managers

Many of the examples we have given in this chapter have focused on an individual making a choice with implications for themselves alone. Now we want to turn to decisions taken by companies and specifically the managers or boards of those companies. Our starting point is to ask what the utility function will look like for business decisions rather than personal ones. The von Neumann–Morgenstern result suggests that there should be an underlying utility function, but what is it? There are a number of issues to consider.

5.4.1   Managers and shareholders

We need to begin by thinking about who makes decisions and what they may be aiming to achieve. It is usual to think about management as acting in the best interest of shareholders, who are the owners of the company, but in reality we have complex systems of corporate governance with a board of directors given the responsibility of hiring and monitoring top management in order to safeguard the interests of the shareholders.

Most shareholders have the opportunity to diversify their holdings across multiple firms. There are exceptions to this when shareholders wish to maintain control or influence by holding a significant fraction of the shares, for example in companies controlled by family interests. Usually, however, the majority of shares are owned by institutions or individual investors who are well diversified and consequently are likely to see only small proportional changes in their overall wealth from changes in the value of a single firm. From this we can deduce that shareholders will be risk neutral in their view of the actions of an individual firm.

To see why this is so, consider a firm that can pay \$$k$ million to take a risky action that delivers either nothing or \$1 million, each with probability 0.5. If a shareholder with a proportion δ of the firm's equity was to make the decision on what is a fair value of k, then the shareholder with current wealth W and utility function u should compare $u(W)$ (manager does nothing) with $0.5u(W - \delta k) + 0.5u(W + \delta(1 - k))$ (manager takes risky action). The shareholder operating on an expected utility basis would want the manager to take the risk provided that

$$0.5u(W - \delta k) + 0.5u(W + \delta(1 - k)) > u(W).$$

But for small δ we can approximate the left-hand side of this expression using the derivative of u at W (which we write as $u'(W)$) to get

$$0.5u(W) - 0.5\delta ku'(W) + 0.5u(W) + 0.5\delta(1 - k)u'(W)$$

$$= u(W) + 0.5\delta(1 - 2k)u'(W).$$

Hence, for any k less than 0.5 this is a worthwhile investment from the shareholder perspective, but it is not worthwhile if k is greater than 0.5. We end up with the risk-neutral value put on the investment.

As an aside, we need to observe that this argument should not be seen as suggesting that investors are completely indifferent to risk. The capital asset pricing model (CAPM) explores how the component of an asset's variance that cannot be diversified away (its β value or systematic risk) is reflected in the price of the asset. But the type of management decision we are considering here, which is idiosyncratic to this particular firm, would not appear in β.

But if diversification makes shareholders risk neutral, the same cannot be said for the managers. A manager is committed to the firm in a way that the investor is not, and the success of a manager's career is tied to the performance of the company. Moreover, a senior manager may well hold stock options which also give her a direct interest in the company share price. This can be expected to lead to risk-averse behavior by a manager and this differs from the risk-neutral behavior that would be preferred by investors. So there is a potential *agency problem* where managers who, in theory, act as agents of the owners, are, in fact, subject to different incentives.

5.4.2   A single company-wide view of risk

In many cases it is convenient to view a company as a single entity; perhaps this is connected to the legal fiction of a company as an individual. But in practice it is clear that there will be differences between the type of actions and choices made by one manager compared to another within the same company. This corresponds to the different personalities involved: one manager may be particularly risk averse by nature, whereas her colleague at the next desk is a natural gambler. When this happens there is a danger of inefficiency, since one manager might pay a risk premium to achieve some measure of certainty, only for this to be negated at the firm level by the relatively risky behavior of a second manager. The Dennis Weatherstone approach at J.P. Morgan (discussed at the start of Chapter 3) certainly attempts to bring the entire company under a single risk umbrella. The same motivation also lies behind the idea that firms need to determine an appropriate risk appetite for the firm as a whole (one of the tenets of Enterprise Risk Management), which implies that a common approach to risk can be discussed and agreed within the top management team. In practice, however, it is not so easy to obtain a uniform approach to risk across the whole of a company.

One problem with the attempt to have a single level of risk appetite for the firm as a whole is that if levels of risk appetite are related to the shape of the utility function, then they may depend on factors like the size of the 'bet' and the current company cash reserves. This explains why it is so hard to give a simple definition of risk appetite: it cannot just be seen as a point on a scale moving from risk averse through to risk seeking.

In Chapter 1 we commented on the way that those involved in trading operations often take a different view of risk than the back office, and it is common to have different levels of risk preference at different levels in a hierarchy, or in different departments. At some level this can be traced back not only to different individual risk preferences (as we see in the caricature of a trader as being a fast-living young man burning out on the adrenaline rush of millions of dollars riding on instant decisions) but also to differences in reward structures, either explicitly through bonuses or implicitly through what is valued within the culture of a work group.

All of this will make us wary of assigning a single level of risk appetite to a company as a whole. It is rarely as simple as that. The complications that arise when dealing with this issue make the 'quick and dirty' approach of measuring value at risk and using this on a company-wide basis seem more attractive.

5.4.3   Risk of insolvency

Whether or not managers are risk averse for most of the time, they certainly become so if there is a threat of insolvency. This would suggest that there is a concave utility function dropping sharply as the solvency of the company becomes an issue. In fact, once a company enters what is called the *zone of insolvency* (i.e. when insolvency is a real possibility), then the board will (or should) change the way that it behaves and the decision processes that are used. Company law may involve individuals on the board becoming personally responsible if the business fails, so it is not surprising that if insolvency is a possibility, the board will act promptly. One aspect of this is that if a company is insolvent, then the creditors have a higher claim on the company assets than the shareholders, and so, for a company that is in the zone of insolvency, the directors should not act in a way that would prejudice the interests of the creditors over against the equity holders if insolvency occurs.

What will the utility look like as a function of total net assets (assets minus liabilities) for a company that faces the possibility of insolvency? The discussion here suggests that the utility function might look something like the curve shown in Figure 5.11. Once the firm goes out of business then the degree of insolvency (the extent of the company debts) will determine how much the creditors receive. Managers will have some interest in seeing creditors given a reasonable deal, but the slope of the utility function in this region will be relatively flat. At first sight, a utility function as shown in the figure suggests the possibility of risk-seeking behavior. To take a simplified example, suppose that a company with current net assets of just $100 000 if allowed to continue trading produces a 10% chance of net assets moving to $-\$0.5$ million, a 70% chance of net assets remaining at $100 000 and a 20% chance of net assets increasing to $0.4 million. Thus, the expected value of the net assets stays at $100 000, since

$$0.1(-0.5) + 0.7(0.1) + 0.2(0.4) = 0.1.$$

From an expected utility viewpoint, this trading outcome is positive (given the utility behavior shown in Figure 5.11). Since $u(0.1) = 0$, the expected utility is

$$0.1u(-0.5) + 0.2u(0.4),$$

which, from the figure, is easily seen to be positive. So the company has an incentive to continue to trade even though this involves the risk of insolvency. In fact, if a more aggressive trading strategy could increase the size of both the gains and the losses while leaving the expected value unchanged, then this would be preferred. Specifically, suppose that an aggressive strategy gives a 10% chance of net assets moving to $-\$1.1$ million, a 70% chance of net assets remaining at $100 000 and a 20% chance of net assets increasing to $0.7 million. Given the shape of the utility function shown in Figure 5.11, the aggressive trading strategy is easily seen to have an even higher expected utility, since the gain in utility in moving from $0.4 to $0.7 million exceeds the loss in utility in moving from $-\$0.5$ to $-\$1.1$ million and occurs with greater probability. Overall, we can see that the corner in the utility function produces a region of convexity and hence risk-seeking behavior.

In practice, however, the scenario sketched above is unlikely to happen, since it supposes that the aggressive strategy can allow a much higher net deficit to

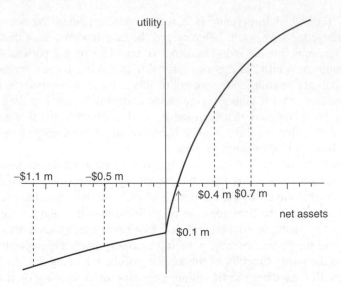

Figure 5.11   The utility function for a company facing insolvency.

be created, whereas the company should cease trading as soon as the net assets become zero. Moreover, the directors would carry significant personal risk that the aggressive trading strategy would be found to be improper, whereupon they would individually carry some liability for the debts.

Notes

The book by Peter Bernstein gives more information about the development of the idea of utility by Daniel Bernoulli and others. The decision tree ideas that we present are quite standard and can be found in any textbook on Decision Theory.

An excellent book which goes quite deeply into the different frameworks that underlie the use of Expected Utility Theory is *Theory of Decision under Uncertainty* by Itzhak Gilboa. This is a book which gives an accessible introduction to some of the important philosophical and conceptual underpinnings of decision theory. Note, though, that Gilboa gives a slightly different form of the continuity axiom in describing the von Neumann–Morgenstern theory. This is arguably a weaker assumption, but slightly increases the complexity of understanding what is going on.

The discussion of utility functions like that shown in Figure 5.6 as an explanation of the coexistence of both gambling on lotteries and insurance (as discussed in Worked Example 5.3) goes back to a paper by Friedman and Savage in 1948. However, a different and more convincing explanation can be found in the ideas of Prospect Theory that we present in the next chapter.

The discussion of Savage's theorem is drawn from Gilboa's book. This theory is designed around a situation with a rich state space (infinite with every individual element null). There are alternative approaches that need a rich outcome space, but a simpler state space, and Wakker (2010) gives such a development.

We have given short proofs of the stochastic dominance results based on an assumption of continuous random variables defined over a finite range, and differentiable utility functions (see the paper by Hadar and Russell, 1971 for a similar approach). It requires a bit more work to prove these results in more generality. Sheldon Ross (2011) covers some of this material in his book on introductory mathematical finance.

Our discussion of risk decisions and utility at the firm level is a simplified view of a complex topic. For example, we have taken no account of corporate taxes (that will often have the effect of making after-tax income a concave function of pre-tax income), nor do we consider the transaction costs of bankruptcy. Smith and Stulz (1985) give a more detailed discussion of these issues.

References

Bernstein, P. (1996) *Against the Gods: The Remarkable Story of Risk*. John Wiley & Sons.

Friedman, M. and Savage, L. (1948) The utility analysis of choices involving risk. *Journal of Political Economy*, **56**, 279–304.

Gilboa, I. (2009) *Theory of Decision under Uncertainty*. Cambridge University Press.

Hadar, J. and Russell, W. (1971) Stochastic dominance and diversification. *Journal of Economic Theory*, **3**, 288–305.

Ross, R. (2011) *An Elementary Introduction to Mathematical Finance*, 3rd edition. Cambridge University Press.

Smith, C. and Stulz, R. (1985) The determinants of firms' hedging policies. *Journal of Financial and Quantitative Analysis*, **20**, 391–405.

Wakker, P. (2010) *Prospect Theory for Risk and Ambiguity*. Cambridge University Press.

Exercises

5.1 Making the right bid

Xsteel is bidding to supply corrugated steel sheet for a major construction project. It knows that Yco is likely to be the only other serious bidder. There is a single price variable and the lower price bidder will win, and if both prices are the same then other factors will determine the winning bid, with Xsteel and Yco equally likely to win. Xsteel believes that Yco will make an offer of $800 or $810 or $820 or $830 per ton, with each of these possibilities equally likely. Xsteel has a production cost of $790 per ton, and only bids at multiples of $10 per ton are possible. What price bid will maximize Xsteel's expected profit?

5.2 EUT and a business venture

James has $1000 which he wants to invest for a year. He can put the money into a savings account, which pays an interest rate of 4%. His friend Kate asks him to invest the money in a business venture for which she needs exactly $1000. However, Kate's business will fail with probability 0.3 and if this happens James will get nothing. On the other hand, if the business succeeds it will make $2000 and this money can be used to repay the loan to James. The time taken to repay the money if the venture succeeds will be one year.

(a) Using EUT and assuming that James is risk neutral, how much would Kate have to repay James in order to convince James to lend her the money?

(b) Assume James is risk averse with concave utility function $u(x) = \sqrt{x}$. How much would Kate have to repay James if the business venture succeeds in order to convince James to lend her the money?

5.3 Calculating utilities from choices

A researcher is trying to understand the utility function of an entrepreneur with a company worth $1 million. He does this by describing various potential ventures and asking the manager whether she would take on this venture under the terms described. By changing the probabilities assigned to success and failure, he finds three ventures where the manager is indifferent between taking them or not. The probabilities on different outcomes are given in Table 5.3.

Use this information to estimate the utilities on the different firm values: $1.5 million, $2 million and $2.5 million, assuming that the utility for firm value $0.5 million is 1 and the utility for firm value $1 million is 2. (These two values can be set arbitrarily because utilities are only defined from choices up to a positive linear transformation: see Theorem 5.1). Sketch a possible utility function for the entrepreneur.

Table 5.3   Outcomes for different ventures in Exercise 5.3.

Outcome:	Lose $0.5 million	Gain $0.5 million	Gain $1 million	Gain $1.5 million
		Probabilities:		
Venture A	0.4	0.6	0	0
Venture B	0.6	0	0.4	0
Venture C	0.7	0	0	0.3

5.4   Stochastic dominance and negative prospects

Show that if a prospect $A = (x_1, p_1; x_2, p_2; \ldots; x_n, p_n)$ stochastically dominates the prospect $B = (y_1, q_1; y_2, q_2; \ldots; y_n, q_n)$, then the prospect $-B = (-y_1, q_1; -y_2, q_2; \ldots; -y_n, q_n)$ stochastically dominates the prospect $-A = (-x_1, p_1; -x_2, p_2; \ldots; -x_n, p_n)$.

5.5   Stochastic dominance and normal distributions

The profits from sales of a product depend on demand, which follows a normal distribution. The demand in week 1 has a distribution with mean 1000 and standard deviation 100. The demand in week 2 has mean 1010.

(a) Suppose that the standard deviation of demand in week 2 is 100. Explain why the profit in week 2 stochastically dominates the profit in week 1, and this result does not depend on the exact relationship between profit and sales.

(b) Show that if demand in week 2 has standard deviation of 105, then the profit in week 2 will not stochastically dominate the profit in week 1.

(c) Show that if demand in week 2 has standard deviation of 95, then the profit in week 2 will not stochastically dominate the profit in week 1.

5.6   Failure of stochastic dominance

Consider the following two prospects: $q = (\$100, 0.1; \$300, 0.2; \$400, 0.2; \$700, 0.3; \$900, 0.2)$ and $r = (\$300, 0.3; \$500, 0.3; \$700, 0.2; \$900, 0.1; \$1000, 0.1)$. Show that neither stochastically dominates the other and find a set of utility assignments where q is preferred and a set of utility assignments where r is preferred. Your utility values should satisfy the requirement that having more money gives strictly higher utility.

5.7   Second order stochastic dominance

Given the two prospects defined in Exercise 5.6, show that r second order stochastically dominates q.

6

Understanding risk behavior

The economics of extended warranties

A consumer who buys either an electronic item (like a TV or laptop) or a white goods item (like a washing machine) will inevitably be offered an extended warranty. The technical term for this is an 'Extended Service Contract' or ESC. This will take the manufacturer's warranty of perhaps one year and extend it to, say, three years from the date of purchase. The consumer pays some additional amount for the peace of mind of knowing that they will not have to face an expensive repair. The sums of money are not small; for example, an ESC on a laptop costing $600 could cost $100. This is an enormous business worth billions of dollars a year and it can be very profitable for retailers who charge a generous margin on top of the cost that they pay to the warranty provider. There are reports that margins can be 50% or more, and that electronics retailers can earn half of their total profits from extended warranty sales.

The consumer is facing a decision where it is hard to estimate the probabilities involved and also hard to estimate the costs that may be incurred. There is some chance of a problem that can be fixed quite cheaply, but it is also possible that the item will fail completely. But, if there is so much money being made by the suppliers of the warranty, then it suggests that the ESC is a bad idea on an expected cost basis. Nevertheless, Expected Utility Theory could explain this as a transaction involving a risk-averse consumer paying for the insurance provided by a more risk-neutral service provider.

Many of the experts, and every advice column, say that buying an ESC is a bad choice on the part of the consumer. In 2006 there was a full-page advert placed in many newspapers by *Consumer Reports*, a respected publication, saying simply, 'Dear shopper, Despite what the salesperson says, you don't need an extended warranty. Yours truly, Consumer Reports'. The argument is made that,

Business Risk Management: Models and Analysis, First Edition. Edward J. Anderson.
© 2014 John Wiley & Sons, Ltd. Published 2014 by John Wiley & Sons, Ltd.
Companion website: www.wiley.com/go/business_risk_management

even assuming a relatively high level of risk aversion on the part of the consumer, it is still hard to justify the high cost of the ESC for many products.

On this basis one might expect that customers who had purchased extended warranties would be unhappy, but not a bit of it. A survey reported in *Warranty Week* (an online newsletter for people working in this area) in March 2012 asked those who had bought extended warranties in the past whether they would do so again: 49% said 'Yes', 48% said 'Perhaps, depending on product and pricing' and only 3% said 'No'. Consumers continue to purchase extended warranties and seem happy to do so.

6.1   Why decision theory fails

In the previous chapter we discussed Expected Utility Theory and how it can be used to make decisions in a risky or uncertain environment. We showed how working with utilities on outcomes and simply making choices that maximize the expected utility value is a strong normative decision theory. It describes the way that individuals should make decisions. In fact, the case for EUT seems unassailable. It can be derived from axioms which seem entirely reasonable. It has all the right properties − such as always preferring a choice that stochastically dominates another. It also enables an elegant understanding of risk-averse or risk-seeking behavior depending on the concavity or convexity of the utility function. However, there is now a great body of evidence to show that EUT does not do well in predicting the *actual* choices that people make. In fact, as we will show, individuals deviate from EUT in consistent ways.

In understanding what might be wrong with Expected Utility Theory as a predictor of individual choice, we need to question the individual components of the theory.

6.1.1   The meaning of utility

Perhaps the most fundamental idea in the decision theory we have presented is that utility is defined simply by the outcomes for an individual. This allows us to say that A is preferred to B, and imply by this that A is always preferred to B. One problem here is that I may make one choice today and another tomorrow, so that there is inconsistency in an individual's choices. This might simply be because I am more or less indifferent between the choices available. Which cereal do I choose for breakfast? Where in the train do I choose to sit? A lack of consistency here is no real problem for the theory; but what if I am not indifferent between two choices, and have a clear preference for one over the other? Does it therefore follow that I will make the same choice on another occasion? Perhaps not, because every decision is made in a particular context. For example, a choice to buy a product or not will depend on the way that the product is described (psychologists talk of this as *framing*) and this can make a difference −perhaps the product has a particularly persuasive salesman. Moreover, the decision I make on this occasion cannot be divorced from the other things that are going on in my life: a fright that I receive while driving my car today will make me more inclined to purchase life insurance tomorrow than I was yesterday.

A separate problem that we should address is the connection between the utility of an outcome and the satisfaction that I achieve from it. In crude terms, I can be expected to choose the action that will bring me the greatest enjoyment, but we have only to make this statement to recognize that it is a dramatic over-simplification.

- I may be altruistic and choose an action that benefits a friend, family member or society at large. So, at the least, we can say that satisfaction or contentment is complex and does not just involve our own pleasures.

- I may postpone doing something that I know I will enjoy. From the point of view of the decision I face right now, the cup of coffee will be well worth its price, but I know that I can enjoy the cup of coffee later in the morning and so decide to wait. Or perhaps I am saving for my retirement and I decide not to purchase the expensive holiday that I want because it will eat into those savings. Part of what is going on in these examples is that there is enjoyment to be had simply in the anticipation of pleasure.

- The action I choose may be determined by who I perceive myself to be, and how I wish to be seen by others. For example, I want to be seen as responsible and restrained and so, tempting though it may be, I do not buy the convertible sports car that I would really enjoy.

These observations all demonstrate that we cannot simply define utility on the basis of immediate personal enjoyment.

In addition to these complexities, we can also observe that the utility of an outcome is often determined in part by a comparison with other outcomes. There are three important aspects to this.

- When assessing an outcome, we often compare it with *the outcomes achieved by others*. For example, an employee will judge her salary not only by its size but also by how it compares with her colleagues and peers. If others do better than me then I will feel worse. As Gore Vidal strikingly put it, 'It is not enough to succeed. Others must fail.' This idea is important when we think about the concept of a Pareto improving outcome, in which all parties do at least as well as they did before, and some people do strictly better. It may seem as though that must be a good thing, but life is more complicated. A boss who arbitrarily gave a bonus of $500 to half of his employees and nothing to the others may produce a Pareto improving outcome for the workforce, but there will certainly be some unhappy people (and imagine what would happen if male employees got the bonus and the women did not!).

- When assessing an outcome, we often compare it with *how things were before*. This means that the utility of an outcome can depend on the route by which it arrived. Two people go into a casino and spend the night gambling: one begins by losing some money and then, over the course of the evening, wins it back to emerge $500 ahead, while the other has some early successes and at one point has made a profit of $5000 before continuing to gamble less successfully, losing 90% of his winnings to finish with a profit of $500. Then, of the two, the person who has slipped from a potential profit of $5000 to a profit of $500 is likely to feel much less happy about the outcome.

• When assessing an outcome, we often compare it with *our expectations*. In a similar way to that in which people compare their current state with their previous state, the expectations of outcomes also play a role. An employee who last year had a bonus of $5000 and expects a similar bonus this year, will feel much less positive on receiving this bonus than an employee who expects a bonus of $2000 and instead receives $5000.

Our discussion so far has demonstrated that utility cannot be understood without taking account of many different factors: our happiness depends both on context and comparison, and the choices we make are not just about our own immediate pleasure. But we can still rescue the idea of a definite utility for each possible outcome, we just need to understand this utility more broadly. For example, we can say that we receive utility from seeing friends do well; from believing that we are thought well of by others; from thinking that we have done better than others; from experiencing an improvement in our circumstances; and from the pleasure of a surprise in relation to our expectations.

The von Neumann–Morgenstern Theorem deduces utility from decisions, not the other way around. And so nothing about the difficulties of constructing utilities from looking at the properties of outcomes necessarily undercuts this theory.

6.1.2 Bounded rationality

If Expected Utility Theory holds, then it suggests that good decision makers (who make consistent and thoughtful choices) should be investing in estimating both the utilities of different outcomes and the probabilities that they may occur. When this has been done, a rational individual will carry out a computation of the expected value of different choices in order to decide between them. Though we can construct artificial examples in which these calculations are relatively easy to carry out, to do this in practice is far more difficult. In most cases there are enough potential outcomes that even listing them would be a challenge, let alone evaluating utilities for them all. And what methods can be used to estimate probabilities for these outcomes? Finally, there is a non-trivial computation of expectations to be carried out. This also does not chime well with what we know in practice, since we make most decisions without recourse to a spreadsheet or calculator.

Herbert Simon called into question whether we can expect so much from decision makers and called this *bounded rationality*. There is not only the question of computational capability, but also whether the time and expense involved is likely to be compensated for by a better final decision. But what is the alternative for making decisions? If a full-scale computation of expected utilities does not take place, then we need to investigate the mental processes that occur instead.

There has been a large amount of academic research into the processes that people use to make decisions. There are a variety of different heuristics and biases that come into play, for the most part subconsciously. Daniel Kahneman describes two systems of decision making or cognition, which we will call *reasoning* and *intuition*. Reasoning is carried out deliberately and requires effort, whereas intuition takes place spontaneously and requires no effort. Most of the time we operate intuitively, with our reasoning capacity acting as a monitor or restraint on our intuitive actions and choices. The reasoning component in decisions requires effort, of which there is only a limited overall capacity, and for that reason it can be interrupted if another reasoning task arises. Kahneman points out that this is what happens when a driver halts a conversation in order to carry out a difficult manoeuvre: the driving decisions temporarily move from being intuitive to requiring reasoning.

But even when decision making is done within a reasoning mode, simplifying heuristics and biases will come into play. For example:

- When faced with a complex choice involving many alternatives, decision makers tend to eliminate some choices quickly using relatively small amounts of information, and only when the choice set is reduced to a small size (maybe two only) does the decision maker attempt to compare on the basis of all available information.

- Decision options or outcomes that have greater *saliency* or greater *accessibility* will be given greater weight. Accessibility here refers to the ease with which something can be brought to mind and may be determined by recent experience or the description of the outcome or option. Saliency refers to the extent that an item is distinctive or prominent. These heuristics imply that the framing of a decision will have a significant effect on the choice made.

- Decision makers will usually accept the formulation that is given relatively passively, and are unlikely to construct their own framework of evaluation for different options. In particular, whatever choice is presented as the default option has a greater likelihood of being selected.

6.1.3   Inconsistent choices under uncertainty

In our discussion so far we have reduced the range of circumstances in which we can expect that Expected Utility Theory will apply. We need to assume that outcomes are simple so that utilities can be evaluated easily. We need to ensure that the decision is taken through reasoning (and applying mental effort) rather than being carried out in an intuitive fashion. Finally, we must have a simple arrangement with well-defined probabilities in order to avoid the constraints of bounded rationality.

It is remarkable that, even working within these limitations, we can find examples where decision makers make choices that are not consistent with *any* choice of utility function, and hence demonstrate a deviation from Expected Utility Theory. The first such observation was made by Maurice Allais in a 1953 article in *Econometrica* ('The behavior of rational man in risky situations – A critique of the axioms and postulates of the American School'). Allais demonstrates that the axiom of independence may not hold in practice. Or, to put it another way, the axiom may not hold in a descriptive rather than normative theory of decision making.

Next we give three examples from amongst many that could be given. In each case people are asked in an experiment to make a choice between two options (or, to be more precise, to say which of two options they would prefer if it was offered). This is Decision 1. Then the same individuals are asked to make a choice between a different pair of options (Decision 2). The choices are constructed in such a way that certain pairs of decisions are inconsistent with any set of utility values. The experiments are repeated with many individuals to demonstrate a consistent pattern in the way that people make decisions. Taken together these examples provide a very convincing case that an alternative to Expected Utility Theory is needed if we want to do a good job of explaining how individuals actually make choices when faced with uncertainty.

Example 6.1   Preference for sure gain: version 1

We give a version of the Allais paradox. Consider the following two experiments. In Decision 1 participants are asked to choose between the two prospects A1 and B1 described as follows:

A1:   gain $2500 with probability 0.33;   gain $2400 with probability 0.66;
        0 with probability 0.01.
B1:   gain $2400 with certainty.

The experiment shows that more than 80% of people choose B1. Under the assumptions of EUT we can convert this into a statement about the utilities of the various sums of money involved. We deduce that, for most people,

$$u(2400) > 0.33u(2500) + 0.66u(2400)$$

since we can assume $u(0) = 0$. This can be simplified to

$$0.34u(2400) > 0.33u(2500).$$

In Decision 2 participants are asked to choose between the two prospects C1 and D1 described as follows

C1:   gain $2500 with probability 0.33;   0 with probability 0.67,
D1:   gain $2400 with probability 0.34;   0 with probability 0.66,

and in this case more than 80% of people choose C1. But, again using $u(0) = 0$, this implies that

$$0.33u(2500) > 0.34u(2400)$$

i.e. the exact reverse inequality to that we just derived.

Try checking what your own choices would be in these two different decisions. For most people, having the inconsistency pointed out to them does not alter the choices they would make. In the Allais paradox, prospects C1 and D1 are obtained from prospects A1 and B1 simply by eliminating a 0.66 chance of winning $2400 from both prospects. This change produces a greater reduction in desirability when it turns a sure gain to a probable one, rather than when both the original and the reduced prospects are uncertain. There is something about the sure gain of prospect B1 that makes it particularly attractive, and this leads to a violation of the independence axiom.   □

Example 6.2   Preference for sure gain: version 2

The same type of phenomenon appears in the following two experiments. In Decision 1 participants are asked to choose between the two prospects A2 and B2 described as follows

    A2:   gain of $4000 with probability 0.8;   0 with probability 0.2,

    B2:   gain of $3000 with certainty.

The majority of people (80%) choose B2. For these people we can deduce that $u(3000) > 0.8u(4000)$. In Decision 2 the two prospects are

    C2:   gain of $4000 with probability 0.2;   0 with probability 0.8,

    D2:   gain of $3000 with probability 0.25;   0 with probability 0.75.

Then the majority of people (65%) choose C2. We can deduce that for these people $0.25u(3000) < 0.2u(4000)$, which is the reverse of the inequality derived from the first experiment. This is another example of people preferring certainty.

In this example, the choice of B2 in preference to A2 is an example of risk aversion. If individuals were risk neutral, then A2 with a higher expected value (of $3200) would be preferred. The preference for certainty here reflects a concave utility function. The problem for EUT is that the two different decisions imply different amounts of concavity (a greater degree of concavity for Decision 1 than for Decision 2).   □

Example 6.3   Dislike of sure loss

The exact opposite of these results is found when losses are involved rather than gains. In this case, in Decision 1 participants are asked to choose between the two prospects A3 and B3 described as follows

A3: loss of $4000 with probability 0.8;   0 with probability 0.2,

B3: loss of $3000 with certainty.

The great majority of people (90%) choose A3. For these people we can deduce that $u(-3000) < 0.8u(4000)$. In Decision 2 the two prospects are

C3: loss of $4000 with probability 0.2;   0 with probability 0.8,

D3: loss of $3000 with probability 0.25;   0 with probability 0.75.

Then, the majority of people (58%) choose D3, leading to the inequality: $0.25u(-3000) < 0.2u(4000)$, which amounts to the exact opposite of the deduction from Decision 1.

Since the great majority of people choose A3 in preference to B3, even though the expected loss under A3 is greater (at $3200), we can deduce that people are risk seeking over negative gains (i.e. losses), where they will gamble to give themselves a chance of avoiding a loss. This implies a utility function that is convex in this area. The difficulty in this example is that there is a greater degree of convexity in Decision 1 than in Decision 2.   □

6.1.4   Problems from scaling utility functions

Our final example of the way that Expected Utility Theory can fail is taken from Rabin and Thaler (2001). It demonstrates that problems arise unless we deal with changes in wealth rather than absolute values. Suppose that you are offered a choice to gamble with a 50% chance of winning $100 and a 50% chance of losing $90. Most people would reject this bet independently of the size of their bank balance. Following Expected Utility Theory, if W is their current wealth then this implies that they prefer W to equal chances of being at $(W - 90)$ or $(W + 100)$. With a utility function u this gives

$$u(W) > 0.5u(W - 90) + 0.5u(W + 100).$$

Hence (multiplying through by 2 and rearranging)

$$u(W) - u(W - 90) > u(W + 100) - u(W). \qquad (6.1)$$

This shows that u is concave over the interval, but we can be more specific. Looking carefully at Figure 6.1 shows that the derivative of u at $W + 100$ is less than the slope of the straight line joining the points on the curve at W and $W + 100$, i.e.

$$u'(W + 100) < (u(W + 100) - u(W))/100.$$

In the same way, the derivative of u at $W - 90$ is more than the slope of the straight line joining the points on the curve at $W - 90$ and W, i.e.

$$u'(W - 90) > (u(W) - u(W - 90))/90.$$

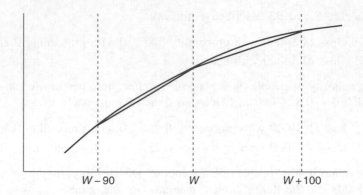

Figure 6.1   Comparing straight line segments to the utility function.

Putting these observations together with Inequality (6.1) shows that

$$u'(W - 90) > (10/9)u'(W + 100).$$

Now, since the W in this inequality is arbitrary, we can deduce that

$$u'(W) > (10/9)u'(W + 190)$$
$$> (10/9)^2 u'(W + 380)$$
$$> (10/9)^n u'(W + 190n).$$

What has happened here is that, in effect, we have applied the inequality arising from not gambling at different points along the curve and then stitched the inequalities together to say something about the way that the slope decreases over a much longer interval. When $n = 50$ this gives

$$u'(W) > 194u'(W + 9500).$$

The slope of the utility function simply tells us what an extra dollar would be worth to us, and so this inequality says that the value of an extra dollar is almost 200 times less if you are $9500 dollars wealthier. This does not seem believable: it is quite reasonable to suppose that increasing wealth makes someone value additional wealth less, but this cannot happen to the extent that this calculation predicts. From this we can see that uniformly applying the consequences of rejecting a small gamble gives results that seem wrong when scaled up.

6.2   Prospect Theory

Many researchers have worked on different ways of explaining deviations from Expected Utility Theory. In this chapter we will describe just one of these theories, called Prospect Theory, developed by Daniel Kahneman and Amos Tversky.

This theory is built on many of the ideas that preceded it and even if some would argue for a different approach, the main ingredients of these theories are similar. So, we will not lose much by concentrating on a version of Prospect Theory. Moreover, Kahneman and Tversky's work (summarized in their papers of 1979 and 1992) is the single most important contribution in this area and they were awarded the Nobel prize for their work in 2002.

6.2.1   Foundations for behavioral decision theory

Prospect Theory is an example of behavioral decision theory. This is a 'theory' in the sense of having predictive power; it can tell us how people are likely to behave facing different choices involving uncertainty. In developing this theory we will start with three fundamental observations about the way people make decisions.

6.2.1.1   Using a reference point

One weakness in Expected Utility Theory is that, for consistency, it must apply to the total wealth of an individual. And yet there seems little evidence that people take much account of their bank balance or housing equity when considering small-scale financial decisions. Obviously this principle will depend to some extent on individual circumstances: when facing bankruptcy, then indeed the absolute wealth may be the focus of attention. But in the normal course of events (say in deciding whether or not to take up an extended warranty offer that involves paying out money now for additional security in the future), our total wealth is not a big factor.

Instead of thinking about the total value of all their assets and using that in a calculation of utilities, people tend to compare possible outcomes against some benchmark in their mind. We have already said that our feeling about outcomes may depend on how well we do in comparison with others around us, or in comparison with an expectation we have formed. But when decisions are made that may involve gains or losses, then the current position becomes the normal starting point. Decision makers focus on changes rather than on absolute values.

The way in which a reference point is constructed will depend on the exact circumstances of the decision. There is an opportunity for framing to lead a decision maker towards a particular reference point. We will concentrate on simple prospects without looking at any of the contextual factors that can come into play in practice. In these cases the current position will be the reference point, unless we are considering a prospect in which every outcome involves a gain; then it seems that the prospect is evaluated by taking the lowest gain as certain and using this lowest gain as a reference point. In much the same way, in evaluating a prospect where every outcome produces a loss, most people take the smallest loss as certain and evaluate the resulting prospect from that point.

6.2.1.2   Avoiding losses if possible

The existence of a reference point opens up the possibility of different behavior on one side of the reference point than the other, and this is exactly what we find.

People dislike losses in an absolute way rather than in a way that corresponds to simply looking at a reduction in utility as wealth decreases. For example, we can look at gambles in which there are equal chances of losing X or gaining Y. For this to be attractive, most people want Y to be about twice as big as X. Thinking in utility terms would suggest that as X and Y get smaller, decision makers should be more inclined to accept the gamble provided that Y (the gain) is larger than X (the loss). In practice, however, this doesn't seem to happen until X and Y are made so small as to be immaterial.

A good way to describe this is to say that individuals have a value function that is kinked around the reference point, as is indicated in Figure 6.2. We use the term 'value function' here to describe something like a utility function but calculated with respect to a reference point (we will make this more precise later). This makes sense of a lot of observations that we can make about people's behavior. For example, it has often been observed that decision makers seem to prefer the status quo. One explanation is that in considering an alternative to the status quo there is usually a possibility of a loss of wealth when compared with the current situation, as well as some probability that there is a gain in wealth. Then the loss aversion effect means that gains, or the probability of gains, need to be disproportionately high in order to make a change worthwhile.

6.2.1.3   Giving too much weight to small probabilities

Most people are not good at understanding intuitively the properties of very small probabilities. For example, we are likely to have a good idea of what it means to leave home late and risk missing the train. We can readily make decisions like whether or not to go back and pick up a forgotten umbrella, given the chance that this will make us miss the train. In essence, this is a choice

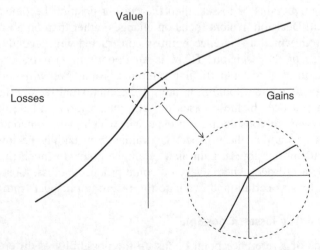

Figure 6.2   The shape of the value function in Prospect Theory.

with different uncertainties (the chance of rain; the chance of missing the train) and different possible outcomes (getting wet; being late), but the probabilities involved are not too tiny (say, greater than 5%). But when we deal with much smaller probabilities, we are less likely to make effective intuitive judgments. For example, if we suddenly recall that there is a back window unlocked, do we then go back and lock it at the risk of missing the train? Here the potential loss if the house is burgled is much greater than just getting wet, but the probability is very small (burglary is attempted; the burglar finds the unlocked window; and the burglar would have gone away if all the windows were secure). Of course, we will make a decision quickly (since there is that train to catch) but the quality of this decision may not be as good.

Experiments show that a small probability of a large gain is given a higher value than we might expect. Suppose that we compare two lotteries: A has one chance in 100 000 of delivering a large prize, and B has one chance in 10 000 of delivering the same large prize. It is easy to see that a ticket for A is worth only one tenth as much as a ticket for B. But since people have a hard time conceptualizing what a chance of one in 100 000 really means, the tickets for A will actually be valued at much more than that.

The same thing happens in reverse with probabilities that are nearly 1, so that the event is nearly certain. Then the chance of the event *not* happening seems to be inflated in people's minds. The result is that a near certainty of a large gain seems less attractive than we might expect. For example, suppose that I am offered a prize of $6000 unless a 6 is thrown twice running with a fair dice (i.e. the $6000 prize is received with a probability 35/36 and nothing is received with a probability 1/36). But there is an alternative, which is to take a prize of $5000 for sure. Most people opt for the certainty of the $5000 prize. This is because we tend to over-weight the small probability of ending with nothing and feeling foolish, and this is the same sort of behavior that is described in Example 6.1 above.

6.2.2   Decision weights and subjective values

Kahneman and Tversky set out their first version of Prospect Theory in 1979 and then updated it later. We will start by discussing 'Version 1' of Prospect Theory before going on to the full-blown cumulative version in the next section. Essentially, Prospect Theory is constructed out of the three observations above.

Since people seem consistently to over-weight small probabilities, it makes sense to define a function π which converts probabilities into decision weights. Thus, $\pi(p)$ is defined for $0 \leq p \leq 1$ and we will assume that $\pi(0) = 0$ and $\pi(1) = 1$. The idea here is that rather than forming an expected value (or utility) based on the probabilities, we will instead use the decision weights given by the π function.

We also need to define a subjective value $v(x)$ for each change in outcome (gain or loss). Here we are using the current wealth as a reference point and we have $v(0) = 0$. The subjective value is a little like a utility function, but it applies to changes in wealth, rather than to absolute values of wealth.

Then, in the simplest case, consider a prospect where a gain of x occurs with probability p and a loss of $-y$ occurs with probability q. Here there is no change with probability $1 - p - q$. We form a value function for the prospect (in comparison with the reference point of no change) by replacing probability with weights according to the decision weight function π and using the values of individual gains and losses in the same way that utilities of outcomes are used in EUT. So we get the prospect value as

$$V(x, p; y, q) = \pi(p)v(x) + \pi(q)v(y). \tag{6.2}$$

Note that since $v(0) = 0$, this does not appear in the expression for V.

However, we also need to capture the observation that if all outcomes are gains, then the lowest gain is treated as certain (and similarly for all losses). Hence, if $x > y > 0$ and $p + q = 1$ then this is perceived as being equivalent to a certain gain y with subjective value $v(y)$ together with a probability p (with decision weight $\pi(p)$) that the gain of y will be replaced by a gain of x. This gives

$$V(x, p; y, q) = v(y) + \pi(p)[v(x) - v(y)]. \tag{6.3}$$

This differs from Equation (6.2) unless $\pi(1 - p) = 1 - \pi(p)$, in which case the two expressions are the same.

The situation for losses is similar. When $x < y < 0$,

$$V(x, p; y, q) = v(y) + \pi(p)[v(x) - v(y)].$$

If the decision weights were equal to probabilities, so that $\pi(p) = p$, and the value function was a utility, then these expressions would just revert to ordinary Expected Utility Theory.

Prospect Theory implies that decisions will generally break the rules for a rational decision maker (with inconsistencies of some sort). Sometimes if a decision maker has these anomalies pointed out, then he will adjust his preferences to avoid being inconsistent. But if the decision maker does not discover that his preferences violate appropriate decision rules, then the anomalies implied by Prospect Theory will occur. Indeed, when there is just one decision to be made (rather than a whole series) and the results are personal to the decision maker, with real gains and losses involved, then a decision maker is unlikely to be concerned about breaking (say) the independence axiom of EUT. In this case, people are likely to follow the predictions of Prospect Theory even when they take time to consider their decisions carefully.

To understand the implications of this theory in more detail, we need to look at the functions v and π. We have already seen in Figure 6.2 roughly how the subjective value function v behaves for many people. One of the observations that emerges from experiments is that, to a first approximation, a prospect that is preferred to another is still preferred if all the amounts of money involved are multiplied by a constant. This is a property that occurs when the subjective value function follows a 'power law,' i.e. we have

$$v(x) = \gamma x^{\alpha}$$

for some γ and α. Since we expect the value function to be concave, we need $\alpha < 1$. This formula works for positive x but cannot be applied to a negative x (e.g. we cannot take the square root of a negative number.) So, when x is negative we need to rearrange this to get

$$v(x) = -\gamma(-x)^{\alpha}.$$

We can also ask how the π function behaves. Again this will depend on the individual, but the general shape is shown in Figure 6.3. Many experiments have been carried out to understand what this function looks like, and they suggest a lot of uncertainty or variation in π at both 0 and 1 (even possible discontinuities). People are simply not very good at evaluating probabilities near zero; either they get ignored, or they are given too much weight, and individuals are inconsistent from one decision to the next.

The shape of the function reflects the characteristics that we discussed above. First, relatively small probabilities are over-weighted. If we know that something happens 10% of the time, we behave much as we 'should' behave under a utility model if this event was much more likely (say 15% or 20%). Secondly, very high probabilities are under-weighted (which ties in with the preference for certainty that we have already observed). We treat an event which occurs 99% of the time, and so is nearly certain to occur, almost as if it happens only 95% of the time.

We can make some deductions from the shape of the functions involved. Notice that both the value function and the decision weight function go through zero and are concave in the region to the right of zero. The decision weight function, $\pi(p)$, is concave in a region approximately from 0 to 0.3, and the

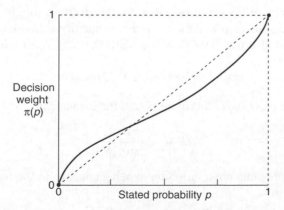

Figure 6.3   A typical decision weight function.

value function $v(x)$ is concave for all $x > 0$. Thus, for $0 < p < q < 0.3$ we know that a straight line joining the origin with the point $(q, \pi(q))$ lies below the π curve and, in particular, below the point $(p, \pi(p))$. Hence

$$\pi(p) > \frac{p}{q}\pi(q)$$

In the same way, if $0 < x < y$, then $v(x) > (x/y)v(y)$. These relationships can be useful in verifying inequalities that imply particular choices will be made under Prospect Theory, as the next example shows.

Worked Example 6.4   Deriving the inequalities for choices made

Suppose that the value function for positive x is given by $v(x) = x^{0.8}$. Show that the choices made by most people in the decisions described in Example 6.2 are consistent with the shape of the decision weight function.

Solution

As the majority of people chose \$3000 with certainty over the option of \$4000 with probability 0.8, we deduce that (\$3000, 1.0) $\succcurlyeq$ (\$4000, 0.8). This is equivalent under Prospect Theory to the inequality

$$v(3000) > \pi(0.8)v(4000).$$

Now

$$\frac{v(3000)}{v(4000)} = \left(\frac{3000}{4000}\right)^{0.8} = 0.794$$

so this inequality can be rewritten

$$\pi(0.8) < 0.794.$$

But looking at Figure 6.3 we can see that, for this decision weight function, $\pi(p)$ is below the diagonal at $p = 0.8$ and so this inequality seems likely to be true.

The example also has (\$4000, 0.2) $\succcurlyeq$ (\$3000, 0.25), from which we deduce that

$$\pi(0.2)v(4000) > \pi(0.25)v(3000).$$

Substituting for $v(3000)/v(4000)$ we obtain the inequality:

$$0.794 < \frac{\pi(0.2)}{\pi(0.25)}. \tag{6.4}$$

But, using our previous observation on π being concave in the region 0 to 0.25, we know that

$$\pi(0.2) > \frac{0.2}{0.25}\pi(0.25),$$

and thus

$$0.8 < \frac{\pi(0.2)}{\pi(0.25)}$$

which makes Inequality (6.4) certain to hold. So, with this value function the choices made in this example are just what we would expect under Prospect Theory.   □

6.2.2.1 Lotteries and insurance

It is instructive to look at the sales of lottery tickets from the perspective of Prospect Theory. Many people are prepared to buy tickets in a fair lottery. A lottery gives a small chance of a large prize. For the lottery to make money for the person running it, the expected value of the winnings needs to be less than the cost of the ticket. But let us consider a lottery where there is no money being made by the organizer. Then, buying a ticket for, say, $5 and getting a one in a thousand chance of winning $5000 is equivalent to preferring the lottery prospect of ($5000, 0.001) to the price of the ticket, which is the prospect ($5, 1). This implies that $\pi(0.001)v(5000) > v(5)$. But if the value function is concave for gains, which is what we would expect, then $v(5) > 0.001v(5000)$. Combining these two inequalities we get $\pi(0.001) > 0.001$. More generally, a willingness to engage in lotteries supports the idea that $\pi(p) > p$ for small p.

Notice that with Expected Utility Theory, the explanation of an individual taking up a lottery has to do with the shape of the utility function (an area where there is risk-seeking behavior). By moving to Prospect Theory we can see that an alternative and more satisfactory explanation has nothing to do with the shape of the value function, and instead is all about over-weighting of very small probabilities.

A full explanation of why individuals are often prepared to gamble in lotteries involves a second important factor, and that is the pleasure in anticipating the possibility of a win even when this does not, in the end, materialize. The lottery ticket is as much about purchasing a daydream as it is about purchasing a small probability of the big prize.

The same underlying idea occurs with insurance. A homeowner insuring his property is essentially preferring the prospect of a certain small cost to a much larger cost which occurs with a small probability. So we can see this as exemplified by saying that the prospect (−$5, 1) is preferred to (−$5000, 0.001). Since we expect the subjective value function to be convex for losses, then we can use the same idea as in our discussion of lottery tickets to show that this also implies $\pi(p) > p$ for small p.

Again there is a second effect that relates to the way that an individual feels over the lifetime of an insurance policy. This can be regarded as 'peace of mind': the knowledge that when an insurance policy is in place we don't need to worry about the possibility of a calamity.

6.3   Cumulative Prospect Theory

The Prospect Theory that we have developed works well when prospects have just two outcomes, but we can get into difficulties if there are more than two outcomes. It turns out that with 'Version 1' of Prospect Theory it is possible for very similar prospects to end up with very different values. For example, compare the prospects:

$$AA : (\$100, 0.05; \$101, 0.05; \$102, 0.05; \$103, 0.05)$$

$$BB : (\$103, 0.2).$$

We expect BB to be preferred since it stochastically dominates AA. Whatever we say about risk aversion or over-weighting of small probabilities, it is hard to imagine a decision maker selecting AA in preference to BB. Now if we just weight values with decision weights, we get

$$V(AA) = \pi(0.05)(v(100) + v(101) + v(102) + v(103))$$

$$\simeq 4\pi(0.05)v(103).$$

Notice that there is a 0.8 probability of getting nothing and so we are not in the position of a sure gain to apply Equation (6.3).

However, because of the properties of concave functions, we know that $\pi(0.05) > 0.25\pi(0.2)$. Thus, $V(AA) > v(103)$, and using Prospect Theory Version 1 will imply that the prospect AA has a substantially larger value than BB.

To fix this problem we need to introduce Cumulative Prospect Theory. A glance at the next couple of pages indicates that this will make the expressions much more cumbersome to write down, but in essence this 'Version 2' of Prospect Theory is only slightly more complicated than 'Version 1'.

We start by defining two different decision weight functions; one will apply to positive outcomes and one to negative. We call these decision weight functions $w^+(p)$ and $w^-(p)$ and they are defined on probabilities p. These weight functions are similar to π. They will have the same general shape as π and we will assume that $w^+(0) = w^-(0) = 0$ and $w^+(1) = w^-(1) = 1$.

We will deal with positive outcomes first. The approach is to order the potential outcomes and apply to each a weight given by the increment in the w^+ function, if $0 < y < x$ then

$$V(x, p; y, q) = [w^+(p + q) - w^+(p)]v(y) + w^+(p)v(x).$$

Notice that with just two outcomes (i.e. no option of zero change) this reverts to the previous version. Thus, if $0 < x_1 < x_2$ and $p_1 + p_2 = 1$, then

$$V = v(x_1)[w^+(p_1 + p_2) - w^+(p_2)] + v(x_2)w^+(p_2)$$

$$= v(x_1) + w^+(p_2)[v(x_2) - v(x_1)],$$

using the fact that $w^+(p_1 + p_2) = w^+(1) = 1$.

More generally, for prospects where outcomes are 0 or x_i with $0 < x_1 < \ldots < x_n$ and there is probability p_i of outcome x_i, then

$$V = \sum \pi_i^+ v(x_i),$$

with

$$\pi_i^+ = w^+(p_i + \cdots + p_n) - w^+(p_{i+1} + \cdots + p_n)$$

and

$$\pi_n^+ = w^+(p_n).$$

Figure 6.4 illustrates the way that this incremental calculation is carried out for a prospect in which there is a 40% chance of a gain of nothing, a 30% chance of a gain of $100, a 20% chance of a gain of $200 and a 10% chance of a gain of $300. Thus, the prospect is ($100, 0.3; $200, 0.2; $300, 0.1). In order to calculate probability weights for each of the outcomes, we divide the probability axis into regions of the appropriate length, starting with the highest gain of $300. Then the π^+ values are read off from the increments in the w^+ function values.

Notice that the weight assigned to the first outcome of $300 is proportionally higher in relation to the probabilities than the weight allocated to a middle outcome of $100 (remember that the worst outcome is $0). In fact, the worst outcome, especially if it has low probability, is also given a higher weighting. These facts follow from the higher slope of the decision weight curve at the two ends of the interval. A key characteristic of Prospect Theory is that it gives higher weights to relatively unlikely extreme outcomes (either large gains or near-zero gains) and this is also true for losses.

The definitions for negative outcomes are similar. Suppose a prospect has outcomes of 0 or x_i with $0 > x_1 > \cdots > x_n$ and there is probability p_i of outcome x_i.

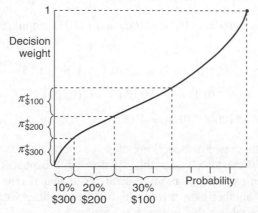

Figure 6.4   Calculating decision weights using the incremental method of Cumulative Prospect Theory.

Then

$$V = \sum \pi_i^- v(x_i),$$

with $\pi_i^- = w^-(p_i + \cdots + p_n) - w^-(p_{i+1} + \cdots + p_n)$ and $\pi_n^- = w^-(p_n)$.

If f is a prospect with both positive and negative outcomes, then we let f^+ be f with all negative elements set to zero, and we let f^- be f with all positive elements set to zero. Then

$$V(f) = V(f^+) + V(f^-)$$

(remember that $v(0) = 0$ so the extra zero value outcomes in f^+ and f^- do not change the value of V).

This is most easily understood by looking at an example: If the outcomes of a prospect are $-\$5$, $-\$3$, $-\$1$, $\$2$, $\$4$, $\$6$, each with probability $1/6$, then

$$f^+ = (\$0, 1/2; \$2, 1/6; \$4, 1/6; \$6, 1/6),$$

$$f^- = (-\$5, 1/6; -\$3, 1/6; -\$1, 1/6; \$0, 1/2).$$

So

$$V(f) = v(2)[w^+(1/2) - w^+(1/3)] + v(4)[w^+(1/3) - w^+(1/6)] + v(6)w^+(1/6)$$
$$+ v(-1)[w^-(1/2) - w^-(1/3)] + v(-3)[w^-(1/3) - w^-(1/6)]$$
$$+ v(-5)w^-(1/6).$$

Now we return to the example we discussed above. Prospect AA will have value

$$V(AA) = v(100)(w^+(0.2) - w^+(0.15)) + v(101)(w^+(0.15) - w^+(0.1))$$
$$+ v(102)(w^+(0.1) - w^+(0.05)) + v(103)w^+(0.05)).$$

Thus, if the values $v(100)$, $v(101)$, $v(102)$ and $v(103)$ are all close to each other, then

$$V(AA) \simeq v(100)(w^+(0.2) - w^+(0.15) + w^+(0.15) - w^+(0.1)$$
$$+ w^+(0.1) - w^+(0.05) + w^+(0.05))$$
$$= v(100)w^+(0.2) = V(BB).$$

So the values of these two prospects are close to each other, as we would expect. In fact, we can show that, under this version of Prospect Theory, $V(BB) > V(AA)$. This is an example of more general result that if one prospect stochastically dominates another then it has a higher value under Cumulative Prospect Theory. We will establish this in the next section.

6.3.1   *More details on Prospect Theory

In this section we will derive two results related to Prospect Theory. First we want to show that a power law for the subjective value function is the only way to achieve a set of preferences that is unaffected by a change in units of money. Secondly, we will show that Cumulative Prospect Theory guarantees that a prospect stochastically dominating another will be preferred.

Suppose that the subjective value function follows a 'power law' where, for $x \geq 1$, $v(x) = \gamma x^\alpha$ for some γ and $0 < \alpha < 1$ (we need this condition to make the value function concave). In this case, if we are indifferent between the prospect (x, p) and the prospect (y, q), then $\pi(p)v(x) = \pi(q)v(y)$ and hence

$$\pi(p)\gamma x^\alpha = \pi(q)\gamma y^\alpha.$$

Thus we have

$$\pi(p)v(kx) = \pi(p)\gamma k^\alpha x^\alpha = \pi(q)\gamma k^\alpha y^\alpha = \pi(q)v(ky),$$

showing that we are still indifferent between these prospects when we have multiplied both the outcomes by a factor of k.

Moreover, we can show that it is only a power law function that has this property. Suppose that multiplying values by k makes no difference to two prospects that are equivalent. Hence, $\pi(p)v(x) = \pi(q)v(y)$ implies $\pi(p)v(kx) = \pi(q)v(ky)$. Thus

$$v(kx)/v(x) = v(ky)/v(y). \tag{6.5}$$

The value y here is arbitrary: for different values of y we simply choose different values of q so that we remain indifferent between (y, q) and (x, p). So we can write h_k for the value of the ratio in Equation (6.5) and $v(x)$ satisfies an equation of the form

$$v(kx) = h_k v(x) \text{ for all } x > 0.$$

Now we define a function g by

$$g(w) = \log(v(e^w))$$

(This is a bit like plotting v on log-log graph paper). Notice that, for any k

$$g(w + \log k) = \log(v(e^w k)) = \log(h_k v(e^w)) = \log(h_k) + g(w),$$

so the function g must have what we can call a 'constant increase' property: in other words, $g(x + A) - g(x)$ depends only on A (and not on x). This implies that g is linear, since if g does not have a constant slope then we can find a point x and distance δ where $g(x) - g(x - \delta) \neq g(x + \delta) - g(x)$ (we just take x somewhere with non-zero second derivative and choose δ small enough). But this contradicts the constant increase property of g. So g must be linear, and we can write it as $g(w) = a + bw$ for some choice of a and b.

Hence, $\log(v(e^w)) = a + bw$ and so

$$v(e^w) = e^a (e^w)^b,$$

which is the form

$$v(z) = \gamma z^\alpha,$$

with $\gamma = e^a$ and $\alpha = b$. Thus, we have established that only power law functions can be used as value functions if we want to preserve ordering of prospects when values are multiplied by a constant.

Now we turn to the question of stochastic dominance. We will just consider prospects with all positive values. $0 < x_1 < \cdots < x_n$ and assume that, for prospect P, there is probability p_i of outcome x_i, and for prospect Q there is probability q_i of outcome x_i. Suppose that P stochastically dominates Q, then

$$\sum_{i=m}^{n} p_i \geq \sum_{i=m}^{n} q_i, \text{ for } m = 2, 3, \ldots, n \tag{6.6}$$

and there is strict inequality for at least one m. We will calculate the π_i^+ values for P and the corresponding values for Q, that we will write as ρ_i^+. Thus

$$\pi_i^+ = w^+(p_i + \cdots + p_n) - w^+(p_{i+1} + \cdots + p_n), \pi_n^+ = w^+(p_n),$$

$$\rho_i^+ = w^+(q_i + \cdots + q_n) - w^+(q_{i+1} + \cdots + q_n), \rho_n^+ = w^+(q_n).$$

We want to show that $V(P) > V(Q)$. Now

$$V(P) = \sum \pi_i^+ v(x_i)$$

$$= \sum_{i=1}^{n-1} \left(w^+(p_i + \cdots + p_n) - w^+(p_{i+1} + \cdots + p_n) \right) v(x_i) + w^+(p_n)v(x_n)$$

$$= v(x_1) + \sum_{i=2}^{n} w^+(p_i + \cdots + p_n)(v(x_i) - v(x_{i-1}))$$

where we have used the fact that $w^+(p_1 + \cdots + p_n) = w^+(1) = 1$ and gathered together the terms with the same w^+ value. In the same way

$$V(Q) = v(x_1) + \sum_{i=2}^{n} w^+(q_i + \cdots + q_n)(v(x_i) - v(x_{i-1})).$$

Because of our ordering for the x_i, we know that $v(x_i) - v(x_{i-1}) > 0$, for $i = 2, \ldots, n$. Thus, we can deduce from Inequality (6.6) that each term in this expansion for $V(P)$ is greater than the corresponding term in $V(Q)$ with strict inequality for at least one term. And hence $V(P) > V(Q)$ as we wanted.

6.3.2   Applying Prospect Theory

Tversky and Kahneman (1992) suggest some functional forms for the functions v, w^+ and w^- and estimate the parameters for these functions for a group of experimental subjects (students). They propose that the value functions for both gains and losses follow a power law and hence

$$v(x) = x^\alpha \text{ if } x \geq 0$$
$$= -\lambda(-x)^\beta \text{ if } x < 0.$$

Note that we can normalize so there is no need of a constant multiplier for the value function for positive x. Moreover, the properties of the power law mean that we don't need to specify the units of money involved here.

The functional forms that Tversky and Kahneman propose for the decision weight functions are:

$$w^+(p) = \frac{p^\gamma}{(p^\gamma + (1-p)^\gamma)^{1/\gamma}},$$

$$w^-(p) = \frac{p^\delta}{(p^\delta + (1-p)^\delta)^{1/\delta}}.$$

Tversky and Kahneman also give estimates for the various parameters:

$$\alpha = \beta = 0.88;$$
$$\lambda = 2.25;$$
$$\gamma = 0.61; \delta = 0.69.$$

We will call these the TK parameter values. They are median values obtained when estimates are made separately for each individual experimental subject.

Figure 6.5 shows the behavior of the value function v with the TK parameters. We can see the way that there is a kink at zero, and also that with an exponent relatively near 1, the value function is pretty much a straight line away from the origin. Surprisingly, the derivative of this curve at zero is infinite, since, for positive z, we have

$$v'(z) = \alpha z^{\alpha-1} = 0.88z^{-0.12}$$

and as $z \to 0$ the value of $z^{-0.12}$ goes to infinity. However, we need to take such tiny values of z to get large values of the slope that this behavior never shows up on this kind of graph (e.g. we have $v'(0.001) = 0.88(0.001)^{-0.12} = 2.02$ and $v'(0.00001) = 3.50$.)

The values implied for w^+ and w^- are given in Table 6.1 and plotted in Figure 6.6. Notice that even though this table gives three decimal places, the numbers arise from parameter choices estimated on the basis of a limited set of decisions made in a laboratory setting and should be taken as only a rough guide to the behavior of any given individual.

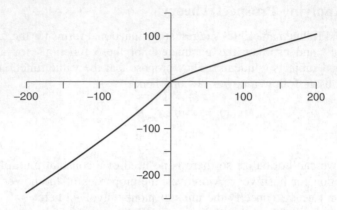

Figure 6.5   Graph of the subjective value function v.

Table 6.1   Median decision weight values found by Tversky and Kahneman.

p	$w^+(p)$	$w^-(p)$	p	$w^+(p)$	$w^-(p)$
0.05	0.132	0.111	0.55	0.447	0.486
0.10	0.186	0.170	0.60	0.474	0.518
0.15	0.227	0.217	0.65	0.503	0.552
0.20	0.261	0.257	0.70	0.534	0.588
0.25	0.291	0.294	0.75	0.568	0.626
0.30	0.318	0.328	0.80	0.607	0.669
0.35	0.345	0.360	0.85	0.654	0.717
0.40	0.370	0.392	0.90	0.712	0.775
0.45	0.395	0.423	0.95	0.793	0.850
0.50	0.421	0.454	1	1	1

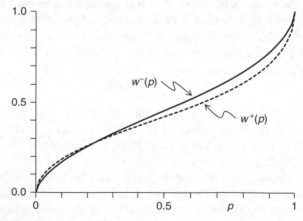

Figure 6.6   Comparison of w^+ and w^- using TK parameters.

Worked Example 6.5   Choosing between three prospects

Use Cumulative Prospect Theory with the TK parameter values to determine which of the following three prospects is preferable:

$$A : (\$100, 0.5; \$200, 0.4; \$300, 0.1),$$

$$B : (-\$100, 0.1; \$300, 0.5; \$800, 0.1),$$

$$C : (\$150, 1).$$

Solution

Prospect A has all gains and we get

$$V(A) = (w^+(1) - w^+(0.5))v(100) + (w^+(0.5) - w^+(0.1))v(200)$$
$$+ w^+(0.1)v(300)$$
$$= (1 - 0.421) \times 100^{0.88} + (0.421 - 0.186) \times 200^{0.88} + 0.186 \times 300^{0.88}$$
$$= 86.35.$$

For prospect B we have one loss and two gains and we get

$$V(B) = w^-(0.1)v(-100) + (w^+(0.6) - w^+(0.1))v(300) + w^+(0.1)v(800)$$
$$= 0.170 \times (-2.25) \times 100^{0.88} + (0.474 - 0.186) \times 300^{0.88}$$
$$+ 0.186 \times 800^{0.88}$$
$$= 88.28.$$

For prospect C there is no uncertainty and we simply have

$$V(C) = v(150) = 150^{0.88} = 82.22.$$

Overall, we see that the highest value is achieved by prospect B, and using these parameters would lead to B being chosen.   □

6.3.3   Why Prospect Theory does not always predict well

Earlier we mentioned the fact that there are many competing theories in this area. And though the general predictions made by Cumulative Prospect Theory are correct, some aspects of the theory are the subject of considerable debate. The primary problem with Prospect Theory as a description of the way that decisions are taken, is that it can be quite hard to calculate the value of a prospect, and it seems hard to imagine that this is a good match with the way that individuals actually make decisions (even in cases where decisions are thought about carefully, with the decision maker operating within a 'reasoning' rather than intuitive framework). An example is the observation we made earlier that in a choice between prospects, outcomes with higher saliency will usually be given greater weight. To some extent this is reflected within Prospect Theory by the use of current wealth as a reference point – extreme changes will be more salient. But it is likely that other aspects of the choice will also have an impact on saliency.

In 'Version 1' of Prospect Theory, as propounded in Kahneman and Tversky's 1979 paper, there was a greater role for heuristics applied by the decision maker in order to simplify the decision to be made (for example, eliminating dominated choices). In moving to Version 2, these preliminary 'editing' steps were dropped; in essence, they were made unnecessary by the use of a rank-based cumulative weighting function. This has the great advantage of making the predicted choice quite definite (once the parameters have been selected) whereas any theory that puts more emphasis on the processes used by a decision maker is likely to lead to cases where the prediction is less clear cut (for example, depending on the order in which some initial editing processes are carried out). However, the price to be paid is that there are many situations in which the neatness of Cumulative Prospect Theory does not seem to capture the whole picture.

There are a number of other reasons why we should be cautious in using Prospect Theory to predict individual behavior:

- Different individuals will have different patterns of behavior, involving different degrees of loss aversion, different degrees of risk aversion, etc. In other words, even accepting the core assumptions of Prospect Theory still leaves the question of what the parameters should be for an individual. Moreover, we should not assume that an individual always operates with a particular decision approach – perhaps our experiences over the past few hours will have an impact on the amount of risk we choose.

- There will be a random element in the way that choices are made, especially when they are perceived as being quite similar in overall value. We recognize this in our own choices sometimes: 'It's six of one and half a dozen of the other'. In an experimental setting it is quite common for the same individual to make different choices between exactly the same pair of prospects when these are presented at different times.

- The decision weight functions w do not appear to work well with probability values that are near zero or near 1. Individuals are particularly poor at making choices when faced with probabilities that are of the order of 0.01 or smaller (or 0.99 or greater). This is where inconsistencies are most likely to arise.

- The particular functional forms chosen for w and v are, to some extent, arbitrary. Different functional forms have been suggested and may result in different predictions.

- Individuals may change their preferences between options over time as a result of repeated exposure to the same choices, or even discussion with colleagues. This issue is complex though, since, if at the outset, a decision maker knows that she will be faced with a whole sequence of similar choices or gambles that are perceived as being grouped together, then her response is likely to be different to the choice made when she is faced with a single gamble.

- Individuals may make different decisions depending on how the choice situation is framed. Since people typically look at changes in wealth with respect to a reference point, framing may occur through the suggestion of a reference point that is different to zero. For example, a choice may be presented as follows: 'Would you rather have $30 for sure or be given a 50% chance of winning $80?' Exactly the same decision problem can be framed by telling someone they have won $30 and then asking 'Do you want to enter a competition where there is a 50% chance of losing $30 and a 50% chance of winning $50?' In this second framing of the choice, the reference point has become +$30. Loss aversion will ensure that fewer people will accept the gamble when they have the $30 as the reference point than in the first framing.

6.4   Decisions with ambiguity

In this section we will discuss the way that individuals take decisions when there is ambiguity in relation to the probabilities involved. The theory of Savage, discussed in Chapter 5, implies that, in many cases, consistent decision makers will be working as though there were a subjective probability associated with any particular event. But at this point we are more interested in describing the way that decisions are made in practice.

The examples we have dealt with so far have all ducked the question of where the probabilities come from. We have assumed that some oracle (the psychologist conducting the experiments) announces for us the probabilities of particular events, allowing us to write a prospect as a string of values and probabilities. Or perhaps the probabilities are generated by throwing dice or tossing coins. This is what Nassim Taleb calls the 'ludic fallacy' – the belief that real decisions are well-represented by the decisions faced when playing simple games of chance. But, as we have pointed out in earlier chapters, actual decisions usually involve probabilities that we can guess at, but not know for sure. This happens when the probability arises because of our uncertainty about the way that the world will change in the future.

For example, suppose we ask 'What is the probability that the price of oil will be above $150 a barrel a year from now?' A business decision may well depend on the likelihood that we assign to this event, so we may be forced to give an answer, either explicitly or implicitly, by the decision that we make.

The prediction of the price of oil is amenable to all sorts of economic analysis, so a decision maker will have some knowledge of this probability, but many business decisions need to be made with very little knowledge of the probabilities involved. For example, we might be interested in the probability that a competing company decides to pull out of a particular marketplace, or the chance that a new pharmaceutical compound under development will prove both safe and effective as a treatment for Alzheimer's disease. In these types of decisions, the extent of our ignorance is greater and we are likely to be forced into looking at statistics for similar cases in the past (if we can find them).

A good example of the way that a lack of knowledge about exact probabilities has an impact on our decisions is provided by the Ellsberg paradox. Suppose that you have two urns each containing 100 balls. In the first urn, there are exactly 50 black balls and 50 white balls, but you have no knowledge of the number of balls of different colors in the second urn. Now you are offered a prize of $100 if you draw out a white ball from one of the two urns, but you only get one attempt and so you need to choose which urn. What would you do? It turns out that a large proportion of people will choose the first urn. In a sense there is less uncertainty associated with the first urn, where we know that there will be a 50% chance of winning; if we choose the second urn then we have no knowledge at all about the probability of winning, which might even be zero if every ball in the second urn is black.

The preference for the first urn remains true for a different problem in which the prize of $100 is given if you can correctly predict the color of the first ball drawn. In this formulation, first the decision maker selects a color and then she chooses one of the two urns, and hence there is no possibility of the composition of the second urn being somehow made unfavorable.

In order to explain the theoretical problem that this creates more clearly, suppose that you are the decision maker presented with two urns and you are offered a prize of $100 if you draw out a white ball from one of the two urns. You are likely to choose the first urn; perhaps you win and perhaps you do not. Then you put the ball back into the urn and shake it up to remix all the balls. Next you are offered a second chance to win a prize of $100, but this time you get the prize if you draw out a black ball. What would you choose? Most people still have a clear preference for the first urn, with the known composition of 50 balls of each color. After all, at this point the second urn is untouched, so it is hard to see why a change of color would change the preference decision.

The first decision is between a prospect ($100, 0.5) and a prospect ($100, p_W) where p_W is the (subjective) probability of a draw of a white ball from the second urn. The preference for the first urn implies that $p_W < 0.5$. This is obvious and is also implied by the prospect valuation, which has $\pi(0.5)v(\$100) > \pi(p_W)v(\$100)$. But if p_W is less than 0.5 then p_B, the subjective probability of a black ball being drawn from the second urn, must be greater than 0.5. Hence, when we reach the second choice it is rational to prefer the second urn. This is the crux of the paradox, which can only be resolved if decision makers were indifferent between the two urns.

This is an example of a situation in which uncertainty about the probabilities in the second urn makes us reluctant to choose it. Sometimes this type of uncertainty is called *ambiguity*, and the effect we see in the Ellsberg paradox is called *ambiguity aversion*. This is an extreme example of a kind of second order uncertainty when the probability is itself a random variable. We will return to thinking about these types of problems in our discussion of robust optimization in Chapter 8.

Much more rarely it is possible to observe an ambiguity preference rather than ambiguity aversion. Ellsberg has suggested an example in which there are two urns: the first has 1000 balls numbered 1 to 1000 and the second has 1000 balls each with a number between 1 and 1000, but numbers may occur more than once and we have no information on which numbers have been used. So, for example, we might have 500 balls marked 17 and the other 500 balls marked with randomly selected numbers between 200 and 800. Now we are asked to write down a number between 1 and 1000 and then draw out a ball from one of the two urns. If the number on the ball matches the number we have written, we win a prize. In this decision scenario people are quite likely to choose the second (ambiguous) urn. The basic structure here is the same as for the Ellsberg paradox but we are dealing with probabilities of 1 in 1000 rather than 1 in 2.

6.5   How managers treat risk

In this final section we will look at the impact of the psychology of risk on management decisions. One of the key observations of Prospect Theory is that individuals make judgments based on a change in their wealth, rather than looking at utilities associated with different values of their total wealth after the decision and its consequences. From a manager's perspective, this is not what shareholders would like; a more rational decision would see maximizing total profit as the real aim, independently of the route taken in getting there.

This is related to what happens when we have multiple opportunities to gamble. For example, the gamble involving gaining $100 or losing $90 might not seem attractive, but if we knew it was offered many times over, then its attractiveness would change. For example, with four repetitions, there is a $1/16$ chance of losing $360, a $1/4$ chance of losing $170, a $3/8$ chance of gaining $20, a $1/4$ chance of gaining $210 and a $1/16$ chance of gaining $400. This still might not be enough to encourage everyone to accept the package of gambles, but it is certainly closer to being attractive than a gamble played just once.

Or we might take another example for which many people are more or less indifferent to accepting the gamble, which is when the loss is roughly twice as large as the gain; say we have a half chance of losing $100 and a half chance of gaining $200. But with two repetitions of this gamble we end with a $1/4$ chance of gaining $400, a $1/2$ chance of gaining $100 and a $1/4$ chance of losing $200. Faced with this prospect, most people give it a positive value. Multiple opportunities to gamble with a positive expected outcome lead to a greater and greater chance of a good outcome.

It seems, however, that as decision makers we are hard-wired to look just at the immediate outcome of the choice facing us, rather than seeing this as one element in a sequence of choices. As a result, we pay attention just to the change arising from the current decision, more or less independent of the results of previous decisions. We can say that decision makers are *myopic* (a technical term for being 'short-sighted' in a decision-making sense).

This leads to relatively high risk aversion for small gambles as well as for large gambles. This is not rational because small gambles are likely to recur. They may not recur in exactly the same form, but over a period of time there are almost certain to be risky opportunities available to a decision maker, which correspond to small gambles. A decision maker who consistently takes these small gambles where they have a definite positive expected value will end up ahead over time.

Much of our discussion so far has been framed around individual decisions on prospects–we might suppose (or hope) that managers' decisions are in some way 'better'. Managers make decisions in contexts that often involve a whole team of people and that are subject to significant scrutiny. However, there seems little evidence that managers do better in making corporate decisions than individuals do in making personal decisions.

The first observation to make is that a manager's decisions are taken within a personal context. A manager's actions are not simply about achieving the best result for the company. In addition, a manager will be asking herself, 'What will this do for my career?' or 'Will this be good for my stock options?'

A second observation is that managers' own ideas about their roles have an impact on their behavior. We may see managers as dealing with uncontrollable risks in a way that accepts these (negative) possible outcomes because they are compensated by significant chances of gain. But this is not the way that managers themselves view their roles (March and Shapira, 1987). Instead, managers are likely to view risk as a challenge to be overcome by the exercise of skill and choice. They may accept the possibility of failure in the abstract, but tend to see themselves not as gamblers, but as careful and determined agents, exercising a good measure of control both over people and events. The net result is that managers are often much more risk averse than we would expect. Consciously or not, many managers believe that risks should be hedged or avoided if they are doing their jobs well.

What is the remedy for this *narrow framing* that looks only at a single decision and ends up being unnecessarily risk averse? Kahneman and Lovallo (1993) suggest that this bias can be helped by doing more to encourage managers to see individual decisions as one of a sequence (perhaps by doing evaluations less frequently) and also by encouraging an attitude that 'you win a few and you lose a few', because it suggests that the outcomes of a set of separable decisions should be aggregated before evaluation.

Even if an observer may see managers as taking significant risks, managers themselves perceive those risks as small – the difference is explained by the way that managers habitually underestimate the degree of uncertainty that they face. Kahneman and Lovallo (1993) describe this in terms of *bold forecasts*. Why do managers so frequently take an optimistic view and underestimate the potential for negative outcomes?

This is an example of a more general phenomenon which is the near universal tendency to be more confident in our estimates than we should be. Even when we are sophisticated enough to understand that any estimate or forecast is really about a distribution rather than a single number, we still tend towards giving too much weight to our best guess, or, to put it another way, we use distributions that do not have sufficient probability in the tails.

There may be many factors at work, but one important aspect of this bias relates to what Kahneman and Lovallo (1993) describe as the *inside view* (which, to some extent, mirrors the narrow framing bias we mentioned above). When faced with an uncertain future and the need to forecast, our natural instinct is to consider very carefully all the specifics of the situation, bring to bear our understanding of the potential causal chains and then determine what seems the most likely outcome. The problem with this approach is that there are often simply too many possible ways in which events may unfold for us to comprehend them all. It may well be that the chain of events that leads to a project completion on time is the most likely amongst all possibilities, but if there are many thousands of possible reasons for delay, each happening with a small probability, then it may well be the case that significant project delay becomes a near certainty.

The remedy for this situation is to step back from considering the specifics of what may happen in detail, but instead to understand the situation in a more statistical sense. In contrast to an inside view, we could describe this as an *outside view*. Is there a set of roughly equivalent circumstances from which a manager can learn more of the likely range of outcomes? Sometimes this is reasonably straightforward – for example, in predicting the box office takings for a four-person comedy drama playing a short season in New York, it is natural to look to information about similar shows in the past. Often, however, it requires care to find the right group of comparator situations. There is a lot of evidence that adopting an outside view is more likely to lead to good predictions, but it is surprisingly rare in practice. As Kahneman and Lovallo (1993) explain:

> 'The natural way to think about a problem is to bring to bear all one knows about it, with special attention to its unique features. The intellectual detour into the statistics of related cases is seldom chosen spontaneously. Indeed, the relevance of the outside view is sometimes explicitly denied: physicians and lawyers often argue against the application of statistical reasoning to particular cases. In these instances, the preference for the inside view almost bears a moral character. The inside view is valued as a serious attempt to come to grips with the complexities of the unique case at hand, and the outside view is rejected for relying on crude analogy from superficially similar instances.'

The specific issue here is related to well-understood characteristics of optimism. In general, people are optimistic when it comes to evaluating their own abilities (so, for example, a large majority of people regard themselves as above-average drivers); they are optimistic about future events and plans; and finally they are optimistic about their ability to control what happens. In general, this is a positive characteristic, and optimism in this form is associated with mental health (Taylor and Brown, 1988). This does not mean, however, that it is a beneficial characteristic when practiced by managers facing important decisions on the future of their organizations.

Notes

There is an enormous amount that has been written on behavioral decision theory and behavioral economics. The book by Wilkinson (2008) gives an accessible introduction to this field, and the paper by Starmer (2000) gives a more detailed discussion of much of the literature in this area. Peter Wakker's book gives a very thorough and technical treatment of all aspects of Prospect Theory, but this is a difficult read for the non-expert. The material presented in this chapter has drawn heavily on the papers by Kahneman and Tversky (1979), Tversky and Kahneman (1993), and Kahneman and Lovallo (1993). Kahneman's book, 'Thinking, fast and slow' is also an easy introduction to this area (see Kahneman, 2003).

For the discussion of ambiguity in decision making I have drawn on the paper by Einhorn and Hogarth (1986).

References

Allais, M. (1953) The behavior of rational man in risky situations – A critique of the axioms and postulates of the American School. *Econometrica*, **21**, 503–546.

Einhorn, H. and Hogarth, R. (1986) Decision making under ambiguity. *The Journal of Business*, **59**, 225–250.

Kahneman, D. (2011) *Thinking, Fast and Slow*. Farrar, Straus and Giroux.

Kahneman, D. and Lovallo, D. (1993) Timid choices and bold forecasts: A cognitive perspective on risk taking. *Management Science*, **39**, 17–31.

Kahneman, D. and Tversky, A. (1979) Prospect Theory: An analysis of decision under risk. *Econometrica*, **47**, 263–292.

March, J. G. and Shapira, Z. (1987) Managerial perspectives on risk and risk taking. *Management Science*, **33**, 1404–1418.

Rabin, M. and Thaler, R. H. (2001) Anomalies: Risk aversion. *Journal of Economic Perspectives*, **15**, 219–232.

Starmer, C. (2000) Developments in non-expected utility theory: The hunt for a descriptive theory of choice under risk. *Journal of Economic Literature*, **38**, 332–382.

Taylor, S. and Brown, J. (1988) Illusion and well-being: A social psychological perspective on mental health. *Psychological Bulletin*, **103**, 193–210.

Tversky, A. and Kahneman, D. (1992) Advances in prospect theory: Cumulative representation of uncertainty. *Journal of Risk and Uncertainty*, **5**, 297–323.

Wakker, P. (2010) *Prospect Theory for Risk and Ambiguity*. Cambridge University Press.

Wilkinson, N. (2008) *An Introduction to Behavioural Economics*. Palgrave Macmillan.

Exercises

6.1 Explanation of Example 6.1

Worked Example 6.4 shows how the behavior shown in Example 6.2 is exactly what one would expect from Prospect Theory if the value function for positive x is given by $v(x) = x^{0.8}$. Use the same kind of analysis to explain the behavior of Example 6.1.

6.2 Prospect Theory when gains turn to losses

Suppose that $w^+(p) = w^-(p)$ and the value function has the property that

$$v(-x) = -\lambda v(x) \text{ for } x > 0.$$

Show that if a prospect $A = (x_1, p_1; x_2, p_2; \ldots; x_n, p_n)$ is preferred to prospect $B = (y_1, q_1; y_2, q_2; \ldots; y_n, q_n)$, and all the x_i and y_i are positive, then prospect $-B = (-y_1, q_1; -y_2, q_2; \ldots; -y_n, q_n)$ is preferred to $-A = (-x_1, p_1; -x_2, p_2; \ldots; -x_n, p_n)$.

6.3 Probabilistic insurance

Kahneman and Tversky carried out an experiment with 95 Stanford students in which the students were presented with the following problem:

'Suppose you consider the possibility of insuring some property against damage (e.g. fire or theft). After examining the risks and the premium, you find that you have no clear preference between the options of purchasing insurance or leaving the property uninsured. It is then called to your attention that the insurance company offers a new program called probabilistic insurance. In this program you pay half of the regular premium. In the case of damage, there is a 50 per cent chance that you pay the other half of the premium and the insurance company covers all losses; and there is a 50 per cent chance that you get your insurance premium back and suffer all the losses. For example, if the accident falls on an odd day of the month, you pay the other half of the premium and the insurance company covers all losses; but if the accident occurs on an even day of the month, you get your insurance premium back and suffer all the losses.

Remember that the premium is such that you find this insurance is barely worth its cost. Under these circumstances, would you purchase probabilistic insurance?'

In this experiment, 80% of the students answered 'No'. Show that this is inconsistent with Expected Utility Theory with a concave utility function because probabilistic insurance gives a strictly higher utility than standard insurance. You can do this by writing the result of probabilistic insurance in the form

$$(1 - p)u(W - (z/2)) + (p/2)u(W - z) + (p/2)u(W - K),$$

and then use the fact that, for a concave function u, we have

$$u\left(W - \frac{1}{2}z\right) \geq \frac{1}{2}u(W) + \frac{1}{2}u(W - z).$$

6.4 Exponent in power law

Students are asked to decide between two choices, Option A and Option B.

Option A: Get $1 with probability 95% and get $381 with probability 5%
Option B: Get $20 for sure.

Most students prefer option B. Next, the students are presented with the same two choices, but with $300 added to all the outcomes, i.e.

Option C: Get $301 with probability 95% and get $681 with probability 5%
Option D: Get $320 for sure.

Many students then switch and decide that they prefer the risky option. Show that, with the standard TK parameters, Prospect Theory will not predict this switch, but that for individuals where the power law exponent α is reduced to 0.6, we will see a switch under Prospect Theory. (This experiment is reported in Bordalo, P., Gennaioli, N. and Shleifer, A., 2011. *Salience Theory of Choice Under Risk*, NBER Working Paper #16387, which provides an alternative explanation for these observations.)

6.5 Laptop warranties

A company selling laptops offers an extended warranty on its products. For one laptop model costing $900 the standard warranty is for one year and the extended warranty covers a further two years at a cost of $75.

(a) Suppose that the probability of a breakdown in the extended warranty period is 0.1 and that a customer who does not take the warranty faces a cost of $500 if this happens. Suppose that a customer's choices can be described using Prospect Theory with the TK parameters. Determine whether the customer is likely to take up the extended warranty.

(b) Now suppose that if a breakdown occurs and the warranty has not been taken up, then the laptop is equally likely to require replacement at a cost of $900 or a simple repair at a cost of $100 (so the expected cost is $500). Calculate the value of the relevant prospect in this case and hence whether the customer will take up the warranty.

6.6 Splitting prospects can change choices

(a) Use Prospect Theory with $\alpha = \beta = 0.9$; $\lambda = 2$ and the standard decision weight functions given in Table 6.1 to calculate the values given to the following four prospects in order to predict which will be chosen.

$$A : (-\$100, 0.5; \$1000, 0.5)$$

$$B : (\$1000, 0.4)$$

$$C : (\$200, 0.3; \$300, 0.3; \$550, 0.4)$$

$$D : (\$340 \text{ with certainty}).$$

(b) Now suppose that prospects A, B and C are constructed in two stages, with $340 received first and then gambles presented (so A is replaced with D followed by $(-\$440, 0.5; \$660, 0.5)$ and similarly for B and C). Show that none of the second stage gambles will be chosen.

6.7 Risk seeking when value function is concave for losses

For some individuals the general pattern is reversed and instead of the value function v being convex for losses, it is either straight or mildly concave. Nevertheless, the behavior of the decision weight function may still lead to risk-seeking behavior (where a risky option with the same expected value is preferred to a certain outcome). Find an example where this occurs and a certain loss of $100 is less attractive than a gamble having the same expected value. You should use the decision weights w^- given in Table 6.1 and take

$$v(x) = x^{0.9} \text{ for } x > 0$$

$$v(x) = -2(-x)^{1.1} \text{ for } x < 0.$$

7

Stochastic optimization

Maximizing profit from pumped storage

A pumped storage facility operates by pumping water up to a high reservoir when power is cheap and then letting that water flow out of the reservoir through hydroelectric generators in order to generate power when it is expensive. There are inefficiencies, so the water, once pumped up hill, can never deliver as much energy from letting it flow through the turbines as was needed to pump it in the first place. Nevertheless, the overall exercise is worthwhile because the actual cost of electricity at night is so much lower than it is during the day. So, cheap electricity can be used to fill the high reservoir during the night and then that power, can be released during the day when prices are high. With increasing amounts of wind power which often delivers energy peaks at a time when demand is not high, the opportunities and need for pumped storage are greater than ever.

The Raccoon Mountain pumped storage plant is a good example of this sort of operation. It was built in the 1970s and is owned and operated by the Tennessee Valley Authority. The reservoir on Raccoon Mountain is more than 500 acres in size (200 hectares). When water is being pumped up to this reservoir, it can be filled in 28 hours. When electricity demand is high, water is released and can generate up to 1600 MW per hour. If the reservoir is full, it would take 22 hours for it to empty if it was run continuously.

To determine the operation policy for the pumped storage plant, each day is divided into different periods: peak is for seven hours, shoulder for eight hours and off peak for nine hours. Typically, the reservoir is pumped in the off-peak hours during the night and is full at the end of that time, then the reservoir is run down during the seven hours of peak demand (nine hours of pumping will give a volume of water sufficient for seven hours of generation). But when demand and prices are high, the reservoir can be used for a longer time to include some of the shoulder period, so that the water level at the end of the day is significantly

Business Risk Management: Models and Analysis, First Edition. Edward J. Anderson.
© 2014 John Wiley & Sons, Ltd. Published 2014 by John Wiley & Sons, Ltd.
Companion website: www.wiley.com/go/business_risk_management

lower and cannot be completely replenished during the night; this leads to a slow drop in the reservoir level over successive days. Eventually, the reservoir reaches a stage where it completely empties during the day, and in the morning the only water available is that which was pumped up overnight. This sets the constraint on the amount of power available on any day (independently of how high the electricity price might be).

The problem facing the plant operator is whether to take advantage of high prices today by running the generators for longer. And if the price is high enough to make it worthwhile to generate power during the shoulder period, how much power should be generated? The longer the generator is run, the lower the resulting water level will be, and the less opportunity there will be to benefit if high prices occur again tomorrow.

7.1   Introduction to stochastic optimization

In this chapter we will use the methods of optimization to determine how to make good decisions in a stochastic environment. Whereas in the previous chapter we primarily looked at decisions made with a clear-cut set of options, each of which contains a small number of possibilities with associated probabilities and consequences, in this chapter we will consider more complex problems. There will be a need to set up a model of what is happening (containing stochastic elements) and then to analyze this model. Because of this additional complexity, our focus will switch back to the normative, rather than the descriptive, so we will ask what managers should do, rather than what they will do. For most of the models that we deal with in this chapter it is necessary to carry out an analysis using some computational tool like a spreadsheet.

An important difference exists between stochastic optimization problems where we get just one opportunity to make a decision and problems where we have an opportunity to make decisions at different points in time. For example, if we are selling fashion clothing items and the demand is uncertain, then we will need to decide at the start of the season how much of a particular fashion item to make. However, we may also have an opportunity to make more halfway through the season when we have some information on how sales are going: this will then be a second decision point.

Before going on with our discussion of stochastic optimization we need to review some fundamental ideas about optimization problems.

7.1.1   A review of optimization

An optimization problem is one in which we need to choose some *decision variables* in order to maximize (or minimize) an *objective function* subject to some *constraints* on the variables specifying which values are possible. These are the three critical components: variables, objective and constraints. If the variables are *n* real numbers that need to be chosen, then we can think of the

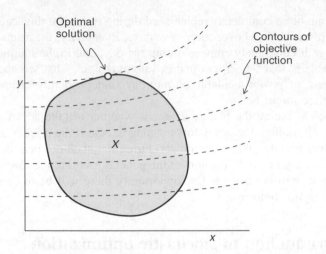

Figure 7.1   A diagram of a maximization problem.

problem in the way that is shown in Figure 7.1. The decision variables here are *x* and *y*, the set *X* is the set of feasible points defined by the constraints of the problem, and the objective function is shown by its contour lines. The optimal point (which we also call the *optimal solution*) is on the boundary of the set *X* and maximizes the objective function (optimization problems may involve either minimizing or maximizing the objective function).

A first categorization of optimization problems distinguishes between problems in which there is only one local optimum and problems for which there are many possible local optima, and we need to compare them to find the best. From the point of view of Figure 7.1, the good property of having only one possible maximum arises from the nature of the objective function and the feasible set. The feasible set is convex (which means that the straight line between two points in *X* can never go outside of *X*) and the objective function is concave (i.e. it has the shape of a hill: a straight line between any two points on the surface defined by the objective function lies below the surface rather than above it). The definition of a concave function can be extended to any number of points, and formally we say that $f(x, y)$ defined on the (x, y) plane is concave if, for any set of *n* points (x_1, y_1), (x_2, y_2), ..., (x_n, y_n), and any set of weights w_j that are non-negative and with $\sum_{j=1}^{n} w_j = 1$, then

$$\sum_{j=1}^{n} w_j f(x_j, y_j) \leq f\left(\sum_{j=1}^{n} w_j x_j, \sum_{j=1}^{n} w_j y_j\right).$$

A similar definition for a concave function works when the function is defined on three or more variables. Also, convex functions have the same definition but with the inequality reversed. With these properties there will just be one local maximum (which is also a global maximum). On the other hand, if we have

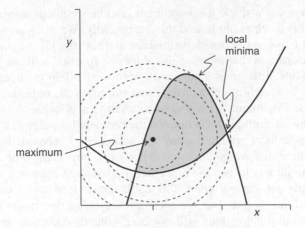

Figure 7.2   This example has a single maximum but more than one local minimum.

a minimization problem rather than a maximization problem, then we want the objective function to be convex, instead of concave.

A convex maximization problem (with a concave objective function and a convex feasible set) may have its optimal solution on the boundary or in the interior of X. Figure 7.2 shows a maximization problem of this sort with an interior maximum.

This maximization problem ($P1$) can be formulated as

$$P1 : \text{maximize } 4 - (x - 1)^2 - (y - 1)^2,$$

subject to the constraints:

$$2y - (x - 1)^2 \geq 1,$$

$$3y + 2(2x - 3)^2 \leq 6.$$

The first constraint defines the lower boundary of the feasible region, and the second defines the upper boundary. The shaded region shows the feasible points. The point which maximizes the objective is $x = 1$, $y = 1$ and this is inside the feasible region X.

But now consider changing the problem so that instead of maximizing the objective function we want to minimize it. The contours of the objective function are shown with dashed lines. From this we can see that there are two local minima on the right of the figure (and one more on the left). A local minimum is a point that is lower than any feasible points close to it. Small changes in the decision variables may lead to an infeasible point if one of the constraints is broken, but at a local minimum small changes that retain feasibility can only make the objective function larger. To find the global minimum we need to compare the objective function values at different local minima. In this example, the higher point on the smooth part of the boundary is the global minimum, as we can see from the contour lines.

If a problem has a single local optimum (and hence this is also a global optimum), then it is much easier to find this numerically. We can start with an initial feasible point (some set of decision variables that satisfies all the constraints) and then consider changes that improve the objective function without breaking any of the constraints. This gives a new and improved feasible point, and we repeat the procedure of searching for a change that improves the objective. Eventually, we reach a local optimum when no further change is possible.

When there are multiple local optima, the problem becomes much harder. We can find a single local optimum using the approach of repeated improvements, but when we have discovered this we will not know how many other local optima there are. One idea is to use the repeated improvement approach again, but to begin at a different starting point: this could either lead us to the same local optima, or a different one. But for a complex problem we might try hundreds of different starting points, and still not be absolutely sure that we have found every local optimum.

To learn about optimization, it is important to try out the ideas in practice. There are a great many pieces of software available that are suited for different types of optimization problem. As a tool for solving simple optimization problems, the Solver add-in for Excel is quite satisfactory. It is a good exercise to solve the problem of minimizing the objective function for $P1$ using Solver. This has been done in the spreadsheet BRMch7-P1.xlsx. Try starting the optimization process at different points: these need not be feasible points, Solver will start by finding a feasible solution before improving it. You will find that the three different local minima can all be obtained depending on what initial values Solver is given.

An important special class of optimization problem, called a *linear program*, occurs when the constraints and the objective function are all linear. This will produce a well-behaved problem, since the set of feasible points is convex and the objective is both convex and concave. At first sight, one might think that solving this sort of problem would be trivial, but in practice when linear programs have a large number of variables and a large number of constraints, finding an optimal solution requires a computer. An example of a linear program is given as problem $P2$ below.

$$P2 : \text{maximize } 6x + 7y + 2,$$

subject to the constraints:

$$2x + 3y \leq 8,$$
$$4x - y \leq 6,$$
$$4y - 5x \leq 1,$$
$$x \geq 0,$$
$$y \geq 0.$$

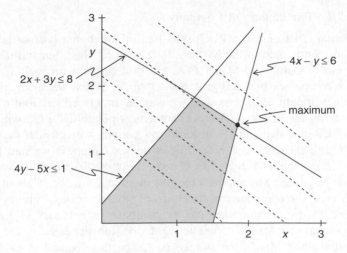

Figure 7.3   The feasible set and the optimal solution for the problem P2.

Figure 7.3 shows the feasible points for $P2$. The dashed lines are contours of the objective function and the maximum occurs at the point shown ($x = 1.857$ and $y = 1.429$).

Linear programs can be solved for very large scale problems using special purpose software. Excel Solver has a setting under options for 'Assume Linear Model' and another setting for 'Assume Non-Negative' and these can be used for small problems like $P2$. The file BRMch7-P2.xlsx gives a spreadsheet for this example.

7.1.2   Two-stage recourse problems

Our starting point in thinking about stochastic optimization is to consider a two-stage problem. In the first stage a decision is made not knowing what will happen in the future, but knowing the probability of different events. After the first stage decision, the random event occurs and uncertainty is resolved. Then, in the second stage, further decisions are made that can depend on what has happened. This is often called a *two-stage stochastic problem with recourse*. The word 'recourse' here refers to what needs to be done at the second stage as a result of the random event. The framework here is very similar to the decision tree analysis we gave in Chapter 5. The main difference is that we will think about decisions where we will choose the value of a continuous decision variable, rather than choosing between discrete options.

In order to illustrate some important ideas and show how we need to take care with our formulation of these problems, we begin by analyzing a simple example.

Example 7.1   Parthenon Oil Company

The Parthenon Oil Company (POC) sells fuel oil for home heating and uses a planning horizon related to the winter selling season. Most demand takes place in the months of October to March. POC has storage tanks for fuel oil and buys it from oil companies that sell at a market price, which fluctuates month by month. In any month, POC can supply customer orders either from its storage or by buying on the market. Customer contracts stipulate that POC will respond quickly to customer demand. To make things simple, we consider the problem facing POC in February, near the end of its season. Suppose we start February with no oil in storage. In February POC buys oil from the market, delivers some to its customers right away and puts the rest in storage for the following month. Then, in March, the company can supply from storage or buy from the market.

We need to decide how much oil to purchase in February (x_1) and how much to purchase in March (x_2). The right decision depends on the price of oil in February and March, the storage cost, and the demand in each period. Suppose that it costs \$5 to store a barrel of oil for a month. With this information, the problem can be modeled as a simple linear optimization problem with the objective to minimize overall cost. In practice, the price and demand in March will be uncertain. Suppose that demand in February is 1000 barrels and the price is \$160. Moreover, we think that March can have one of three equally likely weather scenarios: normal, cold, or very cold. Cold weather means more demand for oil, but at the same time the price that Parthenon pays for the oil will increase. The demand and price data for the three scenarios are given in Table 7.1.

We write d for the demand in March and c for the cost in March. These are the things that are unknown at the point when Parthenon Oil makes a decision on x_1, the amount of oil bought in February. Since demand in February is 1000, the amount in storage at the end of February will be $x_1 - 1000$, for which Parthenon pays \$5 per unit. Thus, we want to minimize total costs:

$$\text{minimize } 160x_1 + 5(x_1 - 1000) + cx_2,$$

subject to the constraints:

$$
\begin{array}{ll}
x_1 \geq 1000 & \text{(there is enough for February demand),} \\
x_1 - 1000 + x_2 \geq d & \text{(there is enough for March demand),} \\
x_2 \geq 0 & \text{(we cannot sell back to the market if we have} \\
& \text{too much).}
\end{array}
$$

Table 7.1   Data for three possible March scenarios for Parthenon Oil.

Scenario	Probability	Oil cost (\$)	Demand (units)
Normal	1/3	160	1000
Cold	1/3	164	1200
Very cold	1/3	174	1400

In this problem we will find out what d and c are before we need to determine x_2, but we need to choose x_1 before we know d and c.

If we know in advance which of the three scenarios will occur, then we can solve a linear program to find the optimal solution (this is shown in the spreadsheet BRMch7-Parthenon1.xlsx). We can calculate that if March is normal, we should take $x_1 = 1000, x_2 = 1000$; if March is cold, we should take $x_1 = 1000$, $x_2 = 1200$; finally, if March is very cold, then it is worthwhile to buy all the oil we need in February, and $x_1 = 2400, x_2 = 0$. But the problem we face involves making a choice of x_1 right now. Since there is a two thirds chance of normal or cold weather, and under both these scenarios a purchase quantity of 1000 is optimal, perhaps that is the best choice of x_1, and we can determine x_2 after we find out the demand in March.

However, a different approach can produce a different answer. One idea is to look at the average behavior, rather than looking at individual scenarios. Often this is coupled with a sensitivity analysis, in which we consider how sensitive the optimal decision is to changes in the parameters. If we find that it is sensitive, then we may consider a range of possible solutions corresponding to the range of values that we expect, but if we find that it is relatively insensitive, then we may simply stick with the 'average' behavior. For the POC problem, each of the three scenarios is equally likely. The average values are $c = 166$ and $d = 1200$. With these values we can solve the linear program and discover that the optimal solution is $x_1 = 2200$ and $x_2 = 0$. So, if we use average figures we should buy ahead in February, leaving 1200 in storage at the end of the month.

Both these approaches are flawed and the proper way to approach this decision is to set up an optimization problem and embed within this problem the correct structure for the decisions we will take. Here we must allow the choice of different values for x_2 depending on the scenario. We call these x_{2A}, x_{2B}, and x_{2C}. Similarly, we use the notation that the scenario demands and costs are given by d_A, d_B, and d_C; c_A, c_B, and c_C. Then we can formulate the problem as

$$\text{minimize } 160x_1 + 5(x_1 - 1000) + (1/3)(c_A x_{2A} + c_B x_{2B} + c_C x_{2C}) \qquad (7.1)$$

subject to the constraints:

$$x_1 \geq 1000 \qquad \text{(there is enough for February demand)}$$
$$x_1 - 1000 + x_{2A} \geq d_A \qquad \text{(there is enough for March demand for}$$
$$\text{each scenario)}$$
$$x_1 - 1000 + x_{2B} \geq d_B$$
$$x_1 - 1000 + x_{2C} \geq d_C$$
$$x_{2A} \geq 0, x_{2B} \geq 0, x_{2C} \geq 0 \quad \text{(purchases are all non-negative)}$$

This is a linear program and we find an optimal solution using Solver in a spreadsheet. The problem has been set up in the spreadsheet BRMch7-Parthenon2.xlsx, and using Solver we find that the optimal solution has $x_1 = 2000$. This means that there are 1000 in store at the start of March; no more are purchased under scenario A, 200 more are purchased under scenario B and 400 more are purchased under scenario C.

The Parthenon Oil Company example should serve as a warning against the two most common approaches adopted by planners faced with uncertainty. One option starts by computing an optimal solution for each scenario separately, then compares these solutions and chooses the decision that is best for as many of these scenarios as possible. The candidate solutions for the POC problem are then to store either 0 or 1400 units of fuel for the next stage. The real optimal policy (as delivered by the stochastic program) is to store 1000 units and this does not correspond to the optimal solution in any of the scenarios. A second common approach is to take the average value as the prediction and solve the problem without a stochastic component, but for the Parthenon Oil Company problem this gives the wrong choice of x_1 and too much oil stored.   □

The Parthenon Oil Company example has the standard structure of a recourse problem. A first stage decision needs to be made, then, as time goes by, the uncertainty in the problem is resolved (for Parthenon the weather in March becomes known) and finally a second decision is made, which will depend on the first decision and the outcome of the uncertain event. There is a kind of nested structure to this problem.

We will show how to put this sort of problem into a general framework. We start with the second stage, where we need to choose a decision variable y (if there is more than one variable involved then y will be a vector). The variable y will be chosen to minimize costs, knowing both the first stage decision, x say, and the outcome of the random event. We will write ξ for the stochastic component in the problem, thus ξ is a random variable whose value is unknown in advance. In the Parthenon example, y is x_2, the amount ordered in March, and ξ is the weather in March that can take just three different values.

We will consider a problem in which the decision maker is risk neutral and aims to maximize expected profit, or (equivalently) minimize expected cost. In general, we can say that costs occur at both the first and second stage. In the first stage they depend only on the first stage decision, so we can write this as $C_1(x)$, but at the second stage costs depend both on first and second stage decisions and also on the random outcome ξ. Thus, we write the second stage costs as a function $C_2(x, y, \xi)$.

The full formal description of the problem will include not only the cost functions C_1 and C_2 but also the feasible set X for the first stage decision x and also the feasible set Y for the second stage decision y. In general, Y might depend on the values of x and ξ, though we don't show this in the notation.

We write $Q(x, \xi)$ for the lowest cost at the second stage, given x and ξ. Thus

$$Q(x, \xi) = \min_{y \in Y}\{C_2(x, y, \xi)\}.$$

The first stage decision is to choose a value of x that minimizes the expected overall costs assuming that the second stage decision is taken optimally. For a particular realization of the random variable ξ, the total cost for a choice x is $C_1(x) + Q(x, \xi)$. Thus, the stochastic problem we wish to solve is

$$\min_{x \in X} \{C_1(x) + E_\xi [Q(x, \xi)]\}, \tag{7.2}$$

where we write $E_\xi[\cdot]$ for the expectation with respect to the random variable ξ.

Example 7.1   (continued) Parthenon Oil Company

We return to the Parthenon Oil Company example in order to put it into this framework. We have a choice as to whether to include the cost of storing oil in the first stage or in the second stage. Suppose that we take it as a first stage cost, then

$$C_1(x_1) = 160x_1 + 5(x_1 - 1000),$$

$$C_2(x_1, x_2, \xi) = c_\xi x_2,$$

where ξ takes the values A, B, or C. Thus

$$Q(x_1, \xi) = \min_{x_2} \{c_\xi x_2 : x_2 \geq d_\xi + x_1 - 1000, x_2 \geq 0\}.$$

Since c_ξ (c_A, c_B or c_C) is positive, the minimum in Q occurs at the lowest possible value of x_2. This is obvious; each x_2 value should be made as small as possible. It will be set so as to just meet the demand in March (allowing for stored oil), or it will be zero if the stored oil is sufficient on its own to meet demand. So

$$x_{2A} = \max(0, d_A - x_1 + 1000),$$

$$x_{2B} = \max(0, d_B - x_1 + 1000),$$

$$x_{2C} = \max(0, d_C - x_1 + 1000).$$

This means that we have

$$Q(x_1, \xi) = c_\xi \max(0, d_\xi - x_1 + 1000).$$

The expectation involves averaging $Q(x_1, \xi)$ over the three values of ξ, and so the expression we need to minimize (over x_1) is

$$160x_1 + 5(x_1 - 1000) + (1/3)c_A \max(0, d_A - x_1 + 1000)$$

$$+ (1/3)c_B \max(0, d_B - x_1 + 1000) + (1/3)c_C \max(0, d_C - x_1 + 1000).$$

Thus we have transformed the problem into one where there is a single decision variable x_1 and a more complex objective function. This can be solved

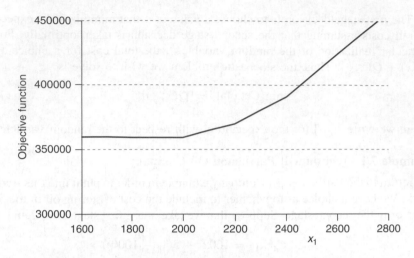

Figure 7.4   Parthenon Oil Company costs as function of x_1.

by evaluating the objective function for different values of the decision variable x_1 and there is no need to solve a linear programming problem. Figure 7.4 shows the way that this function behaves (with d and c values inserted from Table 7.1); each of the $\max(0, \cdot)$ terms corresponds to one of the corners in the function. The initial section slopes slightly down and we can see that we reach the same solution as before, with the minimum achieved at $x_1 = 2000$.   □

We have seen two different ways to solve this problem. In the first approach we use a linear program to do the optimization of both the first and second stage minimizations in one go. In the second approach we figure out what the second stage optimal choices are in order to substitute for second stage costs. Often the first approach, of gathering everything together into a single optimization problem, leads to an easier formulation, and this is the method I would recommend when there are just linear functions involved and the uncertainty relates to only a small set of scenarios.

7.1.3   Ordering with stochastic demand

Now we want to consider a case which involves a continuous form of randomness, as well as continuous decision variables. So, rather than have a finite set of scenarios that may occur, there is instead some continuous random variable and costs will depend on its outcome. One of the most common ways in which we see this happening is when costs depend on the demand for some product that a firm sells, and the demand is a random variable.

Consider a firm that needs to determine a purchase quantity to meet demand, without knowing exactly what the demand will turn out to be. For example, a retailer selling fashion garments may need to order these well in advance of the selling season. If demand is more than the amount ordered then the retailer sells out of that item, but if demand is less than the amount ordered then there is left-over stock at the end of the season. Usually this left-over stock will be marked down to sell, sometimes to below cost price. The right amount to order depends on the distribution of demand and the difference between the amount of money made when selling at full price and the amount of money lost when selling in the end-of-season sale. This is a well-known problem in Operations Management and is usually called the *newsvendor problem*.

We formulate this as a stochastic optimization model. A decision is made by the retailer on how many items of a particular product to buy at a price of $c each. Each item is sold at a price $p during the selling season, and any left over at the end of the season are marked down to a price of $s in order to be sold in an end-of-season sale.

Suppose that x items are purchased by the retailer, then first stage costs are cx. The second stage profits depend both on x and on the stochastic demand D. The retailer will make a profit of px if x is less than D, so that all the products are sold, and a profit of $pD + s(x - D)$ if $x > D$, so that some $x - D$ items are marked down at the end of the selling season.

In this problem there is only a single decision to make. Previously we looked at costs and used a function $Q(x, \xi)$ for the lowest cost at the second stage, given the first stage decision x and a particular value of the random variable ξ. Now we will look at profits and write $Q(x, D)$ for the retailer profit, given an order quantity x (the first stage decision) and a particular value of the random demand D, and we change from minimizing costs to maximizing profits. Then we have:

$$Q(x, D) = px - (p - s)\max(x - D, 0).$$

(You can check that this expression is right for the two cases that x is either less than or more than D). The maximum expected overall profit is given by

$$\max_{x}\{-cx + E_D[Q(x, D)]\},$$

which is the equivalent of the formulation (7.2), but with profits instead of costs.

Suppose that the demand D can take values between 0 and M and has a density function f. We will let $\Pi(x)$ be the expected profit if an amount x is ordered: this is what we will maximize with respect to x. Then, after substituting for Q, we have

$$\Pi(x) = -cx + \int_0^M [px - (p - s)\max(x - z, 0)]f(z)dz$$

$$= (p - c)x - \int_0^M (p - s)\max(x - z, 0)f(z)dz$$

$$= (p - c)x - (p - s)\int_0^x (x - z)f(z)dz. \tag{7.3}$$

Here we have used the fact that $\int_0^M f(z)dz = 1$, and the final step comes from seeing that the integrand is zero for $z > x$.

To find the best choice of x we want to take the derivative of $\Pi(x)$. Here we need a result about how to take a derivative of an integral when both the limits and the integrand depend on the variable of interest. This is called the Leibniz rule and it says that

$$\frac{d}{dx} \int_0^{f(x)} g(x, z)dz = \int_0^{f(x)} \frac{\partial}{\partial x} g(x, z)dz + f'(x)g(x, f(x)).$$

In other words, we can take the derivative inside the integral provided we add a term to correctly allow for the end point that is moving. This extra term needs to take account both of the speed with which the end point $f(x)$ changes and also of the value of the integrand at this end point.

From all this we can deduce that the best choice of x is obtained when the following expression is zero:

$$\frac{d}{dx} \Pi(x) = \frac{d}{dx} \left\{ (p-c)x - (p-s) \int_0^x (x-z)f(z)dz \right\}$$

$$= p - c - (p-s) \int_0^x f(z)dz.$$

The extra (Leibniz) term is zero because the integrand, $(x - z)f(z)$, takes the value zero when $z = x$.

Now we let $F(x)$ be the cumulative distribution function for the demand distribution. Then the condition for a maximum can be written

$$p - c - (p-s)F(x) = 0.$$

Thus, x should be chosen so that

$$F(x) = \frac{p-c}{p-s}. \tag{7.4}$$

Even if you find this sleight of hand with calculus a bit mystifying, we can see what happens quite clearly with an example.

Worked Example 7.2   Troy Fashions

Troy Fashions Ltd. (TFL) sells a type of winter coat for \$200 and works with a clothing manufacturer who can supply at \$100 per item. Winter coats that are not sold at the end of the season will be marked down to \$75. Thus, $c = \$100$, $p = \$200$, and $s = \$75$. Suppose that Troy Fashions is risk neutral and simply wishes to maximize the expected profit. If demand is uniformly distributed between 0 and 200, what is the best choice of order quantity x^* and what is the expected profit if x^* is used?

Solution

Since demand is uniform on $[0, 200]$ we have $F(x) = x/200$ for $0 \leq x \leq 200$. Hence, substituting into Equation (7.4), the optimal choice of order quantity, say x^*, is given by

$$\frac{x^*}{200} = \frac{p - c}{p - s} = \frac{100}{125} = 0.8.$$

Hence, $x^* = 200 \times 0.8 = 160$, and TFL should order 160 of this coat at the beginning of the season.

To check this result and find the optimal expected profit, we consider what happens to the expected profit as a function of x. Using Equation (7.3) we have

$$\Pi(x) = (p - c)x - (p - s) \int_0^x (x - z) f(z) dz$$

$$= (200 - 100)x - (200 - 75) \int_0^x (x - z) \frac{1}{200} dz$$

$$= 100x - \frac{125}{200} \left[\left(xz - \frac{z^2}{2} \right) \right]_0^x$$

$$= 100x - \frac{125}{200} \frac{x^2}{2}.$$

This function is plotted in Figure 7.5 and we can see how 160 is indeed the point at which the maximum occurs. The expected profit when $x = 160$ is given by

$$100 \times 160 - \frac{125}{200} \frac{(160)^2}{2} = 8000.$$

□

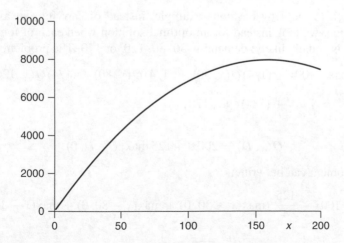

Figure 7.5   The expected profit as a function of x.

7.2   Choosing scenarios

In the two types of stochastic optimization we have looked at so far, we have two different types of uncertainty being played out. For the Parthenon Oil Company example, there is just a small set of possible scenarios, whereas for the newsvendor problem of Troy Fashions the unknown demand is a continuous random variable. The newsvendor problem is unusual in allowing an exact solution to be calculated. For more complex problems, with uncertainty represented by one or more continuous random variables, exact solutions are often not possible. For these problems we need to do something different, and the best approach is to approximate the continuous random variables by generating a set of different scenarios in order to make estimates of expected profit with different decisions. In this section we will introduce Monte Carlo simulation as a way of generating scenarios, and in the next section we will show how these ideas can be applied to complex multistage problems.

To illustrate the idea, we return to the newsvendor problem and see what happens if, instead of there being a distribution of demand given over a range, we suppose that there is a fixed number of different demand scenarios, each with its own probability of occurring. Then the expectation of second stage profits $Q(x, D)$ turns into an average over the different possible scenarios. If there are N demand possibilities $D_1, D_2, \ldots, D_N$, each equally likely, then the expected profit can be written as:

$$\Pi(x) = -cx + E_D[Q(x, D)]$$

$$= -cx + \sum_{i=1}^{N}(1/N)Q(x, D_i).$$

Example 7.2   (continued) Troy Fashions

Going back to the Troy Fashions example, instead of carrying out a complete optimization, we look instead for an optimal solution when each of four demand scenarios is equally likely: demand is 30, 80, 120, or 170. The problem becomes

$$\max_{x} -100x + (1/4)Q(x, 30) + (1/4)Q(x, 80) + (1/4)Q(x, 120)$$

$$+ (1/4)Q(x, 170)$$

with

$$Q(x, D) = 200x - 125\max(x - D, 0).$$

So the problem can be written

$$\max_{x} \left[100x - \frac{125}{4}\left(\max(x - 30, 0) + \max(x - 80, 0) + \max(x - 120, 0) \right.\right.$$

$$\left.\left. + \max(x - 170, 0)\right)\right].$$

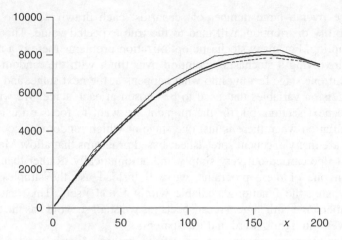

Figure 7.6   Using scenarios to approximate the expected value of profit.

Figure 7.6 shows the exact objective and this approximate objective (the thinner line with five segments).

By adding more scenarios and giving them equal weights we will end up with better and better approximations. One way to generate scenarios is to choose them randomly. (The idea of generating scenarios randomly is called Monte Carlo sampling and we will have more to say about this in the next section.) Here any value of demand between 0 and 200 is supposed to be equally likely. So we choose 15 random numbers between 0 and 200 (rounded to two decimal places): 110.59, 57.42, 168.24, 17.57, 98.77, 190.82, 130.29, 42.81, 188.80, 35.12, 158.13, 24.18, 72.81, 128.20, 62.61. Then we construct the appropriate objective function:

$$100x - \frac{125}{15}(\max(x - 110.59, 0) + \max(x - 57.42, 0) + \ldots$$
$$+ \max(x - 62.61, 0)).$$

This is the dashed line in Figure 7.6. Clearly as we use more and more scenarios we get closer to the exact objective function.   □

7.2.1   How to carry out Monte Carlo simulation

The idea of a Monte Carlo simulation is an important one in modeling risk. Even if we cannot write down explicit expressions to evaluate the expected performance of a decision, we can almost always make an estimate using a simulation. It is worth thinking carefully about the way that a Monte Carlo simulation is carried out.

Each scenario chooses values for the random variables in a way that matches the actual stochastic process through which they will be determined. Then, an average over a large number of scenarios, each drawn with the appropriate probability distribution, will tend to the true expected value. This is Monte Carlo sampling. For many stochastic optimization problems there is a multistage structure involving a stochastic evolution over time, with the random variables occurring at one stage feeding into what happens at the next stage, and there will also be decision variables that need to be chosen at each stage. We will discuss this in the next section, but for the moment we want to focus on using Monte Carlo simulation when there is just one stage at which randomness occurs.

There are many excellent spreadsheet-based programs that allow Monte Carlo sampling to be carried out very simply and automatically. Rather than base this chapter on one of these programs, we will instead use the simplest possible approach using the functions available within Excel itself. This makes things more cumbersome with larger spreadsheets (as we shall see) but has the advantage of being very straightforward and transparent.

In order to create random scenarios from within a spreadsheet we will use the RAND function. This function is written RAND() (and takes no arguments). It returns a uniformly distributed random number between 0 and 1. A new random number is returned every time the worksheet is calculated or F9 pressed. So, to get a demand with a uniform distribution between 0 and 100, we can put the formula =RAND()*100 in a cell.

In practice, it is rare that we want to generate scenarios using a uniform distribution, so what we need is a way of getting from a uniform distribution to some other distribution. This is achieved using the *inverse transform method*, as illustrated in Figure 7.7.

Suppose that we want to draw a sample for a random variable with density function f, and with an associated CDF of F. We start by generating a random number z that is uniform between 0 and 1 and then transform the number z (as shown in the figure) to the point y where $F(y) = z$. If the density f is positive

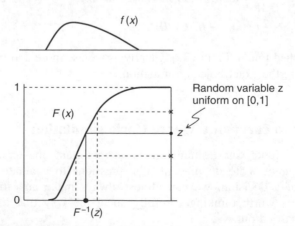

Figure 7.7   A uniform random variable z is transformed into a given distribution through $F^{-1}(z)$.

in its range, then F will be strictly increasing and there will be a single point y defined by this procedure. We can say that y is given by the inverse of the function F applied to the number z and write this as $y = F^{-1}(z)$. (Note that this inverse function is not related to $1/F(z)$!)

Figure 7.7 suggests that this process will be more likely to produce a number where the function F has a large slope, i.e. where the density function has a high value. This is exactly what we want. Now we establish more formally that the inverse transform method will produce a random variable with the right distribution function.

Suppose that X has a uniform distribution on the interval $(0,1)$ and the random variable Y is given by $F^{-1}(X)$. Thus, we get a sample from Y by taking a sample x from the distribution X and then choosing a value y so that $F(y) = x$. To find the distribution of y notice that

$$\Pr(y \le a) = \Pr(F(y) \le F(a))$$
$$= \Pr(x \le F(a))$$
$$= F(a).$$

The first equality is because F is an increasing function; the rest follows from the way that y is chosen and the fact that x is a sample from a uniform distribution. Thus, for any value a, the probability that y is less than a matches what it would be if y had the distribution F. Hence, we have shown that y has the distribution we want.

Worked Example 7.3

We want to generate a scenario in which the demand has a distribution with density $f(x) = 2 - x$ on the range $0.5 \le x \le 1.5$. What formula should be put in the spreadsheet cell to produce a sample from this distribution?

Solution

We need to start by finding the CDF for this density function. The function F is zero below 0.5, F is 1 above 1.5 and between these values we have

$$F(x) = \int_{0.5}^{x} f(z)dz$$

$$= \int_{0.5}^{x} (2 - z)dz = \left[2x - \frac{x^2}{2} \right]_{0.5}^{x}$$

$$= 2x - \frac{x^2}{2} - \left(1 - \frac{1}{8} \right).$$

We can check that this takes the value 1 when $x = 1.5$. We have $2 \times (3/2) - (9/8) - (7/8) = 1$ as required.

Given a value z that is uniformly distributed on [0,1] we generate a new value from the inverse of this function. Hence, we need to find the value of x which solves

$$2x - \frac{x^2}{2} - \frac{7}{8} = z.$$

Thus

$$x^2 - 4x + \frac{7}{4} + 2z = 0$$

and

$$x = 2 \pm \frac{1}{2}\sqrt{9 - 8z}.$$

Now we need to decide whether the higher or lower root is correct; should we take the plus or minus in this expression? We need values of x between 0.5 and 1.5, so we need the minus sign (i.e. the lower root). Hence, we reach the following expression for the spreadsheet cell:

$$\texttt{=2-0.5*SQRT(9-8*RAND())}\qquad\qquad\square$$

Example 7.4   Troy Fashions with normal demand

We return to the example of Troy Fashions, but now we will suppose that the distribution of demand is normal with a mean of $\mu = 100$ and a standard deviation of $\sigma = 20$. We can use Equation (7.4) to determine the optimal order quantity which satisfies

$$F(x) = \frac{p - c}{p - s} = 0.8.$$

The function F here is the cumulative normal distribution, often written Φ. To solve this equation for x we can use tables of a normal distribution but it is simpler to use the spreadsheet function NORMINV, which is designed to find the x value that gives a certain probability of a normal distribution being below it – exactly the problem we need to solve here. The function NORMINV(y, μ, σ) finds the x value for which a normal distribution with specified mean μ and standard deviation σ achieves the probability y of being less than x. Typing the formula =NORMINV(0.8,100,20) into a spreadsheet gives the value 116.832, which we round up to give $x = 117$. This is the optimal choice of x for this problem.

We will illustrate the Monte Carlo simulation approach by using this method to estimate the expected profit when $x = 117$. The idea is to average over different scenarios for the demand, with the scenarios drawn from the desired population. The spreadsheet BRMch7-TroyFashions.xlsx carries out 1000 random draws from a normal distribution for demand in order to estimate the average profit. Each row in the spreadsheet represents a different randomly drawn demand scenario. Have a look at the cell B4 which generates one of the random demands: the cell formula is =NORMINV(RAND(),100,20). As we have already said, the function NORMINV(y, μ, σ) is the inverse of the CDF for a normal distribution and so is the function we need for the inverse transform method to generate demands with a normal distribution.

The average profit is around $9300. Even taking the average over 1000 repetitions, which might be expected to be close to the real expected value, there is still a lot of variation (try pressing F9 repeatedly, which recalculates all the random numbers, and you should see that the average profit jumps around a lot, going from less than $9200 to more than $9400).   □

7.2.2   Alternatives to Monte Carlo

In the Troy Fashions example it is surprising how variable the estimation of expected profit is, even with 1000 scenarios analyzed. The reason for this is related to the randomness of the sample. Essentially, chance will often produce a clustering of the sample in demand regions where there are either higher than average or lower than average profits. There is an alternative approach, which is to spread out the sample points in a regular way. This has been done on the right-hand side of the spreadsheet BRMch7-TroyFashions.xlsx. Instead of making an estimate of the mean by taking 1000 random numbers between 0 and 1 and using an inverse transform of these to generate 1000 demand values, this calculation takes just 98 values 0.01, 0.02, 0.03, ..., 0.98, 0.99 and these values are then used, instead of random numbers, to generate a set of demand values and the expected profit is estimated by averaging these.

By using the inverse transform method in this approach, we generate a set of demand values that is not uniformly distributed; instead the points become closer together at values of the demand with a high likelihood of occurring (i.e. a high density). This gives a far more reliable estimate than a pure Monte Carlo approach, and this approach should always be used when there are just one or two random variables involved in the simulation.

However, the Monte Carlo method really comes into its own when there are many different random variables each with a distribution (or with a joint distribution on them all). Suppose that there are three independent random variables involved in the calculation. The approach of taking evenly spread random numbers now requires 1000 repetitions to get 10 different values for each of the three variables. In essence, we are evaluating the expected profit by averaging the results that occur in a three-dimensional grid, with the spacing of the grid being determined by the density functions for the individual random variables. As the number of variables increases, it gets harder and harder to make a grid-based approach work. For example, if there are 15 different random variables, each of which is uniformly distributed on $(0,1)$, then we might sample these by letting each random variable take three values: say 0.25, 0.5 and 0.75. But with 15 variables there are $3^{15} = 14\,348\,907$ scenarios and this is far too many to allow an exact solution.

Thus, with a high-dimensional problem, we are forced into using a Monte Carlo method, which may be about the only practical way to proceed. In fact, Monte Carlo works reasonably well in most cases, with the accuracy of the solution determined by the number of scenarios independently of the number of random variables.

7.3 Multistage stochastic optimization

Now we turn to the solution of stochastic optimization problems where the decision maker has to make a string of successive decisions with more information becoming available as time goes on. Our aim is to use the Monte Carlo simulation approach in this multistage environment. To give a description of the problem within a general framework can be confusing, so we will make use of a simple example to illustrate our discussion.

Example 7.5   Alexander Patio Supplies

The management team at Alexander Patio Supplies (APS) uses a planning horizon of $T = 3$ months. At the beginning of each month, APS places an order for garden chairs from its supplier and this is delivered at the beginning of the next month. Demand occurs during the month, and if this demand is more than the available inventory then customers will go elsewhere (so that the excess demand is lost). The time line is shown in Figure 7.8.

APS is concerned with the ordering policy for a particular type of garden chair. Demand in each month has a normal distribution with mean 50 and standard deviation 10, and demand in successive months is independent. The chairs are bought at $100 each and sold at $140. To hold a chair over from one month to the next is expensive – it costs $10. Suppose we have in inventory an amount $y_1 = 50$ at the beginning of the first period. The first decision is to choose x_1, the number of chairs ordered in the first month. If demand in the first month is D_1 then we sell $\min(y_1, D_1)$ and hold over an amount $\max(y_1 - D_1, 0)$. We begin the second period with an amount $y_2 = \max(y_1 - D_1, 0) + x_1$ and the whole process repeats. We allow for the cost of holding over inventory at the end of the three months but otherwise do not consider any further costs.

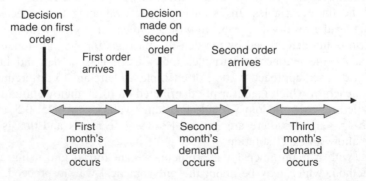

Figure 7.8   Timeline for Alexander Patio Supplies.

A spreadsheet Monte Carlo simulation has been set up in BRMch7-Alexander.xlsx, with each row of the spreadsheet representing a different scenario and each scenario involving three different stochastic demands. There are 1000 different scenarios and the spreadsheet has been set up with an initial inventory of 50 and an order of 55 at the start of week 1 and an order of 40 at the start of week 2.

The contents of cell B6 are =ROUND(NORMINV(RAND(),Mu,Sigma),0). The function NORMINV has the role of converting the uniformly distributed random number RAND() into a normal distribution with mean from the cell with name Mu and standard deviation from the cell with name Sigma. The function ROUND(.,0) rounds this to the nearest integer (customers buy complete chairs not just parts of them!). The same formula is repeated for the other monthly demand in columns E and H. The sales in each month (columns C, F and I) are just given by the minimum of the inventory at the start of the month and the demand. The inventory at the start of month 2 in cell D6 is =Inv_S+Order1-C6, which is the initial inventory (from the cell named Inv_S) plus the month 1 order (from the cell named Order1) minus the month 1 sales from cell C6. Column G contains a similar formula.

Column J gives the costs consisting of the total cost of the products ordered at $100 per chair and the cost of holding stock over at $10 × (starting inventory − sales) for each of the three months. Profits are obtained from three months of $140 × (sales) − total costs.

Notice that the profit figures for different scenarios vary wildly. Even after taking the average over 1000 repetitions, which might be expected to be close to the real expected value, there is still a lot of variation (try pressing F9 and seeing what happens to the average profit.)

Now we ask what are the best values for the month 1 and month 2 orders? This problem is a little similar to the Parthenon Oil Company example and we could set it up as a linear program with decision variables x_1 and x_2 being the two orders made. The easiest thing to do is to use Solver directly to maximize the average profit calculation in BRMch7-Alexander.xlsx. But in order to do so we have to fix the demand values to produce a fixed set of 1000 scenarios against which we will evaluate changes in the order quantity. Using the 'paste values' function, this has been done in the second worksheet in BRMch7-Alexander.xlsx. Try using Solver to find the best choice of the two orders. It turns out that, with this set of scenarios, the best choice is to set $x_1 = 49$ and $x_2 = 36$.   □

In looking at optimal order sizes for the APS problem we need to use functions like min(inventory, demand) to calculate the sales. This introduces corners into the functions (i.e. places where the derivatives jump) and this, in turn, makes it much harder to solve the optimization problem. This difficulty is hidden within what happens 'out of sight' in Solver.

There are ways to set up the problem that avoid these non-smooth functions. In general, an optimization problem in which terms like min(x, y) appear, but that still has a convex feasible set and (if we are maximizing) a concave objective function, can always be replaced by a version of the same thing without the non-smooth functions. We simply replace a constraint of the form $A \leq \min(x, y)$ with two constraints: $A \leq x$ and $A \leq y$. Note that if we have a constraint like $A \geq \min(x, y)$ then the feasible region will no longer be convex and we lose the property of the problem having only one local optimum. If the objective involves maximizing $\min(x, y)$ then we create a new variable v and then we maximize v subject to the original constraints plus two new constraints: $v \leq x$ and $v \leq y$.

In doing these manipulations, it helps to remember the rules of operating with min and max (we will need these rules again in Chapter 9 when we discuss real options).

$$\max(x, y) = -\min(-x, -y)$$

$$a \min(x, y) = \min(ax, ay) \text{ if } a \geq 0$$

$$\min(\min(x, y), z) = \min(x, y, z)$$

$$z + \min(x, y) = \min(z + x, z + y).$$

7.3.1   Non-anticipatory constraints

In our discussion of multistage stochastic optimization so far we have not properly considered the information that becomes available over time. In essence, we have looked at this problem as though we needed to choose all the decision variables at the beginning. This is obviously wrong. For example, in the APS problem, it is important to realize that, at the time when the value of x_2 is chosen (the order placed at the start of month 2), the company already has information on the demand during the first month. If there has been high demand – leading to zero inventory held over – then it makes sense to order more, but if there has been low demand and there are relatively high levels of inventory at the start of month 2, then it will be better to order less.

We need to set up the model of a multistage problem paying careful attention to the exact information that can be used in any decision. A formulation that forces us to choose the decision variables at the start gives too little flexibility but, as we will see, it is easy to make the mistake of a formulation which allows too much flexibility. The model of the decisions should allow us to respond to different scenarios, involving different events. However, we can only respond to what has already happened, and this means including a restriction that stops us using some kind of metaphorical crystal ball in looking ahead to determine the best decision. This is called a *non-anticipatory constraint*. Notice that we naturally want to use whatever knowledge we have of what might happen in the future, for example through understanding the distribution of an unknown random variable. However, what we cannot do is anticipate the exact value that the random variable will take. To explore this idea more thoroughly, we return to the APS example.

Example 7.6   (continued) Alexander Patio Supplies

Consider taking just the first three scenarios that are listed on the second sheet of workbook BRMch7-Alexander.xlsx. Thus, the demand values for the three scenarios chosen are given by the following table:

Scenario:	A	B	C
d_1	60	41	52
d_2	48	58	36
d_3	34	66	53

A natural formulation is to choose different values of x_2 for different scenarios. Scenario A with a high value of the first month demand can then have a higher value of x_2 than scenario B, where the first month's demand is only 41. This is the setup shown in the third sheet of the work book. Try using Solver to find the best choice of the four different variables x_1 and the values of x_2 for the three different scenarios (x_{2A}, x_{2B}, x_{2C}). You should find that the optimal values are $x_1 = 49$, $x_{2A} = 33$, $x_{2B} = 66$, $x_{2C} = 40$. This is surprising: instead of scenario A getting a large order in the second month, it has a small order, with x_{2A} smaller than x_{2B} and x_{2C}. The reason is that the second month order is really only required for the third month's demand. It is the low value of d_3 in scenario A that makes it optimal to order a small amount in the second month.

We can see that by allowing the value of the variable x_2 to depend on the scenario A, B or C then, in effect, we allow x_2 to be affected not only by d_1, but also by d_2 and d_3, which is information not available at the time when the decision is made. In selecting a particular scenario we are making a selection of future variables as well. So, implicit within the procedure we have used, we can say that the decision at the end of month 1 depends on events that have not yet occurred, and this breaks the non-anticipatory constraint. □

A formulation of a stochastic optimization problem might include a specific non-anticipatory constraint forcing decisions that are made with the same information available to be the same. Thus, if two scenarios both start with demand in the first period being 60, say, then the constraint would force the order quantity at the start of the second period to be the same for the two scenarios. However, as we saw in the APS example with just three scenarios, this still leaves the possibility of a wrong formulation through the choice of scenarios that implicitly make information about the future available from an observation of the stochastic demand for the first period. Usually it is safer to build the non-anticipatory constraint more directly into the structure of the problem.

The best approach is to work with a *scenario tree* rather than just a set of scenarios. This allows us to avoid introducing implicit dependence when we don't wish to. In a scenario tree, multiple values are possible for a random variable at stage 2, given what has happened at stage 1, and so on down the tree. When the events at different stages are independent, we can represent this in the scenario

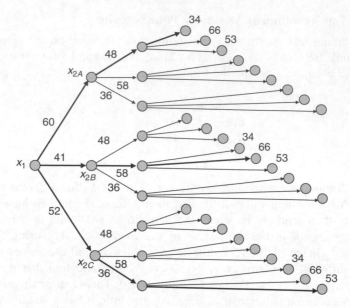

Figure 7.9   An example of a scenario tree.

tree by having the same set of stage 2 outcomes independent of the branch that we are on (i.e. independent of the stage 1 outcome).

We illustrate this in Figure 7.9, for the APS problem. In this figure, the scenarios are built up of demand realizations and, for each month, three possible demands are shown. Of course this dramatically over simplifies what is actually possible, since in each month any demand realization from around 30 to around 70 is quite possible.

The bold arrows in Figure 7.9 show the demand numbers that occur in the scenarios A, B and C that are now just three out of a possible 27 different scenarios. Using this set of 27 scenarios we can construct a more appropriate model in which x_2 is set differently at the three initial nodes according to whether demand has been 60, 41, or 52.

In constructing a scenario tree like this, the more accurately the stochastic component is represented, the more scenario branches there will be at each stage. This can lead to enormous trees and hence great computational difficulties in finding optimal solutions. One option is to reduce the accuracy of the model for steps further into the future by reducing the number of branches at higher levels of the tree.

There has also been a great deal of research on how problems of this sort can effectively be solved numerically. This research is well beyond our scope in this book, but we can sketch a couple of ideas that are useful. One idea is to decompose the problem into separate subproblems for each of the first stage outcomes. These would be the three subtrees in Figure 7.9. If we guess a value for x_1, then it is quite easy to find optimal values for the different values of x_{2A},

x_{2B}, x_{2C}. The solution procedure will also usually generate sensitivity information (especially when the problem is linear), so that for each subtree we can test the consequence of small changes up or down in x_1. This information can be used to find a change in x_1 that produces an overall improvement in expected profit. In the linear case, this idea can be transformed into a process of generating additional constraints in the master problem that can be effective numerically.

Another idea that is important from a practical perspective is to use scenario trees where different scenarios are given different probabilities of occurring rather than being equally weighted. This idea is related to the way that Monte Carlo simulations are generated – it is often possible to get better estimates of the critical quantities by ensuring that the set of samples drawn from the random distributions has particular properties.

There are two other issues that we should mention in relation to multistage stochastic optimization. First we note that it is usually just the first stage decision which is of interest. In practice, we can expect a stochastic optimization problem to be solved on a rolling basis. For example, in the Alexander Patio Supplies problem, a solution includes an order size decision in month 1 as well as decisions made for month 2, conditional on the demand that occurs in month 1. The rolling implementation of this uses this solution to make the order in month 1, but does not commit to what will happen in month 2. Then, at the end of the month when demand for that month is known, the problem can be solved again, but this time pushing one more month out into the future.

The second point to make is that there is a close relationship between this type of problem and that which can be solved using *dynamic programming*. The dynamic programming approach involves looking more closely at what might influence the decisions made at any point in time. In this way the decision is seen as a function of the circumstances at time t. For example, in the APS example the decision on what to order at the beginning of the second month can only depend on the amount of inventory carried over from the first month, if we assume that the demand we observe in each month is independent of the previous demands. This is because at the time the decision is made we can ignore the costs and profits already achieved and look at optimizing the remaining profit. This (optimal) remaining profit can only depend on the state of the system – which just means the current inventory level. If we can find a way to formulate the problem so that at each stage decisions are a function of the state at that stage, then incorporating this into the solution procedure through some type of dynamic programming recursion will usually be worthwhile.

Now we return to the problem we introduced at the start of the chapter, where we asked about the best way to operate a pumped storage facility such as that at Raccoon Mountain. The stochastic element here relates to the uncertain prices, so a scenario tree corresponds to different price trajectories that can occur over time. There will be complex correlations within these price series, since prices depend on demand and demand is critically dependent on temperatures. We want to empty the reservoir when prices are high and fill it when prices are low, so the most natural rules involve two price thresholds: a high threshold above which we generate power, and a low one that signals we use power to fill the reservoir. However, we can expect these thresholds to change with the amount of water (or space) in the reservoir. If the reservoir is nearly empty at the start of the day, we would need a higher price to make it worthwhile to turn on the generators; similarly, if, at the end of the day, there is still a lot of water in the reservoir, then we would wait for a low price that night before starting to pump water. These ideas may enable us to define some policy parameters that could then be optimized using a Monte Carlo simulation of electricity prices.

7.4   Value at risk constraints

So far we have assumed that the problem can be posed as minimizing (or maximizing) the expected value of an objective function. We can use exactly the same approach to deal with a case where the decision maker is risk averse. In this case, we simply define a concave utility function for the decision maker and incorporate the utility function into the objective function. But there are occasions when a different approach is valuable.

We suppose that the decision maker is concerned with risk and, in particular, wishes to avoid bad outcomes. If there is a certain level of loss that is unacceptable then one option is to maximize expected profit as before but to insist that any solution chosen avoids the possibility of a loss greater than the chosen value. So, for example, if we are solving a recourse problem of the following form

$$\min\{C_1(x) + E_\xi[Q(x, \xi)]\},$$

then we could add a constraint

$$C_1(x) + Q(x, \xi) < M \text{ for all } \xi.$$

Since this is a minimization problem, it is large values of the costs given by $Q(x, \xi)$ that are to be avoided. However, in many problems it is not necessary to avoid the possibility of large costs entirely, but just to ensure that it is very unlikely that a large cost occurs. Thus, we end up with a constraint of the form

$$\Pr\{C_1(x) + Q(x, \xi) > M\} \le \alpha,$$

which is called a *chance constraint*.

We will discuss a version of the problem where we maximize the expected profit $E_\xi[\Pi(x, \xi)]$ from a decision x with stochastic behavior described by the random variable ξ. Then, the equivalent chance constraint can be written

$$\Pr\{\Pi(x, \xi) < -M\} \le \alpha.$$

Notice that we are still using the same objective function, but just with an added constraint. So, if $\alpha = 0.01$ and $M = \$1\,000\,000$ then we can express the chance constraint in words by saying that we maximize expected profit subject to the condition that the decision x does not give more than a 1% chance of losing a million or more.

In order to solve this problem effectively we need to have information on the distribution of the profit $\Pi(x, \xi)$. If we know that the CDF for Π is given by the function $F_x(\cdot)$ which depends on x, then we can rewrite the constraint as

$$F_x(-M) \leq \alpha.$$

This is closely related to the value at risk (VaR) measure we discussed in Chapter 3. For example, suppose that a company wishes to maximize expected profit but must operate under risk constraints that impose a limit of $500\,000$ on absolute 95% VaR. Then this can be stated as a problem with a chance constraint that the probability of losses more than $500\,000$ is less than 5%, i.e.

$$\text{maximize} \quad E_\xi[\Pi(x, \xi)]$$

$$\text{subject to} \quad \Pr\{\Pi(x, \xi) < -500\,000\} \leq 0.05.$$

Example 7.7   Portfolio optimization with value at risk constraint

We return to the portfolio optimization problem we introduced in Chapter 2. Suppose there are two investments, both having a normal distribution for the profits after one year. $1000 invested in the first investment returns an expected profit of $1000 with a standard deviation of $400, while the same amount invested in the second investment gives an expected profit of $600 with a standard deviation of $200. A natural stochastic optimization problem with a chance constraint is to suppose that we have $1000 to invest and wish to maximize our expected return subject to the condition that the probability of losing money is less than 0.5%, say. Alternatively, we can express this by saying that the absolute 99.5% value at risk is less than $0.

Suppose we invest an amount $1000w_1$ in the first investment and $1000w_2$ in the second. If we assume that the performance of the investments are independent, then the profits earned follow a normal distribution with mean $1000w_1 + 600w_2$, so the problem can be written

$$\text{maximize} \quad 1000w_1 + 600w_2$$

$$\text{subject to} \quad \Pr\{w_1X_1 + w_2X_2 < 0\} \leq 0.005$$
$$w_1 + w_2 = 1,$$
$$w_1 \geq 0, w_2 \geq 0.$$

where X_1 and X_2 are the random variables giving the individual investment returns. The variance of the total return is given by

$$(400w_1)^2 + (200w_2)^2$$

and the standard deviation is given by the square root of this.

The probability of making a loss can be calculated from the z value giving the number of standard deviations that the mean is above zero. We have

$$z = \frac{1000w_1 + 600w_2}{\sqrt{(400w_1)^2 + (200w_2)^2}}.$$

We can use tables of the normal distribution or the NORMINV function in a spreadsheet to show that we need $z \geq 2.5758$ in order to ensure that the probability of a value less than zero is no more than 0.005. Thus, the constraint becomes

$$\frac{1000w_1 + 600w_2}{\sqrt{(400w_1)^2 + (200w_2)^2}} \leq 2.5758.$$

We can divide through by 100 and square this inequality to show that the problem can be written

$$\text{maximize} \quad 1000w_1 + 600w_2$$

$$\text{subject to} \quad (10w_1 + 6w_2)^2 \geq (2.5758)^2(16w_1^2 + 4w_2^2),$$
$$w_1 + w_2 = 1,$$
$$w_1 \geq 0, w_2 \geq 0.$$

Since the objective is linear, the optimum occurs at a boundary of the feasible region. This means that the inequality constraint will be binding (i.e. hold with equality) and, since we can substitute using $w_2 = 1 - w_1$, we end up with

$$a(16w_1^2 + 4(1 - w_1)^2) - (10w_1 + 6(1 - w_1))^2 = 0$$

where $a = (2.5758)^2 = 6.6349$. Multiplying this out we get

$$(20a - 16)w_1^2 - (48 + 8a)w_1 + 4a - 36 = 0.$$

The maximum is achieved at the higher of the two roots of this quadratic equation. So

$$w_1 = \frac{1}{5a - 4}(a + 6 + \sqrt{a(61 - 4a)})$$

$$= \frac{1}{29.1745}(12.6349 + \sqrt{228.6413})$$

$$= 0.95137$$

giving a split of \$951 in the first investment with the remaining \$49 in the second investment. In this case, any weighting with less than \$951 in the first investment will achieve a 99.5% VaR of less than zero (i.e. less than 0.5% probability of a loss).

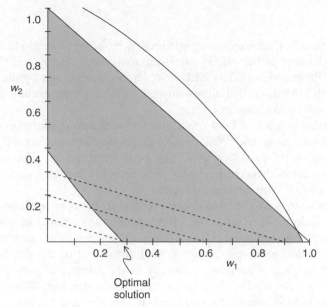

Figure 7.10   Optimizing a three-investment portfolio with a value at risk constraint.

Exactly the same approach can be used when there are more than two stocks. Suppose that we have available a third investment with mean profit of \$1200 and standard deviation 600. The problem becomes

$$\text{maximize} \quad 1000w_1 + 600w_2 + 1200w_3,$$

$$\text{subject to} \quad (10w_1 + 6w_2 + 12w_3)^2 \geq (2.5758)^2 \left(16w_1^2 + 4w_2^2 + 36w_3^2\right),$$
$$w_1 + w_2 + w_3 = 1,$$
$$w_1 \geq 0, w_2 \geq 0, w_3 \geq 0.$$

We can substitute for $w_3 = 1 - w_1 - w_2$ in order to reformulate this as an optimization problem over a two-dimensional region. This is shown in Figure 7.10.

The objective function becomes

$$1000w_1 + 600w_2 + 1200(1 - w_1 - w_2) = 1200 - 200w_1 - 600w_2$$

and the dashed lines in the figure show contours of this. The feasible region is shown shaded. The straight line upper boundary arises from the constraint $w_3 \geq 0$, which translates to $w_1 + w_2 \leq 1$. The curved lines are given by the constraint involving value at risk; in fact, they are part of a large ellipse since this constraint is quadratic in w_1, w_2. The optimal solution is $w_1 = 0.2854$, $w_2 = 0$, $w_3 = 0.7146$.   □

Notes

The Parthenon Oil Company example is loosely based on an example that appears on the wiki pages at the NEOS site (http://wiki.mcs.anl.gov/NEOS/index.php /Stochastic Programming). The NEOS Server is a project run by the University of Wisconsin – Madison that allows anyone to submit optimization problems to state-of-the-art optimization solvers.

There are a number of books that discuss stochastic optimization at various levels of detail. King and Wallace (2012) gives a straightforward discussion concentrating on the modeling area. The book by Birge and Louveaux (2011) gives a more comprehensive treatment looking at the methods that can be used to solve stochastic optimization problems.

The terminology of the classic newsvendor problem in operations management arises from the way it has been formulated in the context of a shop selling newspapers (the first references are actually to the 'newsboy' problem referring to a street vendor). Newspapers left unsold at the end of the day are returned to the publisher, and a decision is needed on the number of newspapers to be ordered by the newsvendor given an uncertain daily demand. The discussion of this problem is based on the tutorial paper by Shapiro and Philpott (2007).

References

Birge, J. and Louveaux, F. (2011) *Introduction to Stochastic Programming*. Springer.

King, A. and Wallace, S. (2012) *Modeling with Stochastic Programming*. Springer.

Shapiro, A. and Philpott, A. (2007). A Tutorial on Stochastic Programming. Manuscript available at http://www2.isye.gatech.edu/ashapiro/publications.html.

Exercises

7.1 Parthenon Oil with risk-averse behavior

Suppose that a utility function $u(x) = \sqrt{x}$ is used by the management of the Parthenon Oil Company. Make the appropriate adjustment to the spreadsheet BRMch7-Parthenon1.xlsx to allow for this utility function. Note

(a) This utility function is defined on the total profit that Parthenon makes (which needs to be positive for this utility function to make sense) – you should assume that all oil is sold at a price of $200 per barrel.

(b) You should assume that POC is maximizing expected utility, so you need to average the utility of the three scenarios. The problem becomes nonlinear, so you need to select the options in Solver appropriately.

Does the optimal choice of the oil purchase in February (x_1) change in any way? (Since the utility function is undefined when POC makes a loss, you will need to ensure that the starting point for the optimization has every scenario profitable.)

7.2 Ajax Lights

Ajax Lights plc sells LED lights and believes that the demand and price are connected linearly. Previously, the price was $10 per bulb and total sales were 50 000 per month. There has been a technical advance, making the cost to produce these bulbs cheaper, at $5 per bulb. There is a three-month time horizon before the market changes radically with a new supplier entering. The firm imports and packs the bulbs but these are obtained from a supplier, who will supply a fixed amount per month with a one-month lead time. Ajax has 100 000 LED lights in stock, and has already set its price at $8 per bulb for the next month. The first decision will be the total amount to be ordered for use during the period of months 2 and 3, which we call Y, and this decision must be made straight away. After discovering the sales in month 1, Ajax will then deduce the two parameters of the demand function (intercept and slope) and set its price for months 2 and 3 so that its entire stock (purchased amount plus remaining stock) is used up by the end of month 3.

Writing Y for the amount ordered and S for the uncertain demand in month 1, show that the problem can be formulated as

$$\max_{Y}(500\,000 + 5Y + E_S(Q(Y, S)))$$

where

$$Q(Y, S) = (100\,000 - S + Y)\left(10 - \frac{S}{5000} + \frac{Y}{50\,000}\right) + 8S.$$

Hence show that Ajax should make an order which matches the optimal policy for the average demand S.

7.3   Generating random variables

Suppose that demand is stochastic and has a density function as follows:

$$f(x) = 0 \text{ for } x \leq 10$$

$$f(x) = (x/2) - 5 \text{ for } x \in (10, 11)$$

$$f(x) = 0.5 \text{ for } x \in [11, 12]$$

$$f(x) = 6.5 - (x/2) \text{ for } x \in (12, 13)$$

$$f(x) = 0 \text{ for } x \geq 13.$$

Use the method discussed in this chapter to generate five random samples from this distribution using the following random samples from the uniform distribution on [0,1]: 0.543, 0.860, 0.172, 0.476, 0.789.

7.4   Non-anticipatory constraints in APS example

Consider the problem solved in the third worksheet in BRMch7-Alexander .xlsx for which $x_{2A} = 33$, $x_{2B} = 66$, and $x_{2C} = 40$. Explain why a non-anticipatory solution to this problem would have $x_{2A} = x_{2C}$.

7.5   Solstice Enterprises

Solstice Enterprises specializes in running bars at sporting events. It needs to make a decision on the size of tent to book and the number of bar staff to provide for an up-coming under-21 women's state cricket final. This is to be played at a venue in Sydney. If the NSW team makes the final, then the crowd is expected to be larger than if NSW does not make it to the final, and the total sales will also depend on the weather on the day (w = wet, d = dry but cool, h = hot). If the NSW team is in the final, the potential sales in dollars are predicted to follow a normal distribution with standard deviation = 2000 and mean $8000, $16 000 or $24 000 according to the weather being w, d or h. If NSW is not in the final, then they will follow a normal distribution with standard deviation 1000 and mean $4000, $8000 or $12 000 according to the weather. The total sales will, however, be limited by the number of people serving, with $5000 per person being the limit and this, in turn, is limited by the size of the tent. There are four tent sizes available suitable for 2, 3, 4, or 5 people. The sequence of events is as follows. First Solstice must decide on a tent size and book this. Then the semi-finals are played, which determines whether the NSW team is in the final. The probability of this happening is 50%. Then, Solstice has to arrange the bar staff for the night, and hence determine the number of people to employ. Finally, the match is played with demand for drinks a random variable that depends on the weather. Solstice does not use weather forecasts in making its decisions. The tent hire cost for a tent to hold x bar staff is $2500x$. Each person on the bar staff is paid $500 for the day's work and the average profit per drink sold is $1.50.

Ten scenarios are generated using a Monte Carlo method and these are shown in Table 7.2.

Table 7.2 Ten scenarios for Solstice Enterprises.

Scenario	NSW in final	Weather	Demand d_i
1	Yes	w	5214
2	No	d	8479
3	No	d	8531
4	Yes	w	7473
5	Yes	h	22 578
6	Yes	d	18 192
7	No	h	11 456
8	Yes	d	16 921
9	No	h	11 439
10	No	d	8477

The model proposed is as follows. Choose x (the tent size: 2, 3, 4 or 5), y_i, (the number of servers in scenario i: 2, 3, 4 or 5) and s_i (the sales in scenario i) to

$$\text{maximize } \frac{1}{10}\sum_{i=1}^{10}(1.5s_i - 500y_i - 2500x)$$

$$\text{subject to } s_i \le d_i,$$

$$y_i \le x,$$

$$s_i \le 5000y_i.$$

Explain why the model does not satisfy non-anticipatory constraints, and hence is wrongly formulated. Reformulate the model in a way that does satisfy non-anticipatory constraints.

7.6 VaR constraints on a portfolio with a bond

An investor has two stocks and a risk-free treasury bond to invest in. Every $1000 invested in the bond will return exactly $1200 in three years' time. The alternatives are stock A, where $1000 will return an average amount of $1250 after three years with a standard deviation of $100, and stock B, where $1000 will return an average amount of $1300 with a standard deviation of $150. The investor wants to find the best portfolio weights to maximize expected profit subject to the constraint that the 99% VaR is zero or less (i.e. there is 1% chance or less of losing money). You may assume that returns are normally distributed over this three-year period and that returns for A and B are independent. Formulate this as an optimization problem and solve it.

8

Robust optimization

Managing risk by gaming

Roger Riley is the CEO of Safety First Corporation (SFC), and certainly takes his company's name to heart. For SFC the key uncertainty relates to demand for its various safety-related products and the uncertainty about manufacturing volume that arises because of the stringent checking that takes place across all product lines. Sometimes they will reject as much as 10% of a week's production on quality grounds. Roger has an unusual approach to risk management, working closely with Wendy Morris as his Chief Risk Officer. A big part of Wendy's role is to dream up possible scenarios in terms of demand and manufacturing yield that will cause difficulties. This is not as easy as it sounds, because manufacturing decisions can be wrong in both directions: producing too much of a product that doesn't sell well will lead to scrap, and making too little of a product that does sell well will mean rush manufacturing orders using expensive overtime and this can also lead to the company losing money.

Roger has always had a pessimistic streak and he sees the risk management process as a game between himself and Wendy. He will come up with a production plan, and then Wendy will play a kind of 'Murphy's law' role to generate a set of demand and yields that is believable, but designed to cause maximum difficulties with the production plan that Roger has chosen. Then Roger and Wendy together use a simple planning spreadsheet to figure out how much money SFC would make (or lose) in this worst case scenario. The next step is for Roger to adjust the manufacturing quantities to try to improve the overall performance of SFC, but each new set of manufacturing decisions is Wendy's cue to redesign the worst case scenario, to try to ensure that SFC does badly. Often, Roger and Wendy go through four or five iterations before Roger decides that he doesn't need to try any more variations.

Business Risk Management: Models and Analysis, First Edition. Edward J. Anderson.
© 2014 John Wiley & Sons, Ltd. Published 2014 by John Wiley & Sons, Ltd.
Companion website: www.wiley.com/go/business_risk_management

Before they set out on this process, Roger and Wendy have jointly to agree the boundaries within which Wendy can choose the relevant numbers, as well as agreeing the estimates of the costs involved to feed into the planning spread-sheet. Now they have this procedure well established and both of them enjoy the challenge of playing the game. Wendy says that it appeals to some malicious instinct in her, and Roger is convinced that the production plans he eventually comes up with are robust: 'These production plans may seem very conservative, but I know that after Wendy has attempted to blow them up, then the plans are not going to be thrown out by an unexpected set of manufacturing and demand data – and that is worth a lot to me'.

8.1   True uncertainty: Beyond probabilities

It is appropriate to look again at a topic that we have discussed briefly in earlier chapters. In most of our discussions of risk we have taken for granted a notion of probability. The risks we take are associated with the losses we may incur and the probabilities associated with those events. Sometimes we can be confident of the probabilities involved ('What is the probability that an Ace is drawn when we choose at random from a full pack of cards?'). Sometimes the probabilities are deduced from looking at the frequency with which something has happened in the past ('What is the probability that a person selected at random in New York city is left-handed?'). And sometimes we make a subjective judgment on the basis of our experience, perhaps putting together what we know from different spheres ('What is the probability that the price of gold will climb over the next two years?').

One of the great cynics of the twentieth century, Frank Knight, would caution us against coming up with a specific number when looking at the probability of a future event. Knight taught in Chicago from 1928 till he died in 1972 at the age of 87. But the idea that he is most remembered for comes from his PhD dissertation of 1916. Knight argues for the existence of a kind of uncertainty that is not amenable to measurement through probabilities:

> 'Uncertainty must be taken in a sense radically distinct from the familiar notion of Risk, from which it has never been properly sep-arated. . . . The essential fact is that "risk" means in some cases a quantity susceptible of measurement, while at other times it is some-thing distinctly not of this character; and there are far-reaching and crucial differences in the bearings of the phenomena depending on which of the two is really present and operating.'

Lord Kelvin made a famous remark about the importance of measurement, claiming that if you cannot measure something then 'your knowledge is of a meagre and unsatisfactory kind'; Knight thought that economists and other social scientists had taken Kelvin's statement to mean 'If you cannot measure,

measure anyhow.' He was scathing about those he saw as trying to turn economics into a science like Physics, based on the rational behavior of all the economic actors.

For Knight, the whole of life was full of examples of individuals making judgments about future events and often the individual could nominate some degree of confidence in this judgment, and yet to talk of the *probability* of a particular judgment being correct is 'meaningless and fatally misleading'.

Writing in 1937 John Maynard Keynes also stressed the difference between what can be calculated as a probability and the uncertainty that prevails over something like the obsolescence of a new invention, 'About these matters there is no scientific basis on which to form any calculable probability whatever. We simply do not know. Nevertheless the necessity for action and for decision compels us as practical men to do our best to overlook this awkward fact...'

Faced with the necessity of making decisions when there is uncertainty, there are two broad approaches: The first is to push hard for at least a subjective assessment of probabilities even under conditions of Knightian uncertainty where we are naturally uncomfortable to provide these. The idea is that, at the point where the decision maker decides between different possible choices, then the decision that is taken will imply some range of values for the missing probabilities. Logically it seems preferable to have our uncertainty translated into a subjective probability of some sort so that it can feed into the decision we need to take, rather than have it emerge as a kind of by-product of the decision we end up making. Looked at from this angle, the question becomes one of finding a way to dig down to the underlying and perhaps unconscious beliefs of an individual regarding the probabilities of different events. There has been much work done on the best way to elicit the beliefs of decision makers, both on the values of different outcomes and on the probabilities of those outcomes.

There is a second approach that seeks to limit the damage from a bad decision rather than fully optimize some specific objective function; it is this idea that lies behind *robust optimization*, which is the topic we will explore in this chapter. One motivation is that we are always inclined to overestimate our certainty and a robust optimization approach will avoid this being too painful. Bernstein quotes G. K. Chesterton as saying that life... 'looks just a little more mathematical and regular than it is; its exactitude is obvious, but its inexactitude is hidden; its wildness lies in wait.' With robust optimization we focus on dealing with the wildness in life.

8.2   Avoiding disaster when there is uncertainty

A robust decision is one which guarantees that the result achieved will not be disastrous. Some aspects of the problem setup are uncertain and there is at least the possibility of a very bad outcome: the idea is to eliminate or minimize this possibility. A focus on the bad results makes it important to know the range of values that some uncertain quantity may take, and these range statements will, in a sense, replace more precise statements about probabilities.

In many cases we are dealing with multiple uncertain variables. So we need to decide whether to specify ranges for each variable independently or whether to look at the combination of values in determining the range.

The first problem we consider is one where a decision maker tries to optimize some objective subject to a set of constraints, but the coefficients in the constraints are uncertain. In this case the bad outcome (that we are trying to avoid) is one of the constraints being violated. The following example illustrates this kind of problem.

Example 8.1   MRB production plan with uncertainty

Suppose that MRB Ltd is a manufacturing company that needs to meet an order for 10 000 units of product A and has two factories that it can use, but there are different costs and efficiencies involved. Factory 1 has higher labor costs of $30 per hour, while factory 2 has labor costs of $26 per hour. However, the machinery in factory 1 is more reliable: in an hour 130 units of product A can be produced in factory 1, but in the same time only 110 units can be produced in factory 2. We can formulate MRB's decision as an optimization problem of minimizing costs subject to meeting the order. If x_1 and x_2 are the hours used in factory 1 and factory 2, respectively, then we want to

$$\text{minimize} \quad 30x_1 + 26x_2$$

$$\text{subject to} \quad 130x_1 + 110x_2 \geq 10\,000$$

$$x_1 \geq 0, x_2 \geq 0.$$

But if we have to determine a schedule in advance, then it is critical that we are able to meet the order, so we may want to be safer. How confident are we that the production rate will be exactly as we have forecast? Machines can break down, personnel can change and obviously the numbers 130 and 110 may not be exactly right. So we would be better to solve a problem where we ask that a constraint like

$$(130 - \Delta_1)x_1 + (110 - \Delta_2)x_2 \geq 10\,000$$

is satisfied for some appropriately chosen values of Δ_1 and Δ_2.

We can easily imagine more complicated versions of the same problem. For example, suppose that MRB also has to meet an order for 5000 of product B and for this product the production rates in factory 1 and 2 are 90 per hour and 80 per hour respectively. Moreover, there is a constraint on the time available, with each factory having only 90 hours available prior to the order delivery deadline. Then, writing y_1 and y_2 for the hours used on product B in the two factories, the overall problem of minimizing costs becomes

$$\text{minimize} \quad 30(x_1 + y_1) + 26(x_2 + y_2)$$

$$\begin{aligned}
\text{subject to} \quad &(130 - \Delta_1)x_1 + (110 - \Delta_2)x_2 \geq 10\,000 \\
&(90 - \Delta_3)y_1 + (80 - \Delta_4)y_2 \geq 5000 \\
&x_1 + y_1 \leq 90 \\
&x_2 + y_2 \leq 90 \\
&x_1 \geq 0, x_2 \geq 0, y_1 \geq 0, y_2 \geq 0.
\end{aligned}$$

Now we have four safety factors Δ_1, Δ_2, Δ_3, Δ_4 to choose. Moreover, what if there is a connection between the production rates for the two different products at factory 1? Clearly things can rapidly become complicated and it would be easy to drown in the detail of this kind of example.   □

Rather than going too quickly into the details, we want to take a step back. Our aim is to optimize an objective subject to meeting a set of constraints for any actual values of the coefficients that may occur. The set of possible coefficient values is clearly critical, and for the moment we suppose that we can identify this set. The form of the general (robust linear programming) problem with just two variables and two constraints is

$$RLP : \quad \text{maximize} \quad c_1 x_1 + c_2 x_2$$

$$\begin{aligned}
\text{subject to} \quad &a_{11}x_1 + a_{12}x_2 \leq b_1 \text{ for all } (a_{11}, a_{12}) \text{ in } A_1 \\
&a_{21}x_1 + a_{22}x_2 \leq b_2 \text{ for all } (a_{21}, a_{22}) \text{ in } A_2 \\
&x_1 \geq 0, x_2 \geq 0,
\end{aligned}$$

and this can obviously be extended to any number of variables and constraints. Notice that the choice to make this a maximization problem with '$\leq$' constraints is fairly arbitrary – we can always convert a maximization problem to a minimization one by looking at the negative of the objective, and we can change the inequalities around by multiplying through by -1.

We will call the set of possible values for the coefficients in a constraint the *uncertainty set* for that constraint. The first thing to notice is that we can determine the uncertainty sets separately for each of the different constraints. There might be some complex interaction between the values of a_{11} and a_{12} for the first constraint and the values of a_{21} and a_{22} for the second constraint, leading to a combined uncertainty set $(a_{11}, a_{12}, a_{21}, a_{22}) \in A$, but since we need to have *both* constraints satisfied for every possible set of parameters, these interactions will not make any difference in the end. The decision variables x_1 and x_2 must satisfy the first constraint for any possible values of a_{11} and a_{12} that appear as a pair in A, and must also satisfy the second constraint for any possible values of a_{21} and a_{22}. Hence, we can split the set A into separate components for each constraint, as has been done in the formulation of *RLP*.

The nature of the solution to this problem depends on the structure of the uncertainty sets involved. Consider a single constraint of the form

$$a_1 x_1 + a_2 x_2 \leq b \text{ for all } (a_1, a_2) \in A.$$

Each element (a_1, a_2) of the set A produces a different constraint and all of them must be satisfied by (x_1, x_2).

The situation is illustrated by Figure 8.1, in which we show the feasible region for the constraints

$$(2 + z_1)x_1 + (3 + z_2)x_2 \leq 3, \text{ for all } (z_1, z_2) \in Z = \{z_1 \geq 0, z_2 \geq 0, z_1 + z_2 \leq 1\}.$$

The set Z here is the set of deviations possible to the base values $a_1 = 2$ and $a_2 = 3$ in order to reach the set of allowable coefficients A.

The figure shows that the overall feasible set is obtained from looking at the constraints generated by the two corner points in Z, i.e. the points where $z_1 = 0, z_2 = 1$ and where $z_1 = 1, z_2 = 0$. The third corner point at $z_1 = z_2 = 0$ gives the base constraint, which is higher than the others and does not contribute to defining the feasible region. All the other points in Z generate constraints which will be satisfied within the feasible (shaded) region. The dashed line is an

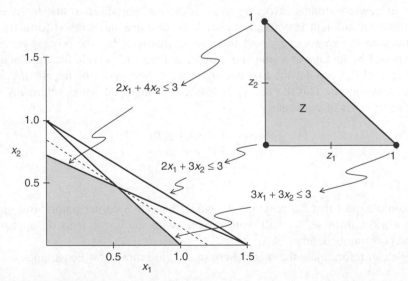

Figure 8.1   The feasible region for the constraint $2x_1 + 3x_2 \leq 3$ when the coefficients are subject to the perturbations in Z.

example of a constraint generated by one such point in Z. The fact that this goes through the point $x_1 = 0.5$, $x_2 = 0.5$ in fact indicates that it comes from a point somewhere along the top boundary of Z where $z_1 + z_2 = 1$.

The corners of the uncertainty set tell us all we need to know, and this property actually holds more generally. If the uncertainty set A associated with a particular constraint is a polytope with a set of k corners (technically called *extreme points* of A), then we can replace the single constraint with k copies defined by the corner points of A, and this will be exactly the same as asking for the constraint to hold at all points in A.

Thus, we can take a general problem like RLP above and simply replace the first constraint by a set of copies derived from the corner points of A_1, and similarly replace the second constraint by a set of copies derived from the corner points of A_2, and so on. This increases the size of the problem, but it retains the same structure and, in the case of a linear program, it is still easy to solve.

We have established the principle that linear programs with polyhedral uncertainty sets for the coefficients remain as linear programs. But to make this a more useful approach in practice it will help to work more directly with the constraints that define the polyhedral set A rather than with the corners of A. When the dimension of the set A increases, the number of corner points can get quite large even for simple constraints.

Thus, we have a second approach to the same underlying problem. Starting with a constraint that must be satisfied whenever the coefficients lie in a set A defined by a set of inequalities, we will show how to rewrite the constraint as a set of new constraints involving some additional variables so that the overall optimization problem is correctly defined. To describe the method formally we would need to express everything in terms of matrices, but the process is easily understood by looking at a simple example, and we will do this for a problem of the form RLP. Consider the first constraint $a_{11}x_1 + a_{12}x_2 \leq b_1$ for all (a_{11}, a_{12}) in A_1 and suppose that the set A_1 is defined by saying that the values a_{11} and a_{12} satisfy a set of constraints:

$$d_{11}a_{11} + d_{12}a_{12} \leq h_1$$
$$d_{21}a_{11} + d_{22}a_{12} \leq h_2$$
$$d_{31}a_{11} + d_{32}a_{12} \leq h_3$$
$$a_{11} \geq 0, a_{12} \geq 0.$$

We could expect that the polytope A_1 would have five corner points. But rather than work with these, we will work directly with the coefficients d_{ij} appearing in the constraints defining A_1.

We can reformulate the original constraint into three new constraints:

$$h_1 y_1 + h_2 y_2 + h_3 y_3 \leq b_1, \tag{8.1}$$

$$d_{11} y_1 + d_{21} y_2 + d_{31} y_3 \geq x_1, \tag{8.2}$$

$$d_{12} y_1 + d_{22} y_2 + d_{32} y_3 \geq x_2. \tag{8.3}$$

We have introduced three new variables y_1, y_2 and y_3 (one for each of the three constraints defining A_1) with $y_1 \geq 0$, $y_2 \geq 0$, $y_3 \geq 0$. The original constraint involving b_1 switches from being about the original variables to being about the new variables. We have added two new constraints in Inequalities (8.2) and (8.3), one for each of the original variables x_1 and x_2. All the coefficients d_{ij} reappear in the new constraints, but transposed. A coefficient like d_{21} that multiplies a_{11} in the second constraint defining A_1, now becomes the multiplier for the second new variable in the constraint associated with x_1. There is a consistency in this pattern that you should check you understand so that you can apply it appropriately.

In general, if the uncertainty set for a constraint is defined by m constraints on the coefficients and there are n coefficients that are uncertain (going with the n original variables) then we will need m new variables and $n + 1$ constraints to represent the original constraint with its uncertainty set.

This reformulation is based on a linear programming duality result that we discuss in more detail in the next section. Applying the rules is reasonably straightforward provided that we begin by putting the problem into a form where the inequalities go the right way round and the set A has all its variables positive. We show how this can work out in practice in the example below.

Worked Example 8.2   Protein constraint in cattle feed

In formulating a feed mix for cattle a firm uses three different grains B, C, and D with different costs. The firm wishes to choose the proportions for these ingredients w_B, w_C, and w_D, to minimize total cost subject to some constraints on the nutritional content of the feed mix. There is a requirement on the protein content in the final mix that can be expressed by saying that

$$bw_B + cw_C + dw_D \geq 2.$$

But the protein content of the individual ingredients varies. All that can be said for sure is that b, c, and d are all positive and the following constraints are satisfied:

$$b + c \geq 3, c + d \geq 4, c - (b+d)/2 \leq 1.$$

(This complex pattern of interdependence might arise because the different grains are sourced from the same growers.) How can the protein constraint be reformulated appropriately?

Solution

We begin by reversing the constraint to get the inequality in the standard direction:

$$b(-w_B) + c(-w_C) + d(-w_D) \leq -2.$$

This way of writing things retains the property that b, c, and d are all positive. We also formulate the constraints on b, c, d with inequalities in the right direction:

$$-b - c \leq -3,$$
$$-c - d \leq -4,$$
$$-(1/2)b + c - (1/2)d \leq 1.$$

Then we have three new variables $y_1 \geq 0$, $y_2 \geq 0$, $y_3 \geq 0$ (one for each of these constraints). We get a total of four constraints as follows

$$-3y_1 - 4y_2 + y_3 \leq -2,$$
$$-y_1 - (1/2)y_3 \geq -w_B,$$
$$-y_1 - y_2 + y_3 \geq -w_C,$$
$$-y_2 - (1/2)y_3 \geq -w_D.$$

Multiplying through each of these constraints by -1 will get rid of many of the negatives and we end with

$$3y_1 + 4y_2 - y_3 \geq 2,$$
$$y_1 + (1/2)y_3 \leq w_B,$$
$$y_1 + y_2 - y_3 \leq w_C,$$
$$y_2 + (1/2)y_3 \leq w_D.$$ $\square$

8.2.1 *More details on constraint reformulation

We have introduced two different methods to take account of the uncertainty set for the coefficients in a constraint. We want to fill in some details for both of these methods. Suppose that the original constraint is of the form

$$a_1 x_1 + a_2 x_2 + \ldots + a_m x_m \leq b,$$

where we know that the vector $(a_1, a_2, \ldots, a_m)$ is in an uncertainty set A.

We start by looking at how we can reformulate this in terms of the corner points of A. Suppose that A has k corner points, and let $\left(a_1^{(j)}, a_2^{(j)}, \ldots, a_m^{(j)}\right)$ be the jth corner point where $j = 1, 2, \ldots, k$. Then any point $(a_1, a_2, \ldots, a_m) \in A$ can be obtained from some set of non-negative weights $w_1, w_2, \ldots, w_k$ with $\sum_{j=1}^{k} w_j = 1$ applied to the corners, and $a_i = \sum_{j=1}^{k} w_j a_i^{(j)}$ for $i = 1, 2, \ldots, m$. In other words, the points in A can be obtained as (convex) combinations of the corner points.

Since a feasible point $(x_1, x_2, \ldots, x_m)$ satisfies the constraints for all $(a_1, a_2, \ldots a_m) \in A$, it must do so at the corners of A. Suppose that is all that we are given, so

$$a_1^{(j)} x_1 + a_2^{(j)} x_2 + \ldots + a_m^{(j)} x_m \leq b \text{ for } j = 1, 2, \ldots, k. \qquad (8.4)$$

Then consider an arbitrary $(a_1, a_2, \ldots, a_m)$ picked from somewhere inside A. Then

$$a_1 x_1 + a_2 x_2 + \ldots + a_m x_m$$

$$= \sum_{j=1}^{k} w_j a_1^{(j)} x_1 + \sum_{j=1}^{k} w_j a_2^{(j)} x_2 + \ldots + \sum_{j=1}^{k} w_j a_m^{(j)} x_m$$

$$= \sum_{j=1}^{k} w_j \left(a_1^{(j)} x_1 + a_2^{(j)} x_2 + \ldots + a_m^{(j)} x_m \right)$$

$$\leq \sum_{j=1}^{k} w_j b = b,$$

and so the point $(x_1, x_2, \ldots, x_m)$ also satisfies the constraint generated by $(a_1, a_2, \ldots, a_m)$. Thus, we have shown that the constraints (8.4) are exactly what is required to represent the whole uncertainty set A.

Next we want to consider the alternative approach when the uncertainty set A is defined by a set of constraints. We have already stated in Inequalities (8.1)–(8.3) how the original constraint can be rewritten to incorporate this information about A. Now we want to show how the new set of inequalities is derived. To do this we need to take a short mathematical detour into the duality theory that is associated with linear programs. The properties of the dual linear program are both surprising and beautiful, and there is no harm in spending a little while looking at this area.

The duality result we need is quite general, but to make it easier to read we will describe the result for a problem with just two variables x_1, x_2 and three constraints. The duality theorem for linear programs states that the value of the (primal) linear program given by

$$LP: \quad \text{maximize} \quad g_1 x_1 + g_2 x_2$$

$$\text{subject to} \quad \begin{aligned} d_{11} x_1 + d_{12} x_2 &\leq h_1, \\ d_{21} x_1 + d_{22} x_2 &\leq h_2, \\ d_{31} x_1 + d_{32} x_2 &\leq h_3, \\ x_1 &\geq 0, x_2 \geq 0, \end{aligned}$$

is the same as the value of the (dual) linear program

$$DLP: \quad \text{minimize} \quad h_1 y_1 + h_2 y_2 + h_3 y_3$$

$$\text{subject to} \quad \begin{aligned} d_{11} y_1 + d_{21} y_2 + d_{31} y_3 &\geq g_1 \\ d_{12} y_1 + d_{22} y_2 + d_{32} y_3 &\geq g_2 \\ y_1 &\geq 0, y_2 \geq 0, y_3 \geq 0. \end{aligned}$$

242   BUSINESS RISK MANAGEMENT

If you have never seen this before you need to stop and look carefully to see what has happened in moving from one problem to the other. The dual linear program has a constraint for each variable in the original LP, and it has a variable for each constraint in the original problem. Also, the coefficients in the objective function get translated into the constraint right-hand sides and vice versa. Not only have the variables and constraints swapped places, but we have changed a maximization problem with '≤' constraints into a minimization problem with '≥' constraints. In both problems all the variables are constrained to be positive. In fact, the duality relation still works if a variable is not constrained to be positive (a 'free' variable), in this case the corresponding constraint has to be an equality. Thus, for example, if the original problem did not have the constraint $x_2 \geq 0$, then the second constraint in the dual would become

$$d_{12}y_1 + d_{22}y_2 + d_{32}y_3 = g_2.$$

Notice what we are saying here: the minimum value of the objective function in the dual *DLP* is equal to the maximum value of the objective in the original *LP*. It is interesting to try and see why these two problems have the same value. You can try, for example, putting actual numbers in instead of all the algebra to check that the result really does hold. But be warned that the reason for the duality result being true is quite deep (it comes down to a separating hyperplane argument; in other words, a generalization to multiple dimensions of the fact that two non-intersecting convex sets in the plane can have a straight line drawn between them). It is easy enough to show that the minimum in DLP is greater than the maximum in LP, but to show that these values are the same is quite a bit harder.

There are other forms of dual that can be written down, but this is the form that is easiest to remember, and what we have said here will be enough for the results we want to derive. What is the connection with our robust optimization problem?

We suppose, as before, that in the problem RLP the uncertainty set A_1 associated with the first constraint is a polytope defined through the following inequalities on the values of a_{11} and a_{12}:

$$d_{11}a_{11} + d_{12}a_{12} \leq h_1,$$
$$d_{21}a_{11} + d_{22}a_{12} \leq h_2,$$
$$d_{31}a_{11} + d_{32}a_{12} \leq h_3,$$
$$a_{11} \geq 0, a_{12} \geq 0.$$

Then we can rewrite the first constraint of RLP that $a_{11}x_1 + a_{12}x_2 \leq b_1$ for all (a_{11}, a_{12}) in A_1 as saying that the maximum value that $a_{11}x_1 + a_{12}x_2$ can take for $(a_{11}, a_{12}) \in A_1$ is less than b_1. And then this can be expressed by saying that the solution of the linear program

$$\text{maximize} \quad a_{11}x_1 + a_{12}x_2$$

$$\text{subject to} \quad \begin{aligned} d_{11}a_{11} + d_{12}a_{12} &\leq h_1 \\ d_{21}a_{11} + d_{22}a_{12} &\leq h_2 \\ d_{31}a_{11} + d_{32}a_{12} &\leq h_3 \\ a_{11} &\geq 0, a_{12} \geq 0 \end{aligned}$$

should be less than or equal to b_1. This is the point where we use linear programming duality. From our duality result this statement is exactly the same as saying that the solution of the dual linear program

$$DLP1: \quad \text{minimize} \quad h_1y_1 + h_2y_2 + h_3y_3$$

$$\text{subject to} \quad \begin{aligned} d_{11}y_1 + d_{21}y_2 + d_{31}y_3 &\geq x_1, \\ d_{12}y_1 + d_{22}y_2 + d_{32}y_3 &\geq x_2, \\ y_1 &\geq 0, y_2 \geq 0, y_3 \geq 0, \end{aligned}$$

should be less than or equal to b_1. Think about this statement carefully and you can see that it amounts to saying that a pair of values x_1 and x_2 will satisfy the first constraint of RLP with its uncertainty set Λ_1 if and only if there are some values $y_1 \geq 0, y_2 \geq 0, y_3 \geq 0$ which satisfy the inequality

$$h_1y_1 + h_2y_2 + h_3y_3 \leq b_1,$$

together with the constraints of $DLP1$. In other words, we want the inequality set (8.1)–(8.3) to be satisfied.

8.2.2   Budget of uncertainty

It is frequently possible to give a range of possible values for an uncertain parameter. So, for a parameter a_1 it often happens that we do not know its exact value but we are confident that it will lie in a certain range. It is convenient to take the midpoint of the range as a kind of base value $\bar{a}_1$ and then define δ_1 as the distance from $\bar{a}_1$ to the two bounds. Hence, the uncertainty set for a_1 is given by $\bar{a}_1 - \delta_1 \leq a_1 \leq \bar{a}_1 + \delta_1$.

When a constraint contains a number of different uncertain parameters $a_1, a_2, \ldots, a_n$, say, then this will determine a combined uncertainty set A by simply asking for each a_i to satisfy a range constraint $\bar{a}_i - \delta_i \leq a_i \leq \bar{a}_i + \delta_i$. With this arrangement the set A becomes an n-dimensional rectangular box centered on $(\bar{a}_1, \bar{a}_2, \ldots, \bar{a}_n)$.

In practice, this results in an extremely conservative uncertainty set, since we allow all the uncertain parameters to take their extreme values at the same time.

Unless they are highly correlated variables it makes sense to be more conservative in the choice of the individual interval lengths δ_i and compensate for this by being less conservative in the points where a lot of parameters are close to their extreme values at the same time.

This leads to defining a *budget of uncertainty B*. If there are 10 variables and we have a budget of uncertainty of 5, this would mean that the sum of the ratios $|a_i - \overline{a}_i|/\delta_i$ is less than 5. This might be achieved, for example, by having five variables at $\overline{a}_i + \delta_i$ and the other five at $\overline{a}_i$, or by having five variables at $\overline{a}_1 + \delta_1/2$ and five variables at $\overline{a}_1 - \delta_1/2$. To see what this looks like for a specific example, Figure 8.2 shows the situation when there are three uncertain parameters and compares what happens when $B = 2.5$ and $B = 1.5$. The solid shapes are centered on the point given by the vector of base values $(\overline{a}_1, \overline{a}_2, \overline{a}_3)$.

Whenever the uncertainty in different parameters is independent, then the assumption that we can cut off the corners in this way is pretty safe. In fact, we will show in the next section that if each of n uncertain parameters is symmetric and independent, then the probability that the optimal solution to a problem of this sort turns out to be infeasible is less than $1 - \Phi\left(B/\sqrt{n}\right)$ for large n. For example, if $n = 36$ then the largest possible value of B is 36, but when B is half this size, so $B = 18$, then the probability of the optimal solution being infeasible is

$$1 - \Phi\left(18/\sqrt{36}\right) = 1 - \Phi(3) = 0.00135,$$

i.e. less than 2 chances in 1000.

Next we investigate the nature of the adjusted linear program that we need to solve when there is an uncertainty set of this form. We start by looking at a simple case where there is a constraint $a_1 x_1 + a_2 x_2 \leq b$ for all $(a_1, a_2) \in A$, where A is defined as the a_1, a_2 satisfying

$$a_1 = \overline{a}_1 + z_1 \delta_1, a_2 = \overline{a}_2 + z_2 \delta_2,$$

with $|z_1| \leq 1, |z_2| \leq 1$, and $|z_1| + |z_2| \leq B$.

Because of the modulus signs, this amounts to a total of eight constraints on the values of (a_1, a_2). If we were to follow our previous discussion, then we

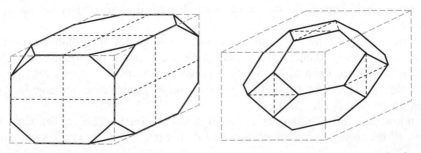

Figure 8.2   There are three variables: the left-hand diagram shows a budget of uncertainty of 2.5, and the right-hand diagram shows a budget of uncertainty of 1.5.

would replace the single constraint with three, but the new constraints would involve eight new variables. There is a way of simplifying this and ending up with three inequalities and just three new variables, as follows:

$$\overline{a}_1 x_1 + \overline{a}_2 x_2 + w_1 + w_2 + Bt \le b \qquad (8.5)$$

$$w_1 + t \ge \delta_1 |x_1|, \qquad (8.6)$$

$$w_2 + t \ge \delta_2 |x_2|, \qquad (8.7)$$

$$w_i \ge 0, t \ge 0.$$

(We will show in the next section how this particular set of inequalities is derived.) This gives a clue as to how a general problem with budget of uncertainty can be formulated. Suppose that there are n decision variables $x_1, x_2,...x_n$ and the original problem is to maximize $c_1 x_1 + ... + c_n x_n$ subject to some constraints, one of which has the form $a_1 x_1 + a_2 x_2 + ... + a_n x_n \le b$ for all coefficients $(a_1, a_2,...a_n) \in A$ where

$$A = \{(\overline{a}_1 + z_1 \delta_1, \overline{a}_2 + z_2 \delta_2, \ldots, \overline{a}_n + z_n \delta_n)\}$$

for $|z_i| \le 1, i = 1, 2, \ldots, n$ and $|z_1| + |z_2| + \ldots |z_n| \le B$.

Then, $n + 1$ new variables $t, w_1, w_2, \ldots, w_n$ are added to the problem, each new variable being constrained to be non-negative and the constraint in question can be replaced by $n + 1$ new constraints

$$\overline{a}_1 x_1 + \overline{a}_2 x_2 + \ldots + \overline{a}_n x_n + w_1 + w_2 + \ldots + w_n + Bt \le b, \qquad (8.8)$$

$$w_i + t \ge \delta_i |x_i|, i = 1, 2, \ldots, n. \qquad (8.9)$$

To get back to a linear program we simply replace each of the constraints $w_i + t \ge \delta_i |x_i|$ with two constraints $w_i + t \ge \delta_i x_i$ and $w_i + t \ge -\delta_i x_i$.

It is easy to check that if there are variables x_i, w_i and t satisfying these conditions, then the original inequality with b is satisfied for all $(a_1, a_2, \ldots a_n) \in A$. We choose a point in A with $a_i = \overline{a}_i + z_i \delta_i$ where $|z_i| \le 1$ and $|z_1| + |z_2| + \ldots |z_n| \le B$. Then

$$a_1 x_1 + a_2 x_2 + \ldots + a_n x_n$$

$$= \overline{a}_1 x_1 + \overline{a}_2 x_2 + \ldots + \overline{a}_n x_n + z_1 \delta_1 x_1 + \ldots + z_n \delta_n x_n$$

$$\le \overline{a}_1 x_1 + \overline{a}_2 x_2 + \ldots + \overline{a}_n x_n + |z_1|(w_1 + t) + \ldots + |z_n|(w_n + t)$$

$$\le \overline{a}_1 x_1 + \overline{a}_2 x_2 + \ldots + \overline{a}_n x_n + w_1 + \ldots + w_n + Bt$$

$$\le b.$$

Here we used inequalities like $z_1 \delta_1 x_1 \le |z_1| \delta_1 |x_1|$ and also made use of the fact that $|z_i| \le 1$ and $\sum |z_i| \le B$. This is one direction of the duality argument, but it is much harder to go the other way around and show that the new set of constraints is no more restrictive than the original set.

Worked Example 8.3   Avignon Imports

Avignon Imports has to determine the order to place for products A, B and C. The entire order for the three products will require delivery together and transport constraints imply that the total weight of the shipment is less than 5000 kg. Products A, B and C all weigh 5 kg per unit but there is uncertainty about the way that the products will be packed and hence the weight of packaging that the suppliers will use. For A and C this is estimated at 0.2 kg per unit, but this is a guess and it is thought that figures between 0.1 kg and 0.3 kg are possible. Product B is more complicated and the packaging is estimated at 0.5 kg per unit, with figures between 0.2 kg and 0.8 kg being possible. All of the items are supplied at a cost of \$100 per unit. After importing them, Avignon Imports will auction the products. The expected price to be achieved by selling product A is \$200 per unit, with a possible variation up and down of \$50. The expected price for product B is \$205, with a possible variation up or down of \$60 per item, and the expected price for product C is \$195, with a possible variation of \$70. There is a requirement that the company makes a profit of at least \$50 000 from the transaction. Avignon Imports wishes to maximize its expected profit subject to the constraints on transport weight and minimum profit achieved. Formulate this as a robust optimization problem using a budget of uncertainty of $B = 2$ for both the constraints and solve the problem in a spreadsheet.

Solution

Let x_A, x_B, and x_C be the amounts ordered for the three products. Write a_A, a_B, and a_C for the weight per unit and s_A, s_B, and s_C for the sale price per unit. The expected profit is \$100, \$105 and \$95 for the three products and so we have a robust optimization problem of

$$\text{maximize} \quad 100x_A + 105x_B + 95x_C$$

$$\text{subject to} \quad a_A x_A + a_B x_B + a_C x_C \leq 5000 \text{ for } (a_A, a_B, a_C) \in A$$
$$-s_A x_A - s_B x_B - s_C x_C \leq -50\,000 \text{ for } (s_A, s_B, s_C) \in S$$
$$x_A \geq 0, x_B \geq 0, x_C \geq 0,$$

with uncertainty sets given by:

$$A = \{(a_A, a_B, a_C) : a_A = 5.2 + 0.1 z_A, a_B = 5.5 + 0.3 z_B, a_C = 5.2 + 0.1 z_C$$
$$\text{for } (z_A, z_B, z_C) \in Z\},$$

$$S = \{(s_A, s_B, s_C) : s_A = 100 + 50 q_A, s_B = 105 + 60 q_B, s_C = 95 + 70 q_C$$
$$\text{for } (q_A, q_B, q_C) \in Z\},$$

where

$$Z = \{(z_A, z_B, z_C) : |z_A| \leq 1, |z_B| \leq 1, |z_C| \leq 1, |z_A| + |z_B| + |z_C| \leq 2\}.$$

Notice that the S constraint on minimum profit has been multiplied by -1 to bring it into standard form. The fact that both budgets of uncertainty are the same means we can use a single set Z for the two different constraints. Now we can use the rules we developed earlier to add constraints as in Inequalities (8.8) and (8.9); the resulting formulation has each of the existing constraints replaced by four new ones (together with four new variables). This gives

maximize $\quad 100x_A + 105x_B + 95x_C$

subject to $\quad 5.2x_A + 5.5x_B + 5.2x_C + w_A + w_B + w_C + 2t_1 \leq 5000$

$\qquad\qquad w_A + t_1 \geq 0.1x_A$

$\qquad\qquad w_B + t_1 \geq 0.3x_B$

$\qquad\qquad w_C + t_1 \geq 0.1x_C$

$\qquad\qquad -100x_A - 105x_B - 95x_C + u_A + u_B + u_C + 2t_2 \leq -50\,000$

$\qquad\qquad u_A + t_2 \geq 50x_A$

$\qquad\qquad u_B + t_2 \geq 60x_B$

$\qquad\qquad u_C + t_2 \geq 70x_C$

and all variables non-negative.

In this formulation we have been able to change $|x_A|$, $|x_B|$ and $|x_C|$ into x_A, x_B and x_C since these are all positive.

The solution to this problem is given in the spreadsheet BRMch8-Avignon.xlsx. We obtain

$$x_A = 785.38, x_B = 81.65, x_C = 69.98,$$

$$w_A = 54.04, t_1 = 24.49, u_A = 34\,370.14, t_2 = 4898.91,$$

and all other variables are zero. In practice, we would need to round the variables to whole numbers (or, better, use an optimization procedure that searches amongst integer solutions). $\qquad\qquad\qquad\qquad\qquad\qquad\qquad\qquad\qquad\qquad\qquad\quad\Box$

8.2.3 *More details on budgets of uncertainty

In this section we want to do two things: first to show how the constraints (8.5) – (8.7) arise from a duality argument, and second to derive the bound we gave on the probability that an optimal solution lies outside the budget of uncertainty (assuming independent symmetric errors).

We begin by considering a constraint $a_1x_1 + a_2x_2 \leq b$ where the values of (a_1, a_2) have base values $(\bar{a}_1, \bar{a}_2)$ and maximum deviations δ_1, δ_2, and also satisfy a budget of uncertainty B. Thus, A is given by

$$a_1 = \bar{a}_1 + z_1\delta_1, a_2 = \bar{a}_2 + z_2\delta_2,$$

$$\text{with } |z_1| \leq 1, |z_2| \leq 1, \text{ and } |z_1| + |z_2| \leq B.$$

If the original constraint holds for all $(a_1, a_2) \in A$, then we can reformulate the constraint by saying that the following linear program with decision variables a_1

and a_2 has a value no greater than b:

$$\text{minimize} \quad a_1 x_1 + a_2 x_2$$

$$\begin{aligned}
\text{subject to} \quad & a_i - \delta_i u_i + \delta_i v_i = \bar{a}_i, i = 1, 2 \\
& u_i + v_i \le 1, i = 1, 2 \\
& (u_1 + v_1) + (u_2 + v_2) \le B \\
& u_1 \ge 0, u_2 \ge 0, v_1 \ge 0, v_2 \ge 0.
\end{aligned}$$

Here we have written u_1 and v_1 for the positive and negative parts of z_1, i.e. $u_1 = \max(z_1, 0)$ and $v_1 = \max(-z_1, 0)$. (Similarly, u_2 and v_2 are the positive and negative parts of z_2). This means that $z_1 = u_1 - v_1$ and $|z_1| = u_1 + v_1$. There is a trick here, since defining the variables in this way means that only one of them is non-zero, whereas the linear program as formulated could have both $u_1 > 0$ and $v_1 > 0$. However, any solution in which both variables are non-zero can be replaced by one in which the same quantity is subtracted from both u_1 and v_1 to make the smaller of the two equal to zero. The equality constraint will still be satisfied and the inequality constraints also still work, since $u_1 + v_1$ is reduced.

Thus, using the duality theorem for linear programs, the constraints imply that the following linear program has a value no greater than b:

$$\text{minimize} \quad \bar{a}_1 y_1 + \bar{a}_2 y_2 + w_1 + w_2 + Bt$$

$$\begin{aligned}
\text{subject to} \quad & y_1 = x_1 \\
& y_2 = x_2 \\
& -\delta_1 y_1 + w_1 + t \ge 0 \\
& -\delta_2 y_2 + w_2 + t \ge 0 \\
& \delta_1 y_1 + w_1 + t \ge 0 \\
& \delta_2 y_2 + w_2 + t \ge 0 \\
& w_i \ge 0, t \ge 0.
\end{aligned}$$

Note that the two equalities $y_i = x_i$ occur (rather than inequalities) because there are no constraints that $a_i \ge 0$. We can use the first two constraints here to substitute x_1 and x_2 for y_1 and y_2. We can also combine the two constraints involving w_1: if $w_1 + t$ is greater than both $\delta_1 x_1$ and $-\delta_1 x_1$ then we have $w_1 + t \ge \delta_1 |x_1|$. Thus, we reach the following optimization problem, which has a value $\le b$.

$$\text{minimize} \quad \bar{a}_1 x_1 + \bar{a}_2 x_2 + w_1 + w_2 + Bt$$

$$\begin{aligned}
\text{subject to} \quad & w_1 + t \ge \delta_1 |x_1| \\
& w_2 + t \ge \delta_2 |x_2| \\
& w_i \ge 0, t \ge 0.
\end{aligned}$$

Hence, we have shown that the original constraint can be written in the form of the constraints (8.5) – (8.7).

Now we want to investigate how likely we are to exceed the budget of uncertainty when each uncertain parameter is symmetric and independent. Rather than simply asking about the probability that the budget of uncertainty is exceeded, we are interested in the probability that the optimal solution that we reach fails to satisfy the constraints of the problem. To be more precise, we suppose that the value of B is used to define an uncertainty set A and then we want to know what is the probability that the optimal solution using this uncertainty set will turn out to be infeasible, assuming that the original ranges are accurate. It is possible that the actual values of the uncertain parameters lie in those parts of the n-dimensional cube cut off by the budget of uncertainty constraint, and this is a question about how likely it is that this will lead to infeasibility.

Suppose that x^* is an optimal solution to the problem with the budget of uncertainty in place. We know that the solution x^* satisfies the constraint with any choice of z_i satisfying the budget of uncertainty, and we will make a specific choice for z.

We do this by first reordering the variables so that the highest values of $\delta_i \left| x_i^* \right|$ come first and then we choose $z_i = 1$ or -1 according to the sign of x_i^* for the first $L = \lfloor B \rfloor$ of these variables, and $z_{L+1} = B - L$ or $-B + L$ for the next one (again in order to match the sign of x_{L+1}^*). The remaining z_i are all set to zero. With this choice of z_i we will have $\sum_{i=1}^{n} \left| z_i \right| = B$ and the coefficients $a_1, a_2, \ldots, a_n$ will lie in the defined uncertainty set A. Hence, the constraint will be satisfied and we can deduce

$$\sum \overline{a}_i x_i^* + \sum_{i=1}^{L} \delta_i \left| x_i^* \right| + (B - L)\delta_{L+1} \left| x_{L+1}^* \right| \leq b. \tag{8.10}$$

Now suppose that the constraint does not hold at the optimal solution x^* for some set of a_i values in the ranges given. Thus, there is a set of z_i values for which

$$\sum \overline{a}_i x_i^* + \sum z_i \delta_i \left| x_i^* \right| > b,$$

and hence, from Inequality (8.10),

$$\sum \overline{a}_i x_i^* + \sum z_i \delta_i \left| x_i^* \right| > \sum \overline{a}_i x_i^* + \sum_{i=1}^{L} \delta_i \left| x_i^* \right| + (B - L)\delta_{L+1} \left| x_{L+1}^* \right|.$$

This inequality can be rewritten

$$\sum_{i=L+1}^{n} z_i \delta_i \left| x_i^* \right| > \sum_{i=1}^{L} (1 - z_i)\delta_i \left| x_i^* \right| + (B - L)\delta_{L+1} \left| x_{L+1}^* \right|.$$

Because of the ordering of the $\delta_i \left| x_i^* \right|$ (and using the fact that $1 - z_i \geq 0$) this implies

$$\sum_{i=L+1}^{n} z_i \delta_i \left| x_i^* \right| \geq \delta_{L+1} \left| x_{L+1}^* \right| \left(\sum_{i=1}^{L} (1 - z_i) + (B - L) \right).$$

So

$$\sum_{i=1}^{L} z_i + \sum_{i=L+1}^{n} z_i \frac{\delta_i \left|x_i^*\right|}{\delta_{L+1} \left|x_{L+1}^*\right|} > B.$$

This inequality has the form

$$\sum_{i=1}^{n} z_i h_i > B \qquad (8.11)$$

with $0 \leq h_i \leq 1$ for all i.

Now we ask what is the probability that the z_i values make the constraint not satisfied (if each z_i is chosen in a way that is independent and symmetric around 0)? Since Inequality (8.11) is satisfied if the constraint is broken, the probability of this inequality holding must be greater than the probability that the constraint is broken.

We can use the central limit theorem to produce a bound on this probability for large n. The random variable $\sum_{i=1}^{n} z_i h_i$ has mean zero (since each z_i has zero mean) and variance $V = \sum_{i=1}^{n} h_i^2 V_i$ where V_i is the variance of the variable z_i. Since z_i lies in the range -1 to 1, it cannot have variance larger than 1 and h_i^2 is also less than 1. Hence, the variance of $V \leq n$. Finally we have

$$\Pr \left(\sum_{i=1}^{n} z_i h_i > B \right) \approx \Pr(N(0, \sqrt{V}) > B) \leq 1 - \Phi \left(\frac{B}{\sqrt{n}} \right).$$

Thus, we have established that for large n and any symmetric distribution of coefficient errors around the base levels, provided these errors are independent, using a budget of uncertainty B in solving the optimization problem will give a probability of the constraint being broken at the optimal solution of no more than $1 - \Phi \left(B / \sqrt{n} \right)$.

8.3   Robust optimization and the minimax approach

An important type of uncertainty relates to the objective function in the minimization. In this context the classical stochastic optimization would look at the expected value of the objective under some model describing the probabilities of different parameters in the objective function. But we are interested in an environment in which there is no known distribution for these parameters, at the most we simply have a range of possible values. The objective function parameters belong to an uncertainty set A. In this context it is natural to consider a *minimax* approach which assumes the worst and makes decisions in order that the worst will not be too bad.

To apply this approach we will begin by rewriting the problem with an extra variable representing the objective function. Thus, a standard optimization problem may be written

maximize $f(x)$ subject to $x \in X$,

where f is the objective function and X is the feasible set defined by the constraints. This standard problem can be rewritten adding an unconstrained (scalar) variable v as:

$$\text{maximize} \quad v$$

$$\text{subject to} \quad v \leq f(x)$$
$$x \in X.$$

The advantage of this rearrangement is that an uncertainty in the objective function is translated into an uncertainty in the constraints, as was dealt with in the previous section.

Now consider the case that the objective function is linear, so we have $f(x) = c_1 x_1 + c_2 x_2 + \ldots + c_n x_n$ and suppose we know that $(c_1, c_2 \ldots c_n)$ lies in a given uncertainty set A. Then the problem becomes

PZ: maximize v

$$\text{subject to} \quad v - c_1 x_1 - c_2 x_2 - \ldots - c_n x_n \leq 0 \quad \text{for all } (c_1, c_2, \ldots, c_n) \in A$$
$$x \in X.$$

This formulation allows all the machinery introduced in the previous section to be applied. Note that by asking for the constraint to apply for all choices of coefficient in the set A, we end up with a value v that is equal to the smallest value of the objective for possible $(c_1, c_2, \ldots, c_n) \in A$. Thus, the formulation PZ is equivalent to

$$\max_{x \in X} \left(\min_{(c_1, c_2, \ldots, c_n) \in A} \{c_1 x_1 + c_2 x_2 + \ldots + c_n x_n\} \right).$$

We can see that with this formulation we are maximizing the objective subject to the most pessimistic assumptions on the values of the uncertain parameters. We can think of this as a game between us and an opponent, just the sort of game that we saw Roger Riley playing in the scenario at the start of this chapter. We choose the values of the decision variables $x_1, x_2, \ldots, x_n$ and then the other player chooses the values of $c_1, c_2, \ldots, c_n$. But our opponent here is simply malicious: their aim is to give us the worst possible outcome. Sometimes people talk of playing against nature, though this implies a rather paranoid view of the world! We will simply regard this problem as one of guaranteeing a reasonable outcome whatever nature throws at us.

There is a link to the idea of reverse stress testing introduced in Chapter 3. However, reverse stress testing simply carries out the inner minimization (finding the worst possible scenario); whereas here we are choosing decision variables as well. Thus, we are not only finding the worst result possible, but also working to avoid this being too bad.

More generally, we can think of a profit function Π that depends not only on our actions x but also the values of some uncertain parameters given by the vector a and the only information we have is that $a \in A$, the 'uncertainty set' for the problem. Then the best we can do in guaranteeing a certain level of profit is to maximize the minimum value of the profit for $a \in A$, i.e. we solve

$$\max_{x \in X} \left\{ \min_{a \in A} \Pi(x, a) \right\}. \tag{8.12}$$

An important case of this problem is when the profit function Π is a concave function of a for each value of x and A is a polytope with corners $a^{(1)}, a^{(2)}, \ldots, a^{(k)}$. Each of these corners is itself a vector, so we can write $\left(a_1^{(j)}, a_2^{(j)}, \ldots, a_m^{(j)} \right)$ for the jth corner point of A where $j = 1, 2, \ldots, k$.

In this case the equivalent to the formulation PZ becomes

PZ1:   maximize v

subject to $v - \Pi(x, a) \leq 0$ for all $(a_1, a_2, \ldots, a_m) \in A$
$x \in X$.

Using the same approach we used before, the next step is to replace the single constraint with k copies: one for each of the extreme points of A. We get the following

PZ2:   maximize v

subject to $v - \Pi(x, a^{(1)}) \leq 0$
$v - \Pi(x, a^{(2)}) \leq 0$
$\ldots$
$v - \Pi(x, a^{(k)}) \leq 0$
$x \in X$.

Clearly if v and x are feasible for PZ1 they must satisfy the constraints for all the corner points of A and hence are feasible for PZ2. We can use our assumption on the concavity of Π to show that if each of the constraints of PZ2 is satisfied, then the constraint of PZ1 will also be satisfied. The argument is the same as that given earlier in respect to Inequality (8.4) and Exercise 8.5 asks you to repeat this. So, the end result is that when the profit function is concave in the uncertain parameter and the uncertainty set is a polytope, we can replace the problem (8.12) with the optimization problem PZ2.

Worked Example 8.4   Sentinel Enterprises

Sentinel Enterprises sell tablet computers and e-readers. They have a new product launch of the 'FlexReader' for which there has been considerable advertising. They have advance orders for 5000 FlexReaders. The advance order customers have been given a secure code which they can use to make an online purchase on the launch date, which will be two weeks before the FlexReaders are available at retail stores. FlexReaders come in two screen sizes (large and small) – and, due to a mistake, the advance order customers were not asked which of these they wanted. To make matters worse, the manufacturers have been experiencing problems with meeting the launch date and the result will be an extra cost for

FlexReaders available for advance purchase. Sentinel will pay $550 for the large-screen format and $520 for the small-screen format for FlexReaders available in time for advance purchase, while FlexReaders delivered two weeks later will cost $70 less. The large FlexReaders sell for $640, the small ones for $590. Customers who have placed an advance order but cannot get their preferred format are likely not to purchase at all. FlexReaders that Sentinel gets delivered early and which are not needed for the advance order customers will simply be sold later. How many of the two different readers should Sentinel order for the launch date?

Solution

The aim is to use a robust approach to deal with the uncertainty over how many of the advance order customers will want the large format and how many will want the small format. This will have the effect of maximizing Sentinal's profit assuming the worst case for the split of demands.

We want to determine the order size x_L for large FlexReaders and x_S for small FlexReaders to be delivered at launch date. If the total advance orders are split as d_L of large and d_S of small, then the advance order sales are $z_L = \min(x_L, d_L)$ and $z_S = \min(x_S, d_S)$.

Let w_L and w_S be the normal sales of the two sizes. The purchase costs for the large size format are

$$550x_L + 480(w_L - x_L + z_L),$$

where the second term arises because there are $x_L - z_L$ left-over FlexReaders from the advance purchase. Similarly, the purchase costs for the small format FlexReaders is

$$520x_S + 450(w_S - x_S + z_S).$$

Thus, the final profit is

$$\Pi = 640(z_L + w_L) + 590(z_S + w_S) - 550x_L - 480(w_L - x_L + z_L)$$
$$-520x_S - 450(w_S - x_S + z_S)$$
$$= 160z_L + 140z_S - 70(x_L + x_S) + 160w_L + 140w_L.$$

To maximize the profit we want to choose x_L and x_S to

$$\text{maximize } 160\min(x_L, d_L) + 140 \ \min(x_S, d_S) - 70(x_L + x_S),$$

but nature will choose d_L and d_S from the uncertainty set

$$A = \{(d_L, d_S) : d_L \geq 0, d_S \geq 0, d_L + d_S = 5000\}.$$

We can formulate this as the optimization problem of maximizing v, subject to the constraints

$$v - 160\min(x_L, d_L) - 140\min(x_S, d_S) + 70(x_L + x_S) \leq 0 \text{ for all } (d_L, d_S) \in A,$$
$$x_L \geq 0, x_S \geq 0.$$

Since the minimum operators are concave, this satisfies the conditions we need to replace the uncertainty set with copies of the constraint at the two extreme points of A; these are $d_L = 5000, d_S = 0$ and $d_L = 0, d_S = 5000$.

We can assume that $x_L \leq 5000$ and $x_S \leq 5000$, since there can be no reason to order more than the maximum demand. Then, the constraints at the two extreme points can be simplified and we obtain

$$\text{maximize} \quad v$$

$$\text{subject to} \quad v - 160x_L + 70(x_L + x_S) \leq 0$$
$$v - 140x_S + 70(x_L + x_S) \leq 0$$
$$0 \leq x_L \leq 5000, \quad 0 \leq x_S \leq 5000.$$

The spreadsheet BRMch8-Sentinel.xlsx is set up for the solution of this problem. The solution turns out to be $x_L = 4375$ and $x_S = 5000$, which gives a v value of \$43 750. In other words, with these values of x_L and x_S a profit of at least \$43 750 will be made and this is the best 'guaranteed' profit there can be.   □

8.3.1 *Distributionally robust optimization

Up to this point we have been considering a situation where the uncertainty relates to particular numbers within the problem statement. Rather than assume that we know the distribution of those parameters, we have assumed simply that we know an uncertainty set to which they belong. There are many cases, however, in which we know more than the range of values that a parameter may take but less than its complete distribution. When this happens it makes sense to use a *distributionally robust* model in which we specify, not a set of points, but a set of distributions as the uncertainty set. An example occurs if we are confident that, for a specific commodity, tomorrow's price follows a normal distribution with a mean the same as today's price, but we are uncertain about the standard deviation.

In this case we will write the uncertainty set using a script 'A' ($\mathcal{A}$) to remind us that it is a set of distributions. If ξ is the parameter in question and $\Pi(x, \xi)$ is the profit using decision variables ξ and we know the distribution of ξ, then (if we are risk neutral) we want to maximize the expectation of $\Pi(x, \xi)$. Since we will be considering changes of distribution, it is helpful to write this expectation in terms of the distribution. We use the notation $E_F[\Pi(x, \xi)]$ to mean the expectation of $\Pi(x, \xi)$ when ξ has the distribution F.

We will concentrate on the problem where the uncertainty occurs in the objective rather than in the constraints. Then the distributionally robust optimization problem is to find the best expected profit that is guaranteed if we only know that the distribution of ξ lies in an uncertainty set $\mathcal{A}$. We can write this as

$$\max_x \left\{ \min_{F \in \mathcal{A}} E_F \left[\Pi(x, \xi) \right] \right\}.$$

If the uncertainty in the distribution can be represented by a small set of defined parameters, then this problem can be brought back to a 'point-based' robust formulation, by working out the value of the expectation in terms of the parameter values. In the example about tomorrow's commodity price, we might bound the possible standard deviations between $\sigma_{\min}$ and $\sigma_{\max}$ and then $\mathcal{A} = \{N(0, \sigma) : \sigma_{\min} \leq \sigma \leq \sigma_{\max}\}$. Now suppose that we can calculate the expected profit achieved for a given value of standard deviation σ, and decision variables x, say this is $\overline{\Pi}(x, \sigma)$. Then we can rewrite the distributionally robust optimization problem in the form

$$\max_x \left\{ \min_{\sigma \in A} \overline{\Pi}(x, \sigma) \right\},$$

where $A = \{\sigma : \sigma_{\min} \leq \sigma \leq \sigma_{\max}\}$.

In the distributionally robust optimization problem, if the uncertainty set $\mathcal{A}$ includes distributions that have the extreme behavior of putting all the probability weight on a single value of ξ, then we can assume that nature will choose one of these extreme distributions. The reason is simple, the expected value of Π under the distribution F must be greater than the minimum value that it could take; in other words, if the density of F is f and this is non-zero on the range $[a, b]$, then

$$E_F \left[\Pi(x, \xi) \right] = \int_a^b \Pi(x, s) f(s) ds$$

$$\geq \int_a^b \left(\min_{a \leq z \leq b} \Pi(x, z) \right) f(s) ds = \min_{a \leq z \leq b} \Pi(x, z).$$

We write δ_z for the distribution that puts all its weight at the single point z (sometimes this is called a Dirac delta distribution). Also we write $R(\mathcal{A})$ for the set of all values that may occur under a distribution in $\mathcal{A}$. Then, in the special case that for every value of $z \in R(\mathcal{A})$ then δ_z is also in $\mathcal{A}$, we can deduce that

$$\max_x \left\{ \min_{F \in \mathcal{A}} E_F \left[\Pi(x, \xi) \right] \right\} = \max_x \left\{ \min_{z \in R(\mathcal{A})} \Pi(x, z) \right\}$$

so we are back to a pointwise robust optimization problem.

In some contexts it is natural to consider an uncertainty set $\mathcal{A}$ consisting of all distributions which are unimodal – the densities increase to a maximum and then decrease. For example, when considering the distribution of demand for a product that has an uncertain relationship to the weather (at sufficient distance into the future that weather forecasts are not much use), we may be comfortable restricting the distribution to a unimodal one – even though almost nothing else is known.

We will consider an uncertainty set which consists of all distributions with unimodal density functions defined on a range, but we add the condition that the mode is known in advance. We can transform the problem so that the mode is zero. For example, we may judge that the most likely value for the price of oil in a week's time is the price today, and we regard this price as stochastic with a distribution which is unimodal. We can see that the distribution is obtained by taking a unimodal distribution with mean zero and adding to it today's oil price.

A key observation is that any unimodal distribution that has support in a range $[-a, b]$ and mode 0 can be obtained by first choosing a number from a particular distribution G on $[-a, b]$ and then multiplying by a number drawn randomly from the interval $[0, 1]$. We can say that any unimodal distribution can be obtained as the product of two independent samples, one from G and one from $U(0, 1)$, the uniform distribution on $(0, 1)$. This result is called Khintchine's Theorem. Another way to put this is to say that we can obtain any unimodal distribution from a mixture between uniform distributions, each of which has a range either of the form $[0, x]$ or of the form $[-x, 0]$. Figure 8.3 demonstrates this by considering a distribution where the density function is an increasing step function for $x < 0$ and a decreasing step function for $x > 0$. We have shown how the density splits into horizontal rectangles; each represents a separate uniform distribution between the horizontal endpoints of that block (either a range $[-x, 0]$ or a range $[0, x]$). Suppose that we select each rectangle with a probability equal to its area (these sum to 1 since they equal the integral of the original density function f) and then sample from within the given rectangle uniformly within its horizontal range. It is not hard to see that the probability of ending up at any point matches that from the original distribution.

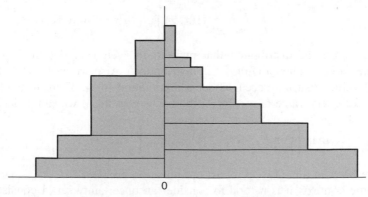

Figure 8.3   An example of a unimodal distribution as a mixture between uniform distributions.

Suppose that we are considering the inner minimization. This is nature's problem: given a choice of x made by the decision maker, how should the distribution of ξ be chosen to minimize the expected value of $\Pi(x, \xi)$? The available distributions are unimodal (with mode 0). Whatever distribution is chosen, an alternative to get the same result is to split this into its uniform (horizontal rectangle) components, as in Figure 8.3, and then choose one of these with the appropriate probability. Thus, for example, if the distribution F was composed from three uniform distributions U_A, U_B and U_C with probabilities p_A, p_B and p_C then

$$E_F[\Pi(x, \xi)] = p_A E_{U_A}[\Pi(x, \xi)] + p_B E_{U_B}[\Pi(x, \xi)] + p_C E_{U_C}[\Pi(x, \xi)].$$

This is a convex combination of the expectations taken over the three uniform distributions, and so one of the three must have a value of $E_F[\Pi(x, \xi)]$ or lower. If all three had values greater than $E_F[\Pi(x, \xi)]$ then the result of choosing between them with certain probabilities would also be greater than $E_F[\Pi(x, \xi)]$.

This is an example of the kind of argument we have seen already: when minimizing a linear function over a polytope, we can just consider the extreme points of the polytope. In the same way, if we are minimizing an expectation over a set of distributions $\mathcal{A}$, we can just consider the extreme points of the set $\mathcal{A}$.

In fact, we can make this whole argument in a more abstract and general way. The set $\mathcal{A}$ of unimodal distributions with mode 0 and support in the range $-a$ to b is itself a convex set, since if we take a convex combination of two such distributions, the result is still a unimodal distribution with mode 0. Moreover, the uniform distributions on $[0, y]$ for $0 \leq y \leq b$, or $[-y, 0]$ for $0 \leq y \leq a$ are the extreme points of $\mathcal{A}$ (since they cannot be obtained through a convex combination of two other distributions in $\mathcal{A}$). When minimizing a function that is linear on distributions, we only need to consider the extreme points (i.e. these uniform distributions).

So, to find a unimodal distribution for ξ (with mode 0) that minimizes $E[\Pi(x, \xi)]$ we need only consider uniform distributions, either on $(-y, 0)$ or on $(0, y)$. We can go further in the case that $\Pi(x, \xi)$ is a concave function of ξ. In this case we can say that the minimum of $E[\Pi(x, \xi)]$ over distributions $F \in \mathcal{A} = \{$unimodal distributions with support in a range $[-a, b]$ and mode 0$\}$ is attained either when F is uniform on $(-a, 0)$ or when F is uniform on $(0, b)$. We will give a sketch proof of this result by showing that one or other of these two uniform distributions gives a smaller expected value for $\Pi(x, \xi)$ than the uniform distribution over $(0, y)$ for $0 < y < b$. (Exactly the same method can be used to show that one or other has a smaller value than the uniform distribution over $(-y, 0)$ for $0 < y < a$.)

The value of x in $\Pi(x, \xi)$ plays no part in this discussion, so we just write $\Pi(\xi)$ for $\Pi(x, \xi)$. Note that when ξ has a uniform distribution on $(-a, 0)$ then

$$E(\Pi(\xi)) = \frac{1}{a} \int_{-a}^{0} \Pi(u) du$$

i.e. the average value of Π over $(-a, 0)$, and there are similar expressions for the expectation of Π when ξ has a uniform distribution on $(0, y)$ or $(0, b)$. We choose an arbitrary y with $0 < y < b$: our aim is to show that if

$$\frac{1}{a} \int_{-a}^{0} \Pi(u)du > \frac{1}{y} \int_{0}^{y} \Pi(u)du,$$

then

$$\frac{1}{y} \int_{0}^{y} \Pi(u)du > \frac{1}{b} \int_{0}^{b} \Pi(u)du.$$

Since Π is concave, we can draw the straight line between the points on the graph of Π at 0 and y and the values of Π are above this line between 0 and y and below this line outside this range. This is illustrated in Figure 8.4. Now if this straight line were horizontal at a height h, we could immediately deduce that the average value of Π on $(0, a)$ would be more than the average on $(-a, 0)$. Since we assume that the average value of Π over $(-a, 0)$ is greater than in the range $(0, y)$, it is clear that the line must slope downwards. Hence, we can see that the value of Π in the range (y, b) is lower than the values in $(0, y)$. Thus

$$\frac{1}{b} \int_{0}^{b} \Pi(u)du = \frac{1}{b} \int_{0}^{y} \Pi(u)du + \frac{1}{b} \int_{y}^{b} \Pi(u)du$$

$$< \frac{1}{b} \int_{0}^{y} \Pi(u)du + \frac{(b-y)}{b}\Pi(y)$$

$$< \frac{1}{b} \int_{0}^{y} \Pi(u)du + \frac{(b-y)}{b}\frac{1}{y} \int_{0}^{y} \Pi(u)du = \frac{1}{y} \int_{0}^{y} \Pi(u)du.$$

This completes what we need to establish: one or other of $(1/a) \int_{-a}^{0} \Pi(u)du$ or $(1/b) \int_{0}^{b} \Pi(u)du$ must be less than or equal to $(1/y) \int_{0}^{y} \Pi(u)du$. And so we have established that the minimum occurs at one of these values.

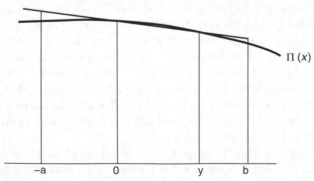

Figure 8.4   Diagram to illustrate the argument about averages of a concave function Π.

We can replace the mode zero with an arbitrary value w_0. So finally we have established that when we wish to minimize the expected value of a concave function $\Pi(x, \xi)$ over choices of distribution for ξ which are unimodal, have support in (w_L, w_U) and have their mode at w_0, then we need only consider two options: the uniform distribution on (w_L, w_0) and the uniform distribution on (w_0, w_U). We show how to use this result to calculate a robust solution in the worked example below.

Worked Example 8.5   Toulouse Medical Devices

Toulouse Medical Devices (TMD) needs to order heart monitors to meet an uncertain demand. TMD is uncertain about the distribution of demand but believes that the distribution is unimodal, with the lowest level of demand being zero and the highest being 200. The most likely value of demand is 100. Units cost \$4000, and are sold at \$10 000. There are costs associated both with having unsold units and with not being able to meet the demand. TMD estimates that if it orders x and demand is d, then these 'mismatch' costs are $50(x - d)^2$. Thus, the profit function (in \$1000s) is

$$\Pi(x, d) = 10 \min(x, d) - 4x - 0.05(x - d)^2. \qquad (8.13)$$

TMD realizes that in some cases it may make a loss, but wishes to maximize its expected profit given the worst possible distribution of demand.

Solution

The profit function in Equation (8.13) is a concave function of d once x is fixed, since the three components are all concave. From the result above, we need only to evaluate the expected profit for two particular distributions of d: either d is uniform on $(0, 100)$ or d is uniform on $(100, 200)$.

$$E_{U(100,200)}[\Pi(x, d)] = \frac{1}{100} \int_{100}^{200} (10 \min(x, u) - 4x - \frac{1}{20}(x - u)^2) du.$$

Clearly the value of this integral depends on the value of x. If $x \leq 100$

$$E_{U(100,200)}[\Pi(x, d)] = \frac{1}{100} \int_{100}^{200} (6x - \frac{1}{20}(x - u)^2) du$$

$$= 6x + \frac{1}{100} \left[\frac{1}{20}(x - u)^3 / 3 \right]_{100}^{200}$$

$$= 6x + \frac{1}{6000}((x - 200)^3 - (x - 100)^3).$$

If $x > 100$

$$E_{U(100,200)}[\Pi(x,d)]$$

$$= \frac{1}{100}\int_{100}^{x}(10u - 4x)du + \frac{1}{100}\int_{x}^{200}6x\,du - \frac{1}{100}\int_{100}^{200}\frac{1}{20}(x-u)^2du$$

$$= -4x\frac{x-100}{100} + \frac{1}{100}\left[5u^2\right]_{100}^{x} + 6x\frac{200-x}{100} + \frac{1}{2000}\left[(x-u)^3/3\right]_{x}^{200}$$

$$= \frac{x}{10}(160 - x) + \frac{1}{20}(x^2 - 100^2) + \frac{1}{6000}\left((x-200)^3 - (x-100)^3\right).$$

The other distribution can be evaluated similarly

$$E_{U(0,100)}[\Pi(x,d)] = \frac{1}{100}\int_{0}^{100}(10\min(x,u) - 4x - \frac{1}{20}(x-u)^2)du.$$

If $x \le 100$

$$E_{U(0,100)}[\Pi(x,d)]$$

$$= \frac{1}{100}\int_{0}^{x}(10u - 4x)du + \frac{1}{100}\int_{x}^{100}6x\,du - \frac{1}{2000}\int_{0}^{100}(x-u)^2\,du$$

$$= -4x\frac{x}{100} + \frac{1}{100}\left[5u^2\right]_{0}^{x} + 6x\frac{100-x}{100} + \frac{1}{2000}\left[(x-u)^3/3\right]_{0}^{100}$$

$$= \frac{1}{10}x(60 - x) + \frac{x^2}{20} + \frac{1}{6000}\left((x-100)^3 - x^3\right),$$

and finally if $x > 100$

$$E_{U(0,100)}[\Pi(x,d)] = \frac{1}{100}\int_{0}^{100}(10u - 4x)du - \frac{1}{100}\int_{0}^{100}\frac{1}{20}(x-u)^2du$$

$$= -4x + \frac{1}{100}\left[5u^2\right]_{0}^{100} + \frac{1}{2000}\left[(x-u)^3/3\right]_{0}^{100}$$

$$= -4x + 500 + \frac{1}{6000}\left((x-100)^3 - x^3\right).$$

The spreadsheet BRMch8-Toulouse.xlsx shows the values of $E_{U(0,100)}[\Pi(x,d)]$ and $E_{U(100,200)}[\Pi(x,d)]$ as x varies, and these are also shown in Figure 8.5. The robust optimum value for x is the one that maximizes the minimum profit. The optimum (integer) order size is $x = 73$, which guarantees a minimum expected profit of \$99 883. We can see that for many x values a negative expected profit is possible if nature deals us a bad hand in the choice of demand distribution. □

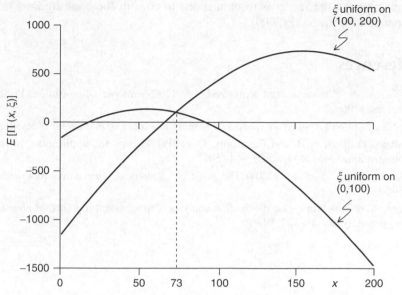

Figure 8.5   The expected profits for TMD as a function of x for the two extreme uniform distributions of demand.

Notes

The material on Knightian uncertainty is mainly taken from the book by Bernstein. The whole area of robust optimization has excited a great deal of interest in the last few years and there are many papers that deal with different aspects of robust optimization. The review article by Bertsimas, Brown and Caramanis (2011) gives a good introduction to the very extensive literature on robust optimization. Our treatment here has been elementary and focused on relatively small-scale problems with simple uncertainty sets. The discussion in Section 8.2 on budgets of uncertainty arises from work by Dimitris Bertsimas and co authors (Bertsimas and Sim, 2004). In that section we use duality properties to establish the exact problem to solve – this approach through duality can be extended to a whole variety of more complex robust optimization problems.

In the same way, our treatment of distributionally robust optimization can be extended in many ways. The uncertainty set we have looked at in most detail, unimodal functions with a known mode, is particularly simple to analyze. Some related theory in a more general context of multidimensional distributions can be found in Shapiro (2006).

There has also been a great deal of work that looks at efficient computation of robust optimal solutions for a variety of problems – for example, problems with dynamic characteristics matching the discussion of stochastic optimization that we gave in Chapter 7. For a much more comprehensive discussion of all of this material, the reader is recommended to consult the book by Ben-Tal, El Ghaoui and Nemirovski (2009).

References

Ben-Tal, A., El Ghaoui, L. and Nemirovski, A. (2009) *Robust Optimization*. Princeton University Press.

Bernstein, P. (1996) *Against the Gods: The Remarkable Story of Risk*. John Wiley & Sons.

Bertsimas, D., Brown, D. and Caramanis, C. (2011) Theory and applications of robust optimization. *SIAM Review*, **53**, 464–501.

Bertsimas, D. and Sim, M. (2004) The price of robustness. *Operations Research*, **52**, 35–53.

Shapiro, A. (2006) Worst-case distribution analysis of stochastic programs. *Mathematical Programming Ser. B*, **107**, 91–96.

Exercises

8.1 Rewriting constraints

Consider a problem of choosing a production quantity x_i for products $i = 1, 2, 3, 4$ where the failure rates are high and uncertain. The time required to produce one unit of any product is 1 minute. But because of the failure rates, to produce an amount x_i of product i requires a time $a_i x_i$ minutes, where $a_i \geq 1$. The production rates a_i are uncertain. Suppose that we are confident that $a_1 \leq 1.1$, $a_3 \leq 1.1$. Moreover, the failure rates are known to be correlated in a complex way. The expert view is that the following additional constraints can be assumed:

$$a_1 + a_2 + a_3 + a_4 \leq 4.3,$$
$$a_1 - a_2 + a_3 - a_4 \leq 0.1.$$

Formulate the constraint on the production quantities for the four products if it is necessary to guarantee that these amounts can be made in less than 40 hours in total (i.e. 2400 minutes).

8.2 Coefficients are unrelated

A robust optimization problem has a constraint

$$a_1 x_1 + a_2 x_2 + a_3 x_3 \leq b$$

which must hold for all $(a_1, a_2, a_3) \in A$ where A is the set where $a_i \in (\bar{a}_i - \delta_i, \bar{a}_i + \delta_i)$. There are no connections between the a_i values. Show that by using absolute values $|x_1|$, $|x_2|$, $|x_3|$ this can be rewritten with a single constraint.

8.3 Impact of the budget of uncertainty

(a) Solve the Avignon Imports example with a budget of uncertainty of $B = 1.5$ and compare the objective functions with the $B = 2$ case to see how much the expected profit is increased by taking the less conservative approach.

(b) Show that if there is no budget of uncertainty (and each variable can take any value in its range), which is equivalent to setting $B = 3$, then there is no feasible solution.

8.4 Robust optimization for Sentinel

In the Sentinel example, use the spreadsheet to show that if the selling prices are reduced to \$600 and \$550 for the large and small formats, then the robust optimal solution has $x_L = x_S = 0$. Explain why this happens.

8.5 Uncertainty sets with concave functions

Suppose A is a polytope with corners $a^{(1)}, a^{(2)}, \ldots, a^{(k)}$ (each of these is a vector in R^n). If $\Pi(x, a)$ is a concave function of a for each fixed x and $v \leq \Pi(x, a^{(i)})$ for $i = 1, 2, \ldots, k$ (so v is less than $\Pi(x, a^{(i)})$ at each corner point), then show that $v \leq \Pi(x, a)$ for all $a \in A$.

8.6 Provence Rentals

Provence Rentals has a fleet of 100 cars that it rents by the day. It is considering investing in GPS systems for these cars and will charge a premium of \$4 per day for hire of the GPS systems. Each of the GPS systems will cost \$500 to purchase and also requires the fitting of a secure holder with a cost of \$250 per car. Provence sells its cars after 500 days and this is also the effective lifetime of the GPS system. So, a car with a system installed has no extra value after this period in comparison with a car without the system installed. Once Provence Rentals advertises this service, it will be expensive in terms of goodwill not to be able to provide it; Provence Rentals reckons that there is a cost of \$10 when a customer is not able to have the system and requests it.

(a) Set this up as a robust optimization problem with the assumptions: (A) all cars are rented every day. (B) Provence Rentals has to make a decision at the start of the 500 days on how many GPS systems to install and cannot install any more over the course of the 500-day period. (C) The same proportion of customers requests the GPS system each day (but Provence Rentals has no way of predicting this proportion).

(b) Show that if Provence Rentals makes a decision on how many systems to install and then the value of p, the proportion of customers wanting a GPS system, is chosen so that Provence Rentals makes least money, then either p will be chosen at 0 or at 1.

(c) Use the observation in (b) to solve the robust optimization problem.

8.7 Toulouse Medical Devices

In the Toulouse Medical Devices example, suppose that TMD knows that the demand distribution is unimodal but is not able to say what the value of the mode is (with other aspects of the problem staying the same). Explain why the decision reduces to an ordinary (pointwise) robust optimization problem and calculate the best choice of order x.

9

Real options

Commitment anxiety: good or bad?

'Why be in such a rush?' is Darae's comment when she hears what her boss Kandice is proposing. Kandice is the CEO of Analytics Enterprises, an animation company working primarily on short advertisements and music videos, and Darae is the Financial Director. For more than a year they have been looking for an opportunity to start a new division of their company working in the computer games area. Now a small company, Eckmet Ltd, specializing in computer games has approached them seeking an injection of capital. Eckmet's main asset, apart from its 10 employees, is the rights to 20% of the sales revenue from a new computer game that the company has been working on, due to launch in three months' time. The first royalty payments made to Eckmet under this contract are due in six months, but there is great uncertainty as to how successful the game will be and hence how much will be received in royalties.

Both Kandice and Darae agree that linking up with Eckmet is a good idea. Kandice is all for buying a controlling interest in Eckmet straight away. But Darae has been looking at the possibility of buying a smaller stake in the company, with the intention of taking a larger stake only if the new computer game does well. Sometimes a new game can take a while to catch on, so Analytics would want to have an option on further share purchases for at least a two-year period. Kandice has argued that this is even more expensive – a 20% stake in Eckmet would cost $2 million (with an option for a further 40% to be purchased at any time in the next two years at a cost of $4 million), while $4.8 million would give them a 60% stake right away.

Kandice cannot see the sense in this: why set yourself up to pay a total of $6 million for something that you can get right away for $4.8 million? But Darae is concerned about betting such a large sum of money on the success of one computer game; if that goes badly the investment in Eckmet will not be

Business Risk Management: Models and Analysis, First Edition. Edward J. Anderson.
© 2014 John Wiley & Sons, Ltd. Published 2014 by John Wiley & Sons, Ltd.
Companion website: www.wiley.com/go/business_risk_management

worth much. There must be a benefit in delaying making this commitment, but is it worth the extra price that they will end up paying?

9.1 Introduction to real options

In most cases increased uncertainty in outcomes is regarded as undesirable. Our discussion in Chapter 6 of the way that individuals make decisions confirms this to be the case for individuals when dealing with uncertain gains: we would rather have $80 000 for sure than an 80% chance of $100 000 and a 20% chance of nothing. We also met the same idea in Chapter 2 discussing portfolio theory, where we expect higher returns when the risk (i.e. the variance) is greater.

However, there are circumstances where higher volatility or higher variance is beneficial. This happens whenever there is an implied option, with the effect of limiting any downside associated with the variation. This beneficial effect is not as the result of a particular preference structure used by the decision maker: it happens when the decision maker is risk neutral.

At first sight it seems odd that an increase in variance might make things better, and a good way to understand this is through looking at an example.

Example 9.1   Investing in biotech companies

Suppose that you are considering investing in one of two biotechnology companies, both involved in similar research and development work. Company A typically produces one new design idea every three months and, on average, the potential profitability of these product ideas, if they were to be put into production, has a mean of $30 000 per year with a standard deviation of $20 000 a year. (There is the possibility of losing money if the less successful product designs were ever put into production, so these designs are simply shelved). The second company, Company B, also produces one new design idea every three months, with mean product profitability of $30 000, but the standard deviation is higher at $40 000 a year. Which is the better company to invest in?

We naturally wish to minimize risk, and these two companies are the same except for the variance of the returns on new design ideas. So, at first sight, we might expect that Company A with smaller variance would provide the better investment. But actually it is Company B, with the *higher variance* of returns, that will give a higher expected return and represents the better investment. Product ideas that lose money will never be put into production. So the less successful designs have value zero, rather than a negative value. Hence, there is a gain from the larger positive variations in profitability that is not offset by the larger negative variations. The idea is shown by Figure 9.1, in which a distribution of product profitability for the two companies is given with the assumption that unprofitable products have value zero. The bar at zero represents the probability of either of the profits taking a value zero. The mean value for the product profitability in the low variance case is $31 052 and for the high variance (dashed line) is $33 991. The greater the proportion of the distribution that is cut off, the higher the expected value for profit. And if we

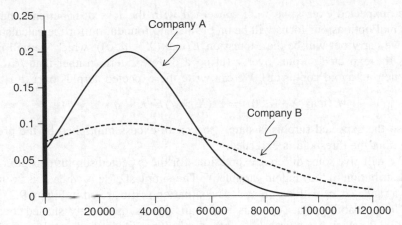

Figure 9.1   Profit distribution truncated at zero gives different results depending on the variance.

were to consider an even smaller standard deviation then we would find almost no probability of a negative profitability – leading to an expected profit equal to the mean of $30 000, a lower amount even than Company A. ☐

This example demonstrates how, when the bottom tail is cut off, making the variance smaller will make the expected profit worse. This is the underlying phenomenon that we explore in this chapter. It is important for managers to have a good understanding of why holding an option (in this case the option not to develop a product) makes increasing variability suddenly valuable. There are simple tools that will help us with the calculations needed to put a dollar value on the 'optionality' here, and this is what we turn to next.

9.2   Calculating values with real options

Suppose that the outcome of some venture or investment is uncertain but there is a guarantee that the result will not be lower than a given value a. Perhaps we have the option of not going ahead with the venture, in which case the outcome can never be less than zero, or perhaps we can always sell our investment for a given amount a since we hold a put option for this amount (we will come back to a fuller discussion of financial options later). In any case, if X is a random variable giving the profit in the absence of the option, then our expected profit, with the option to always make an amount a, will be given by

$$E(\max(X, a)) = E(\max(X - a, 0)) + a.$$

This expression is reminiscent of our discussion of the excess distribution and the mean excess that we met in Chapter 4, and there is indeed a relation between the two. But the definition of the excess distribution is different because it is *conditional* on $X > a$ (for a threshold value a). Moreover, in our discussion of the expected excess we were concerned with the loss distribution, whereas with real options our focus will be on profit. The foundation for the calculations that we carry out will be the expression $E(\max(X - a, 0))$, which we will refer to as the expected surplus over a (using a different term, rather than 'excess' may help to avoid confusion). We can write the expected surplus over a as

$$E(\max(X - a, 0)) = \Pr(X > a)E(X - a \mid X > a)$$

and so the expected surplus is simply the mean excess multiplied by the probability that the threshold is exceeded.

We will give some different expressions for the expected surplus depending on the distribution of the random variable X. The simplest case is when X is uniform over a range (b, c) with $b < a < c$. The situation is illustrated in Figure 9.2.

The probability that $X - a$ is less than zero is the lightly shaded area in Figure 9.2, which has a probability of $(a - b)/(c - b)$: in the formula $\max(X - a, 0)$ this is the probability mass that is set to zero, indicated by the solid bar at zero. In the event that $X - a$ is not less than zero, then it is equally likely to take any value between 0 and $c - a$. So, in this event the average value is $(c - a)/2$ (this is the mean excess over the threshold of a). Thus, the expected surplus over a is

$$E(\max(X - a, 0)) = \frac{(a - b)}{(c - b)} \times 0 + \left(1 - \frac{(a - b)}{(c - b)}\right) \times \frac{(c - a)}{2}$$

$$= \frac{(c - a)^2}{2(c - b)}$$

We can derive the same expression using integration. We have

$$E(\max(X - a, 0)) = \int_b^c \max(u - a, 0)\frac{1}{c - b}du$$

$$= \frac{1}{c - b}\int_a^c (u - a)du$$

$$= \frac{1}{c - b}\left(\frac{c^2}{2} - ca - \frac{a^2}{2} + a^2\right)$$

$$= \frac{(c - a)^2}{2(c - b)}.$$

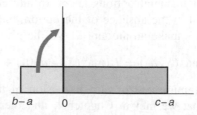

Figure 9.2   Diagram to show calculation of expected surplus over a for uniform distribution on (b, c).

It is also useful to have a formula for $E(\max(a - X, 0))$. This can be derived in a similar way:

$$
\begin{aligned}
E(\max(a - X, 0)) &= \int_b^c \max(a - u, 0) \frac{1}{c - b} du \\
&= \frac{1}{c - b} \int_b^a (a - u) du \\
&= \frac{1}{c - b} \left(a^2 - \frac{a^2}{2} - ba + \frac{b^2}{2} \right) \\
&= \frac{(a - b)^2}{2(c - b)}.
\end{aligned}
$$

Worked Example 9.2   Deep water salvage

A salvage company has been asked to consider carrying out a difficult deep water salvage operation. The salvage value of the boat is $3 million (20% of a $15 million insured value), and there will be a cost of $300 000 to carry out a preliminary investigation which will determine the actual cost of the salvage operation. This cost is estimated to be between $2 million and $4 million and the company regards any amount between these numbers as equally likely. Obviously if the cost is found to be too large, the company will not go ahead with the salvage operation, but it will not recover the $300 000 preliminary cost. What is the expected value of this project to the salvage company?

Solution

Once the $300 000 has been spent it is a question of whether or not to go ahead with the salvage, and this will be worthwhile if the cost is no more than $3 million. Hence, the expected profit in millions is

$$
-0.3 + E(\max(3 - X, 0))
$$

where X is uniform on the interval between 2 and 4 (million). Using the formula above we have $a = 3, b = 2, c = 4$ and an expected profit of

$$
-0.3 + \frac{(3 - 2)^2}{2(4 - 2)} = -0.05.
$$

Hence, this is a project that is not worth going ahead with. We can get the same result without using the formula since the numbers are all quite easy. 50% of the time the costs are more than $3 million and the salvage is not worth doing, giving a loss of 0.3 million. On the other hand, if the salvage is worthwhile (as happens 50% of the time), then the costs are uniformly distributed on the range 2 million to 3 million, giving an average cost of the salvage of $2.5 million, meaning a profit of just 0.2 million after paying the upfront costs. With equal chances of a $300 000 loss or a $200 000 profit, the salvage company should walk away from this deal.   □

Solving problems involving real options almost always requires the same underlying model for what is going on, so it is worth trying to unpack this in more detail.

1. *Identify what is being valued.* The underlying structure of real options calculations is one of valuation. We want to know how much is this project or potential investment worth, or equivalently how much is our expected profit? Knowing the value will tell us the maximum amount we should pay (and when we should walk away from the project if the net value is negative). In the salvage example, we want to value the entire project.

2. *Identify the decision point and the information that will be used.* The underlying idea of real options is that we can make a decision at some point in the future between different choices on the basis of information which will be available then, but is not available when we make the valuation upfront. For simplicity, we assume that there are just two choices being considered (often they are of the 'go' or 'no go' kind). In the salvage example, the decision point occurs after we find out the cost of the salvage operation, and the option is simply to go ahead or not.

3. *Calculate the profits for different choices in terms of the information available.* If we write the information on which our decision will be based as X (which is thought of as a random variable at the outset, when we do the valuation) then we need to compare the two choices in terms of their profits, given X. If the choices are A and B we might write $\Pi_A(X)$ and $\Pi_B(X)$ for the two profit functions. In the salvage example, X is the cost of the salvage operation, and letting A be 'go ahead' and B be 'don't go ahead', the profits are (in millions):

$$\Pi_A(X) = 3 - X - 0.3$$
$$\Pi_B(X) = -0.3.$$

4. *Identify the cutoff point between different choices.* The value of X will determine which option is taken and there will be some cutoff point for X at which we switch choices. So we make choice A if $X < a$ for some threshold a, and we make choice B if $X > a$, and we are indifferent between the choices when $X = a$. In the salvage example, choice A (go ahead) is best if $X < 3$ and choice B (quit) is best if $X > 3$.

5. *Express the project value in terms of* $\max(a - X, 0)$. Next we need to find a way of expressing the overall profit in terms of the $\max(a - X, 0)$ formula (or sometimes it is more convenient to use the $\max(X - a, 0)$ form). At this point we are still working with profit as a function of the 'information' variable X. The overall profit is the greater of the two possibilities, so

$$\Pi(X) = \max(\Pi_A(X), \Pi_B(X))$$
$$= \Pi_B(X) + \max(\Pi_A(X) - \Pi_B(X), 0). \qquad (9.1)$$

Usually we can put the second term into the required form (we know already that $X = a$ is the point where $\Pi_A(X) - \Pi_B(X) = 0$). In the salvage example, Equation (9.1) can be written immediately as:

$$\Pi(X) = -0.3 + \max(3 - X, 0).$$

6. *Use the distribution of* X *to find the expected final value.* The final step is to evaluate the expected profit using the information available on the distribution of X and a suitable formula for $E(\max(a - X, 0))$. In the salvage example, this is the point where we use the formula derived from the fact that X has a uniform distribution.

Now we want to introduce the formula for the expected surplus with a normal distribution. We give the derivations in the next section, but at this point we simply quote the two formulae which are expressed in terms of the density and CDF of a normal distribution. If X is a normal random variable with mean μ and standard deviation σ, then

$$E[\max(a - X, 0)] = (a - \mu)\Phi_{\mu,\sigma}(a) + \sigma^2\varphi_{\mu,\sigma}(a) \qquad (9.2)$$

and

$$E[\max(X - a, 0)] = (\mu - a)(1 - \Phi_{\mu,\sigma}(a)) + \sigma^2\varphi_{\mu,\sigma}(a), \qquad (9.3)$$

where $\Phi_{\mu,\sigma}(\cdot)$ is the CDF for X and $\varphi_{\mu,\sigma}(\cdot)$ is the density function for X, i.e.

$$\varphi_{\mu,\sigma}(x) = \frac{1}{\sqrt{2\pi\sigma^2}} \exp\left(-\frac{(x - \mu)^2}{2\sigma^2}\right)$$

where we use the notation $\exp(x)$ to mean e^x.

Now we can return to Example 9.1 and see where the valuation numbers come from. The expected profit from Company A in \$1000s is given by $E(\max(X, 0))$ when X has a $N(30, 20)$ distribution. Setting $a = 0$ in Equation (9.3) we have

$$E(\max(X, 0)) = 30(1 - \Phi_{30,20}(0)) + 20^2\varphi_{30,20}(0)$$

$$= 30(1 - \Phi_{30,20}(0)) + 400\frac{1}{\sqrt{800\pi}} \exp\left(-\frac{900}{800}\right).$$

The first term can be evaluated from tables of the normal distribution, but it is simplest just to use a spreadsheet to evaluate the whole formula

```
= 30*(1-NORMDIST(0,30,20,1)) + 400*NORMDIST(0,30,20,0).
```

This uses the spreadsheet formula NORMDIST$(x, \mu, \sigma, \cdot)$ which returns either the cumulative distribution function $\Phi_{30,20}(x)$ or the density function $\varphi_{30,20}(x)$ according to whether the final argument is 1 or 0. Using this formula gives the value \$30 586.14. This can then be repeated with $\sigma = 40$ to obtain the profit figure \$35 246.68.

Another way to think about what is going on in this example is to use the language of options. Suppose we are considering the expected benefit from a single product idea for Company B. We can regard the value of a single product idea as an option to purchase the resulting product patent, in other words it gives 'the right but not the obligation' (to use the standard phrase) to purchase the product patent. If, once the idea is fully worked out and becomes an actual product, it turns out that the product is a loss maker, then we will not choose to exercise our option: in other words, we will not take the product to market. The key idea here is that though we pay money up front, there is a point later on when we have more information, and then we will make a decision whether or not to go ahead.

9.2.1   *Deriving the formula for the surplus with a normal distribution

Our aim in this section is to derive Equations (9.2) and (9.3). We will do this by establishing the following two intermediate formulae for integrals involving the normal density function $\varphi_{\mu,\sigma}$:

$$\int_{-\infty}^{a} x\varphi_{\mu,\sigma}(x)dx = \mu\Phi_{\mu,\sigma}(a) - \sigma^2\varphi_{\mu,\sigma}(a), \tag{9.4}$$

$$\int_{a}^{\infty} x\varphi_{\mu,\sigma}(x)dx = \mu(1 - \Phi_{\mu,\sigma}(a)) + \sigma^2\varphi_{\mu,\sigma}(a). \tag{9.5}$$

As before, we write $\Phi_{\mu,\sigma}$ for the CDF of a normal random variable with mean μ and standard deviation σ. Thus, $\Phi_{\mu,\sigma}$ is the integral of $\varphi_{\mu,\sigma}$ and

$$d\Phi_{\mu,\sigma}(x)/dx = \varphi_{\mu,\sigma}(x).$$

Equation (9.4) can be derived by noting

$$d\varphi(x)/dx = \left(-\frac{(x-\mu)}{\sigma^2}\right)\frac{1}{\sqrt{2\pi\sigma^2}}\exp\left(-\frac{(x-\mu)^2}{2\sigma^2}\right) = -\frac{(x-\mu)\varphi(x)}{\sigma^2}$$

and so

$$\varphi(a) = \int_{-\infty}^{a}[d\varphi(x)/dx]dx = \int_{-\infty}^{a} -\frac{(x-\mu)\varphi(x)}{\sigma^2}dx$$

$$= \frac{\mu}{\sigma^2}\Phi(a) - \frac{1}{\sigma^2}\int_{-\infty}^{a} x\varphi(x)dx.$$

We get Equation (9.4) simply by multiplying through by σ^2 and rearranging this equation.

Equation (9.5) comes from

$$\mu = \int_{-\infty}^{\infty} x\varphi(x)dx = \int_{-\infty}^{a} x\varphi(x)dx + \int_{a}^{\infty} x\varphi_{\mu,\sigma}(x)dx$$

and hence

$$\int_a^\infty x\varphi_{\mu,\sigma}(x)dx = \mu - \int_{-\infty}^a x\varphi(x)dx$$

and substituting from Equation (9.4) gives the result we are looking for.

Thus, when the random variable X has a $N(\mu,\sigma)$ distribution, we get

$$E[\max(a - X, 0)] = \int_{-\infty}^a (a - x)\varphi_{\mu,\sigma}(x)dx$$

$$= a\int_{-\infty}^a \varphi_{\mu,\sigma}(x)dx - \int_{-\infty}^a x\varphi_{\mu,\sigma}(x)dx$$

$$= (a - \mu)\Phi_{\mu,\sigma}(a) + \sigma^2\varphi_{\mu,\sigma}(a)$$

as we require.

Similarly, we can derive the other formula:

$$E[\max(X - a, 0)] = \int_a^\infty (x - a)\varphi_{\mu,\sigma}(x)dx$$

$$= \int_a^\infty x\varphi_{\mu,\sigma}(x)dx - a\int_a^\infty \varphi_{\mu,\sigma}(x)dx$$

$$= \int_a^\infty x\varphi_{\mu,\sigma}(x)dx - a(1 - \Phi_{\mu,\sigma}(a))$$

$$= (\mu - a)(1 - \Phi_{\mu,\sigma}(a)) + \sigma^2\varphi_{\mu,\sigma}(a).$$

9.3   Combining real options and net present value

Applying a real option approach to an investment decision will usually involve taking account of a choice that will be made later. Thus, we will have to consider the money flows over time, and in practice we need to do these calculations using an appropriate discounting of returns over time. In other words, we need to add real options into a net present value (NPV) calculation.

We want to compare different choices open to us, and the assumption we make is that different choices lead to different amounts of money being paid to us, but that these amounts are paid at different times, so that we are not simply comparing cash sums but instead are comparing a schedule of cash payments over time (we can call this a *cash flow stream*). To make these comparisons we have to allow for interest rates and doing this leads to the concept of net present value (often called a discounted cash flow analysis).

The starting point is to split time up into periods: most often these are years, but we could use months or quarters. We will assume the existence of an interest rate per period, which we write as r: this is what money would earn if put into a bank account (or we might use some other risk-free instrument). Putting an amount x into the bank at the start of a period allows an amount $x(1 + r)$ to be withdrawn at the end of the period. This amount becomes $x(1+r)^2$ after two periods and so on. To compare two different cash flow streams we can suppose that cash received is put into the bank and held there until the end of the planning horizon. The future value of a cash flow stream will be the cash sum that can be withdrawn at the end of the planning horizon. Whichever cash flow stream has the highest future value will be preferred.

To apply this method we need to take care to track exactly when the cash becomes available. For simplicity we will write x_0 for the money we are paid at time 0 (which we can think of as the start of period 1), and x_1 for the amount we receive at the end of period 1 (or the start of period 2) and so on.

Of course, projects don't just involve receiving money, they also involve making payments. We can deal with this in the same way provided we have an *ideal* bank that does not make money from its depositors and therefore charges the same interest rate r for loans as it gives for deposits. So, if we borrow an amount x to make a payment at time 0 we will have to pay back to the bank an amount $x(1+r)$ at the end of period 1 and, in general, an amount $x(1+r)^k$ at the end of period k. Thus, the future value at the end of period n of a cash flow stream $(x_0, x_1, \ldots, x_n)$ is given by

$$FV = x_0(1+r)^n + x_1(1+r)^{n-1} + \cdots + x_{n-1}(1+r) + x_n.$$

In this expression positive x values correspond to receiving cash and negative ones correspond to making a payment. This gives us a way to compare two different cash streams. If two cash flow streams have the same future value, then we are going to be indifferent between them on this model.

We can also look at a fair value to be put on the cash stream at time 0. This is the amount of money that we could be paid now which would make us indifferent between taking the money upfront or receiving the cash flow stream. In other words, it is the amount paid at time 0 which has the same future value as the cash flow stream. This is the present value (call it PV) and on this argument

$$PV(1+r)^n = x_0(1+r)^n + x_1(1+r)^{n-1} + \cdots + x_{n-1}(1+r) + x_n,$$

which we can rewrite as

$$PV = x_0 + \frac{x_1}{(1+r)} + \frac{x_2}{(1+r)^2} + \cdots + \frac{x_n}{(1+r)^n}.$$

We can see that payments and receipts farther out into the future are discounted by greater amounts. Money received at time 0 is not discounted at all, money received at the end of period 1 is discounted by a factor $(1+r)$, and so on. Often we call the present value the *net present value* to emphasize that we are including both positive and negative cash amounts.

Worked Example 9.3   Two cash flow streams compared

Two investment opportunities are available. Investment A requires $5000 dollars now and will return $2000 a year for the next three years. Investment B requires

$6000 to be paid now and will return $1500 at the end of year 1, $2500 at the end of year 2 and $3200 at the end of year 3. Compare the two NPV amounts using a discount rate of 5%.

Solution

The cash flow stream for A is, in $1000s, $(-5, 2, 2, 2)$, giving a net present value of

$$V_A = -5 + \frac{2}{1.05} + \frac{2}{(1.05)^2} + \frac{2}{(1.05)^3} = 0.4465$$

Doing the same calculation for B gives

$$V_B = -6 + \frac{1.5}{1.05} + \frac{2.5}{(1.05)^2} + \frac{3.2}{(1.05)^3} = 0.4604$$

So, from this calculation both investments have a positive NPV and investment B is slightly preferable.   □

In evaluating net present values it is often necessary to deal with a situation in which there is a constant cash flow each period. Of course it is hard to think of an example where we would truly need to allow for payments being made forever, but it is, in any case, useful to work out the value of such a cash flow, which is called a *perpetual annuity*. The present value of an amount x received a year from now and with the same amount paid to us each year from then on, if discounted at rate r, will be the infinite sum

$$PV = \sum_{k=1}^{\infty} \frac{x}{(1+r)^k} = \frac{1}{r}x.$$

Here we have used a standard formula to evaluate the infinite sum. Notice that this is for payments starting at the end of the first period: if there is a payment of an amount x made to us at time 0 as well, then the formula becomes

$$PV = x + \frac{x}{r} = \frac{1+r}{r}x. \tag{9.6}$$

Now it is time to make use of these NPV ideas in the context of a real option involving cash flows over time. We will do this by looking at an example.

Example 9.4   Foxtrot Developments

Foxtrot Developments is considering the purchase of a block of land for development. If purchased now (time 0) at the start of year 1, then development approvals are expected to be completed after two years (end of year 2) and building will take a full year to complete, so that the property can be let from the beginning of year 4. Currently the building on the land is let and generates an income of $50 000 a year. If the building is developed for commercial purposes, Foxtrot

Table 9.1   Net present value calculation for Foxtrot Developments.

	year 0	year 1	year 2	year 3	year 4	terminal value
cash flow	50	50	−2800	200	200	4200
× discount factor	1	1/1.05	$1/1.05^2$	$1/1.05^3$	$1/1.05^4$	$1/1.05^5$
= present value	50	47.6	−2539.7	172.8	164.5	3290.8

calculates that it will bring an income of $200 000 a year. However, there is a question about the building costs in year 3. Building costs typically vary randomly from year to year. Foxtrot estimates that if the building were constructed now, the costs would be $2 800 000. The question is, how much is the maximum that Foxtrot should pay for this development assuming a discount rate of 5%?

Suppose first that Foxtrot calculates this number by simply guessing that building costs stay the same as they are now. Then the income stream it receives is shown in Table 9.1, where the cash flows in a column 'year k' occur at the end of year k, which is the same as the beginning of year $k + 1$ (all sums are in $1000s and we have worked in constant 2010 dollar values – so income streams are shown as having a constant value even if they could be expected to increase with inflation).

We have chosen to assume that the lease payments to Foxtrot are paid annually in advance and the building costs incurred in year 3 have to be paid upfront at the end of year 2. The terminal value here is the value at the end of year 5 of an income stream of $200 000. Since the first payment is at the end of year 5, we can use Equation (9.6) with a discount rate of $r = 0.05$ to give a present value at the end of year 5 of $(1.05/0.05)x = 21x$. Since $x = 200$ in this case, we get a terminal value of 4200.

We can add the present values together to get a value for the project of

$$50 + 47.6 - 2539.7 + 172.8 + 164.5 + 3290.8 = 1186.$$

This figure needs to be compared with the value if there was no development of the site. In this case the value would be equivalent to an income stream of $50 000 a year, which, under these conditions, has an NPV (in $1000s) of $21 \times 50 = 1050$. The relatively high building costs have canceled out most of the benefit of the additional rental income.

Now we want to carry out a real options valuation, taking account of the uncertainty in building costs, together with the option value of the possibility of not going ahead with the development. We will see that the actual value of the investment to Foxtrot is higher than the $1 186 000 deduced from Table 9.1.

The decision point occurs at the beginning of year 3, when Foxtrot needs to decide whether or not to build. This decision will be made on the basis of information on the cost of the building work. Suppose that Foxtrot obtains a bid of x, then the choice is between continuing with $50 000 a year or paying x this year and then receiving $200 000 a year. Write A for the choice of building and B for the choice of not building. The cash flows with A are $(50, 50, -x, 200, 200, \ldots)$

and the cash flows with B are $(50, 50, 50, 50, \ldots)$. We already calculated $\Pi_B = 1050$ and the profit for A can obtained from Table 9.1 with x instead of the cost 2800. We get

$$\Pi_A = 50 + 47.6 - \frac{x}{1.05^2} + 172.8 + 164.5 + 3290.8 = 3725.7 - \frac{x}{1.05^2}.$$

Next we need to work out the breakeven point for x. This is a value of x so that we are indifferent between the two options – meaning that any higher value for the building costs would lead to Foxtrot not going ahead. We can do this by solving

$$3725.7 - \frac{x}{1.05^2} = 1050.$$

Alternatively, we can look at the position at the beginning of year 3 (valuing everything at that point). Then the choice is between a present value of $21 \times 50 = 1050$ and a present value of

$$-x + 4200/(1.05) = 4000 - x.$$

These are equal when $x = 2950$. This tells us how the decision will be made: Foxtrot will build only if the price is less than \$2 950 000.

Now we want to give a single formula for the project value. We have

$$\Pi = \Pi_B + \max\left(\Pi_A - \Pi_B, 0\right)$$

$$= 1050 + \max(2675.7 - \frac{x}{1.05^2}, 0)$$

$$= 1050 + \frac{1}{1.05^2}\max(2950 - x, 0).$$

It is worth pausing at this point to consider this formula. Once we know that the breakeven point for going ahead is at $x = 2950$, then we know that any higher value of x will just deliver the baseline NPV of 1050. On the other hand, having a lower value of x is just like getting the baseline NPV of 1050 plus a payment in year 3 of the saving \$2950 $- x$.

The remaining piece of the jigsaw puzzle is an estimate of the volatility of building costs. Suppose that Foxtrot believes that building costs are equally likely to move up or down from the current value of \$2.8 million, and the result is normally distributed with a mean of \$2.8 million and a standard deviation of \$200 000. The expected current value of the investment in \$1000s is given by

$$E(\Pi) = 1050 + \frac{1}{1.05^2}E(\max(2950 - x, 0)).$$

Now we can use our earlier formula, Equation (9.2), to show

$$E(\max(2950 - x, 0)) = 150\Phi_{\mu,\sigma}(2950) + 200^2\varphi_{\mu,\sigma}(2950)$$

$$= 176.23.$$

So the final valuation is

$$= 1050 + \frac{1}{1.05^2} 176.23 = 1209.827,$$

or \$1 209 827, which is \$23 827 more than the previous figure. So, in this example, recognizing the real option is worth an additional 2% in the value of the investment. □

9.4   The connection with financial options

In the financial world options come in two varieties: a call option gives the right (but not the obligation) to purchase an underlying financial instrument at some point in the future at a given 'strike' or 'exercise' price. If the date of the exercise of the option is fixed, it is called a European option; if the option can be exercised at any point up to the expiry date of the option, it is called an American option. A put option is similar except that it gives the right (but not the obligation) to sell the underlying financial instrument at a given price.

For example, on 21 March 2013 Apple stock was trading at \$452.73, and an (American) call option with a strike price of \$430 to be exercised by May 17, 2013 was selling for \$35.25; a call option at \$460 was priced at \$18.70; and a put option at \$450 was priced at \$22.00. We can review what those numbers mean: for an outlay of \$35.25 an investor gets the chance to buy the stock for \$430 on or before May 17. If the stock's value on that date is less than \$430 the option is valueless, but if Apple is selling for \$430.50 then the option is worth \$0.50, and if the price is higher, the option will be worth even more. A put option is the reverse: for an outlay of \$22 an investor gets the opportunity to sell the stock at \$450. If the stock price is actually above that level on May 17 then the put option is valueless, but if, for example, the Apple stock price drops to \$420 then the put option will be worth \$30.

As well as buying put or call options at a whole lot of different exercise prices and a number of different expiry dates, there is also the opportunity to sell these options (sometimes described as 'writing' an option). So there are a very large number of different positions that an investor may take, and investors will often decide to hold a portfolio of options in a stock in order to tailor the profile of possible gains or losses that they could experience.

A risk-averse investor may want to purchase a significant shareholding in Apple shares and, at the same time, buy put options at an exercise price of, say, \$410 to provide a type of insurance against a large drop in the price of the shares. If the same investor was to sell a call option at a higher exercise price, say \$480, then they would limit their possibility of a large gain, but the money they receive for the call option could be put towards the cost of buying the put option. Notice, however, that there is nothing to stop an investor from buying and selling the options without actually holding any shares. In fact this is the norm. The option contract may specify physical delivery (of, say, shares in Apple) at the point of settlement, but often this will not actually take place and instead the option will be cancelled out by buying a covering position. Some option contracts (for example, those where the underlying security is an index, like the Dow Jones) specify cash settlement so that payments are made but no stocks delivered. Usually it is best, rather than thinking of a call option on Apple shares in terms of the right to buy Apple stock, to consider the option as a financial contract involving an agreement for the seller to pay the buyer the difference between the Apple stock price and the strike price if this is in the right direction.

Figure 9.3 shows how the value of a put or call option depends on the underlying share price at the time of exercise and the exercise price.

Now we return to the example of the previous section. If we think of building costs as like a stock price, then a low value is good for Foxtrot – the lower the building cost, the more valuable the investment will be, but once the building cost goes above 2.95 million, then it no longer matters what the price is, since the building will not be worthwhile. So the investment has the characteristics of a put option with an exercise price of 2.95 million. Suppose we take as a baseline case the project when the building cost is 2.95 million, which makes Foxtrot indifferent between going ahead with the building or not. Then any lower cost is equivalent to Foxtrot receiving the difference in year 3, but any higher cost leaves things as they are. So this matches the put option where we have the right to sell at the exercise price, so that when the price falls below the exercise price we can buy at one price and sell at a higher price, making a profit of the difference.

In the Foxtrot example, we can see the purchase as having two parts: first we buy the property in its undeveloped state with a certain value and in addition we buy a put option on the building price index. This is a European option to be exercised at the beginning of year 3 with an exercise price of 2.95 million for the building. Obviously if the building price index were on a square meter basis then the exercise price would be divided by the size of the building.

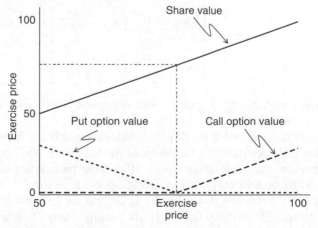

Figure 9.3   The payments for put and call options as a function of the share price.

A final question relates to the size of the option to be purchased. Here we need to look at the slope of the put option line in Figure 9.3 and compare it with the amount of increased profit if the price drops below 2.95 million. In regards to the additional value of the option *at the time of its exercise*, every dollar less is a dollar more earned for Foxtrot. Thus, the equivalence is with a put option for the full value of the building cost. Notice that in these calculations the complexities associated with taking a one-year gap in earnings and then replacing an annual sum of $50 000 with an annual sum of $200 000 are all dealt with within the single number of 2.95 million, which is the exercise price of the option.

We have given a direct approach to valuing an option when the uncertainty can be represented as a normal distribution with a known mean and variance at the date of exercise of the option. However, the most famous approach to valuing options is the Black–Scholes formula. This gives the price of a European option in terms of five quantities:

a. the underlying stock price now, S_0,

b. the volatility in the stock price, σ,

c. the time till the exercise of the option, T,

d. the exercise price, K,

e. the discount rate to be applied (risk-free rate of return) r.

When there is no dividend yield, the Black–Scholes formula gives the price of a European call option (the right to buy the stock at price K at time T) as

$$S_0 \Phi_{0,1}(d_+) - e^{-rT} K \Phi_{0,1}(d_-)$$

where

$$d_+ = \frac{1}{\sigma \sqrt{T}} \left[\log\left(\frac{S_0}{K}\right) + \left(r + \frac{\sigma^2}{2}\right) T \right],$$

$$d_- = \frac{1}{\sigma \sqrt{T}} \left[\log\left(\frac{S_0}{K}\right) + \left(r - \frac{\sigma^2}{2}\right) T \right].$$

The first three variables (S_0, σ and T) will determine the distribution of stock prices at the exercise date.

One big difference between the Black–Scholes approach and the examples we have given for real options is that the financial markets have stock prices moving in a multiplicative way – so, rather than a $100 share price being equally likely to move to $110 or $90 (say), a movement up by a factor 1.1 would imply an equally likely movement down by 1/1.1, so that an increase from $100 to $110 and a decrease from $100 to $90.91 are equally likely. This multiplicative behavior means that it is the log of the share price that is likely to exhibit a normal distribution rather than the share price itself.

However, the Black–Scholes formula is not derived by just looking at an expected value at the exercise time. Instead, the approach is more sophisticated and involves constructing a synthetic risk-free instrument that must then match the risk-free rate of return by a 'no arbitrage' argument. This allows the calculation to take place without consideration of a drift over time: the volatility alone is enough to work things out. The idea is that if the option relates to an asset that is traded, then the price for the asset now is not independent of the expected future behavior of the asset price: if the price now gets out of line with what will happen in the future, then someone will seize the opportunity to buy or sell and make money from the trade. In other words, we cannot separate out the drift in asset value and the risk-free rate of return and have these two things independently chosen.

In more complex situations there is a range of different possible approaches to valuing a real option: we can use the Black–Scholes formula; we can solve a set of stochastic differential equations with appropriate boundary conditions; we can carry out an evaluation using a type of binomial (or trinomial) lattice; or we can use a Monte Carlo simulation approach. The different possibilities are illustrated in Figure 9.4. A stochastic differential equations approach arises from assuming some form of Brownian motion or related stochastic process for the uncertain prices (or returns). The Black–Scholes formula is a special case available for traded assets under certain assumptions. But in any case the stochastic differential equations can be regarded as the limit of a discrete time stochastic process as the time increments get smaller and smaller. Lattice calculations work directly with these discrete time stochastic processes and allow extra flexibility in the modeling by using a discrete state as well. The idea is to calculate the probabilities of being at different states arranged on a two-dimensional lattice with time on one dimension and price on the other. Finally, the Monte Carlo approach replaces a calculation of probabilities in the lattice approach by a simulation. The Monte Carlo approach is the simplest and may be the most useful for valuing real options in practice: we will discuss this method in more detail in the next section.

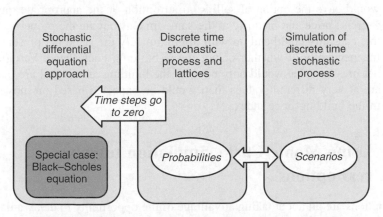

Figure 9.4   Different approaches to valuing real options.

Finally, it is worth commenting on some of the advantages in making a connection between real options and financial options:

- The business world now contains many people with a familiarity with options: there is no difficulty in understanding an option approach and it can be helpful in appreciating the financial structure of a potential investment (as well as explaining why a greater variation can lead to a more profitable outcome).

- With a map from the investment into a set of options it becomes possible to use well-established techniques and software for valuing options. We have already mentioned the Black–Scholes formula that applies for European options when the price movements can be modeled as a multiplicative Brownian. But there are other techniques in common use by those who need to value different types of options. American options are usually valued using some sort of simulation process. In this example, the option was European but different types of real options scenarios can give rise to either European or American options. We will show how the Monte Carlo simulation technique can be used in a real options framework in the next section.

- If the underlying uncertainty relates to a traded instrument (like a commodity price), then the option value may not be something which relies on calculation – perhaps it can be found simply by looking at the prices in the marketplace.

- In the event that there is a market for options that are closely related to the options occurring in the investment, there is also the possibility of buying or selling options to entirely cancel the uncertainty in relation to the exercise of the implicit option (thus giving an exact hedge). There are a number of building price indices available in different parts of the world. To the best of my knowledge there are no traded options in any of these indices. However, for the sake of an illustration suppose there were. Then Foxtrot would have the option of selling a put option of the appropriate quantity, exercise price and maturity at the same time as buying the property. Doing this carefully could end up with approximately the same net present value independently of whether or not the re-building goes ahead. Variations in net present value would only occur if the building quotes that are obtained are at very different values than would be expected based on movements in the building price index.

9.5 Using Monte Carlo simulation to value real options

Real options are all about taking advantage of the opportunity to delay a decision until more information is available, and often the decision is based on something that varies in a stochastic way. It is natural to use a Monte Carlo approach in this context and this will give us the flexibility to represent different aspects of what is going on. An advantage of using this method is that we can easily see the distribution of possible outcomes, so that we can obtain risk measures like value at risk from the same calculations. To illustrate this we will work through a simple example.

Example 9.5   Quickstep Oil

Quickstep Oil is considering the development of a shale oil project with high costs, where the decision whether or not to go ahead is related to the price of oil in the future. The first decision is whether Quickstep should buy rights to the resource. Quickstep will then need to build some substantial additional processing capacity and infrastructure, but this could be delayed until prices are higher than their current value. It will take a year to build the plant (with build costs spread over this period). Once in operation there is a large (fixed) cost per year of operation together with substantial costs that depend on the volume of oil produced.

We suppose that the price of crude oil is modeled as a stochastic process involving an underlying price which follows a geometric Brownian motion with drift upwards, but that in addition, this is subject to fluctuations due to short-term global supply and demand changes that we model through a mean reverting process. Writing w_t for the log of the oil price, then $w_t = y_t + \theta_t$ where y_t represents the underlying level and θ_t is the short-term mean reverting component (θ_t has mean zero). We can define the two components through recursions:

$$y_t = y_{t-1} + \alpha + \varepsilon_t,$$
$$\theta_t = \beta\theta_{t-1} + \delta_t,$$

with constants α (giving the drift) and $\beta < 1$ (determining how quickly θ_t is pulled back to zero). In these equations, ε_t and δ_t are random noise.

Whether or not this is a good model for oil prices, it does illustrate the way that the Monte Carlo methodology is not restricted in the type of price processes that can be considered. In fact, it is not easy to decide on the right model for oil prices over a long horizon: prices seemed to be very stable prior to the early 2000s, then they began to steadily climb over the period since around 2002, reaching more than $130 a barrel in the middle of 2008, before dropping to below $50 at the end of that year, since when they have steadily climbed again.

We assume that the current oil price is $100 per barrel. We will work with a six-month time unit. We have set $\alpha = 0.04$, $\beta = 0.75$ and ε_t is normal with mean zero and standard deviation 0.07, and δ_t is normal with mean zero and standard deviation 0.09. Figure 9.5 shows a sample of 10 price realizations over a 20-year period given these parameters.

The cost of securing the site properly and decommissioning some existing plant (works that need to be carried out immediately on purchase) is $25 million. The cost of building the plant (which will take a year) is $400 million. Once in

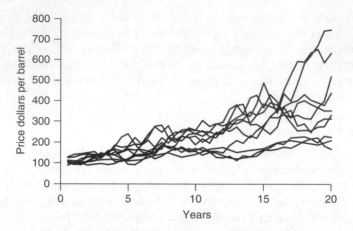

Figure 9.5   A sample of 10 price series for the Quickstep example.

operation the plant can produce 2 million barrels per year. The fixed operating cost per year is $80 million and the (variable) production costs are $70 per barrel. Quickstep uses a discount rate for capital projects of 7% (when dollar amounts are all converted to constant 2013 dollar values). The reserves are sufficient for the plant to run for 20 years and we will ignore what happens at the end of this period (so we do not allow for the residual value of production equipment, nor for the cost of cleaning up the site).

Given this arrangement, the spreadsheet BRMch9-Quickstep.xlsx gives the Monte Carlo simulation. This is a very unwieldy spreadsheet since it includes 1000 simulations, each one over a period of 40 years. Each simulation involves four rows: the two time series y_t and θ_t used to generate the prices, the prices themselves and the cash flow line which involves some messy formulae that have the effect of ensuring that the production plant starts to be built when the price reaches some specified price threshold and then switches off 20 years later.

Because of the delay in building the plant, the earliest that the plant can begin operation is 18 months after purchase. The average price at that point will be above the figure of $110 per barrel that makes the whole thing economic (during 18 months the log price will increase by $3a = 0.12$, implying an increase in the oil price by a factor of $\exp(0.12) = 1.1275$). A starting point is to use the Monte Carlo simulation to calculate the expected net present value of the project given that there is no delay in building the plant – this can be achieved by setting the price threshold to something lower than 100. The result will depend on the specific random numbers that happen to come up in the simulation, and even with 1000 scenarios there is quite a bit of variation (which we can see by pressing 'F9' to get another 'draw' of the random numbers). The overall average net present value is about $3190 million, but a set of 1000 individual scenarios can have an average value anywhere in a range from around $3110 million to around $3250 million.

The next step is to try to take account of the flexibility that is available to Quickstep: if the price of oil happens to drop at the start, then it makes

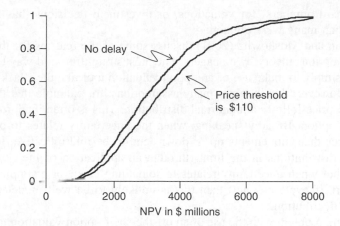

Figure 9.6   Improved NPV risk profile for Quickstep given option to delay.

sense to wait before committing to spend $400 million on building the plant. A conservative approach would be to wait till the price gets above the breakeven level of $110 before starting. This will not be a guarantee against a later price drop, but because there is usually an upward drift in the price process, the chance of losing money is certainly reduced. When using this approach, the Monte Carlo simulation gives an overall average net present value of about $3550 million, but a set of 1000 individual scenarios can have an average value ranging from around $3430 million to $3620 million. The average improvement in NPV arising from the flexibility to delay starting construction is about $365 million in this example. Figure 9.6 shows a comparison between the cumulative distribution of net present value, obtained from two sets of simulations – one with no delay and one with a price threshold of 110. Because these are based on data from multiple runs of the simulation, the CDFs are not smooth.

The threshold of $110 dollars is a little arbitrary here and one could carry out simulations with different thresholds to see what works best. Notice that, in any case, we are not going to be able to calculate the kind of clear-cut breakeven point which we have seen in the other examples of this chapter. Also, notice that this example differs from previous ones because the exercise of the option (to start) can be made at different time points, depending on when the price gets above the threshold value.   □

9.6   Some potential problems with the use of real options

When real options were first discussed there was considerable excitement about their application as a strategic tool in evaluating investments and other management decisions. It would be fair to say that the use of real options theory to produce 'hard numbers' for valuations, or investment decisions, has been less common than many were predicting.

Bowman and Moskowitz (2001) discuss some of the reasons for the limited use of real options theory by managers. The most straightforward way to use real options is simply to make use of an option valuation tool like the Black–Scholes formula. However, this involves many assumptions, for example that the underlying stock price follows a lognormal distribution. This is often inappropriate for a strategic option. Broadly speaking, when the uncertainty relates to prices and dollar values then movements up or down tend to be multiplicative, leading to lognormal distributions in the limit (through an application of the central limit theorem), but when uncertainty relates to something else (for example, sales of a new short-lifecycle product) then this is unlikely to be well modeled using a lognormal distribution.

A second problem with the use of an off-the-shelf option valuation tool is that there is an assumption of tradeable assets where the possibility of arbitrage has a big effect on how prices behave. Thus, for example, with an exchange-traded option, stock prices are easily observable and an option holder can readily buy or sell shares at this price to realize the profit from (or to cut the loss on) an option position. In contrast, for real options, the analogous stock price is often very hard to ascertain and it may also be hard to trade at the price implied.

There are also problems with the time to expiration. For strategic real options, there is often no set time to expiration. For example, a research project could be extended for a longer period of time, and an investment in a new product distribution system indefinitely retains the option to add additional products.

So there are a number of problems with applying a methodology lifted straight from financial option analysis, and it may be better to think of building a more advanced and customized option valuation model. But this brings with it some dangers. Creating such a model is a technical challenge that will take it out of the hands of the managers who will rely on its results. Moreover, the complexity of the options approach can also make it difficult to find errors in the analysis, or to spot overly ambitious assumptions used by optimistic project champions.

This list of difficulties explains why we have given quite a lot of attention to relatively unsophisticated approaches like Monte Carlo simulation. It is important also to recognize that much of the value of an options analysis will be at the stage of project design. By bearing in mind the timing of decisions, particularly those of an options nature (e.g. the decision to go ahead with, or to defer some expenditure) we may well be able to create additional value. In understanding these situations, a very accurate numerical model may not be necessary, since a small model inaccuracy in these situations is unlikely to change the decision we take. As Bowman and Moskowitz (2001) point out: 'Whereas small deviations are worth fortunes in financial markets, they are fairly inconsequential in product markets.'

Notes

Our approach to real options has focused on the fundamentals of how to use flexibility and the calculations that are needed in specific examples to evaluate projects where there is flexibility. Because we think there are problems with its application in a real options environment, we have given rather little attention to the Black–Scholes equation, which has sometimes formed the basis for these valuations (see, for example, Luehrman, 1998). The approach we take is broadly in line with the recommendations in Copeland and Tufano (2004) and also with the approach proposed in the book on flexibility in design by de Neufville and Scholtes (2011).

We have only given a very brief introduction to this area and there is much more to be said. The paper by Smith and McCardle (1999) gives a careful and helpful treatment of some of the issues that need to be dealt with in practice. There are many books dealing with real options, ranging from the original discussion of Dixit and Pindyck (1994) to more recent books by Mun (2005) and Guthrie (2009).

References

Bowman, E. and Moskowitz, G. (2001) Real options analysis and strategic decision making. *Organization Science*, **12**, 772–777.

Copeland, T. and Tufano, P. (2004) A real-world way to manage real options. *Harvard Business Review*, **82**, 90–99.

De Neufville, R. and Scholtes, S. (2011) *Flexibility in Engineering Design*. MIT Press.

Dixit, A. and Pindyck, R. (1994) *Investment under Uncertainty*. Princeton University Press.

Guthrie, G. (2009) *Real Options in Theory and Practice*. Oxford University Press.

Luehrman, T. (1998) Investment opportunities as real options: Getting started on the numbers, *Harvard Business Review*, **76**, 3–15.

Mun, J. (2005) *Real Options Analysis: Tools and techniques for valuing strategic investment decisions*, 2nd edition. John Wiley & Sons.

Smith, J. and McCardle, K. (1999) Options in the real world: Lessons learned in evaluating oil and gas investments. *Operations Research*, **47**, 1–15.

Exercises

9.1 Toothpaste dispenser

A company is launching a new type of toothpaste dispenser in the UK. It will test market the product with only local advertising in South Wales, where it has test marketed other products before. The dispenser sells for a wholesale price of £2.50 and initially the cost of manufacturing is £2.40, giving a profit per unit of 10 pence for the manufacturer. On average, the sales in the UK turn out to be 100 times as large as the sales in the test market. Because it is an unusual product, sales are difficult to predict. Test market sales are expected to be between 500 and 2500 per month, and are thought to be equally likely to take any value in this range. Local advertising costs for the month will be £2000. After one month a decision will be taken as to whether to ramp up to full-scale production, or whether to drop the product. To ramp up production will involve expenditure of £20 000 in installing a new production line, but this will bring the per unit production costs down to £1.50, giving a margin of £1. National advertising costs will be, on average, £5000 a month. The company has already spent £20 000 on bringing the product to a point where it can be launched. Use a real options analysis to calculate the expected profit from this product over a two-year period (ignoring any discounting of future profits).

9.2 Pop concerts

Two investments are available in pop concerts. One of them involves paying an upfront sum of $10 000 and then receiving 10% of the net ticket sales, which are uncertain but are expected to be $120 000 with a standard deviation of $20 000. The other venture is more speculative and will fund a series of three shows. The expected ticket sales are $230 000 with a standard deviation of $80 000. The fixed cost to put on the shows is $190 000. 19 investors have all been asked to put in $10 000 to cover these fixed costs. There is a chance that the shows make less than $171 000, in which case it has been agreed that each investor will receive back $9000 and any final shortfall will be met by the producers of the show. If the shows achieve net ticket sales of more than $171 000, then each of the 19 investors will receive $9000 dollars back plus one twentieth of the profit over and above $171 000 (so at the point where the ticket sales are $191 000, the investors will receive all of their $10 000 stake back). The remaining one twentieth share will be paid to the producers. Calculate the expected value of both investments. Which investment is preferred?

9.3 Gold extraction

Charleston Mining Company is considering buying a licence allowing the extraction of gold from mine tailings for a period of three years. Extraction is an expensive process and only worthwhile if the gold price is high enough. The operation costs an average of $1000 to extract 1 oz of gold. It is expected that the operation will produce a total of 500 oz per month. The price of gold is volatile and is currently $1200 per oz. If the price drops below $1000 per oz the operation will simply be stopped until the price rises above that threshold.

(a) If the price of gold in January next year is estimated to have a normal distribution with mean $1200 and standard deviation $200, what is the expected revenue from the operation for January?

(b) What options purchase would match the cash flows for the operation in January next year? (If this was repeated for each month of the licence it would give a route to valuing the licence without needing to estimate volatility in gold prices.)

9.4   SambaPharm

A company has the option to buy a small firm called SambaPharm whose main asset is a patent on a pharmaceutical product currently undergoing clinical testing. Testing will take a further three years at a cost of $60 000 a year. The long-term profitability of the drug will depend on the results of these clinical trials. The best existing treatments are effective for 40% of patients. The final sales of the product are related to the number of patients for whom it is effective. It will just break even if it is equally as effective as the best current drug. But for every additional 10% of patients for which it is effective, the net annual income (after production costs) will increase by $50 000. The best guess is that the effectiveness is equally likely to be at any value between 0 and 80%. Hence, with a probability of 0.5 the drug will be found to be less effective than the best existing treatment and will not be put into production. Once in production the drug will have an estimated five-year life before the next variety of this pharmaceutical family appears and profits are reduced to zero (or close to zero). But for a period of five years, starting in year 4 immediately after clinical testing, the profitability of the drug is expected to be stable. Taking account of the implied real option, what is the value of the company holding this patent if future profits are discounted at a rate of 8% per year?

9.5   Trade shows

To develop a new product, Tango Electronics must spend $12 million in 2014 and $15 million in 2015. If the firm is the first on the market with this product it will earn $60 million in 2016. If the firm is not first in the market it will do no more than cover its costs. The firm believes it has a 50% chance of being first on the market. From 2017 onwards there will be a number of other firms that enter this market and though Tango may well continue with production, it expects to do no more than cover its costs. The firm's cost of capital is 12% and you should use this figure to discount future earnings.

(a) Should the firm begin developing the product?

(b) Now suppose that there will be a trade show on 1 January 2015, when all the potential market entrants will show their products. After the trade show Tango will make a new estimate of its probability of being first on the market. Assume that at this point it can correctly state its probability of being first in the market, and further assume that this probability estimate is equally likely to take any value between 0 and 1. Should the firm begin developing the product?

9.6   Option prices imply distribution parameters

Suppose that a European call option on Apple stock with a strike price of \$430 sells for \$35.25; and a call option with a strike price of \$460 sells for \$18.70. Assume that at the exercise date the price of Apple stock is a random variable with a normal distribution (rather than a lognormal distribution). Use a spreadsheet and the formula for $E(\max(X - a, 0))$ to estimate the mean and standard deviation of the normal distribution. Use this normal distribution to find the price of a put option with a strike price of \$450.

10

Credit risk

Credit ratings track a firm in hard times

Liz Claiborne was an American fashion designer who was born in Brussels in 1929. In 1976 she was one of the cofounders of Liz Claiborne Inc., which made it into the Fortune 500 in 1986. She was the first woman to be the CEO of a Fortune 500 company, and had a big influence on the way that fashion is sold, insisting that the clothes in her collection be grouped together in the store, rather than Liz Claiborne skirts being put together with other skirts and in a separate place to shirts. Claiborne retired from active management in 1989 and died in 2007.

Liz Claiborne Inc. was generating 2 billion dollars in annual sales in the early 1990s and expanded through acquisitions through the 1990s and early 2000s, buying Lucky Brand Jeans, Mexx and Juicy Couture. But then they hit problems. Perhaps there were too many acquisitions, but there were also difficulties managing relationships with the big retailers Macey's and J.C. Penney. William McComb took over as CEO in 2006, and in early 2007 the stock price peaked at $46.64, but there were clear problems to deal with: an ageing set of customers and a headline brand that was in decline. Things then got much worse, with losses starting in the last quarter of 2007 and continuing right through to 2012. The stock price dropped dramatically, going lower than $2 in 2009 at a point when the company was laying off workers and closing distribution centers. In 2011 the company made a substantial loss (of $172 million) but the share price continued its slow recovery and in 2013 (as this is being written) the share price is around $19.

Meanwhile the company has, for a long time, had a credit rating from Standard & Poor's. Looking at the history since 2000 we can see that Liz Claiborne Inc. was rated at investment grade (BBB in S&P terminology) until 3 June 2008, when it was cut to BB, which is below investment level (commonly called *junk*). At the time, S&P cited significantly higher debt levels and

---

Business Risk Management: Models and Analysis, First Edition. Edward J. Anderson.
© 2014 John Wiley & Sons, Ltd. Published 2014 by John Wiley & Sons, Ltd.
Companion website: www.wiley.com/go/business_risk_management

a challenging retail environment for this regrading. Then, on 17 August 2009 Liz Claiborne's rating was cut again to B, and in March 2010 it was cut to CC. This rating means that S&P regards the company as 'highly vulnerable' and is an enormous red flag. The final indignity occurred on 11 April 2011 when the company was judged to have made a selective default. In some senses when the S&P team chose to regard a particular tender offer refinancing as 'distressed' and equivalent to a selective default, this was a technical decision, and a long way from a standard default on debt. In fact, the very next day the company was re-rated back to a B. Since then the company (which has now rebranded as Fifth & Pacific Companies) is in much better shape though challenges remain; some brands have been sold to pay back debt and the company has concentrated on its three main brands of Kate Spade, Juicy Couture and Lucky Brand.

For a company like Liz Claiborne that needs to borrow money, credit ratings are essential. The aim of the credit rating agencies is to give an indication of the chances that the company will be unable to meet repayments that are due. But to what extent does a credit rating give information that is different to the stock market valuation? And how reliable are the ratings?

10.1   Introduction to credit risk

In this chapter we will give an introduction to credit risk, including a discussion of consumer credit (a topic likely to be relevant to managers working outside the financial sector). Of course we may well have an interest in credit risk as it affects us as individuals, since our own circumstances and credit history will influence our ability to obtain credit and the interest rates that we pay.

As we discussed in Chapter 1, credit risk refers to the possibility that a legally enforceable contract may become worthless (or at least substantially reduced in value) because the counterparty defaults and goes out of business. In the case of an individual, an outstanding debt may be uncollectable even without the individual concerned becoming bankrupt, and this too would be classified as credit risk.

Two aspects of credit risk will drive our discussion in this chapter. First, credit risk has a yes–no characteristic, with either a default or not. Detailed analysis of the tails of distributions and a range of possible outcomes is no longer very relevant. Second, credit risk is about what happens to another party and so we have less information than for our own businesses. We need to work harder to get the maximum value out of whatever information is available to us. We will want to look at what happens over time (is a company moving to a less stable situation?) and we will want to make use of any indirect pointers that are available (is the fact that this individual has taken out four new credit cards over the last six months a bad sign?).

The Basel II framework is designed for banks. In this context, credit risk will relate to a large number of different kinds of loans – both loans to businesses and loans to individuals. A large component of the lending to individuals is through mortgages. In this case the value of the property itself provides some security for the bank, but there can still be a significant credit risk if property prices fall sufficiently far that the outstanding debt becomes larger than the house value (as happened in the US with sub-prime mortgages).

From a corporate perspective, credit risk is related to credit ratings that are given by one of three major credit rating agencies: Standard & Poor's, Moody's, and Fitch. When entering into a contract with, and especially when lending money to, a firm that has a lower rating and hence a higher risk of default, it will be appropriate to pause and think carefully. At the very least it will be wise to ask for a higher rate of interest on loans to firms where there is a higher perceived risk. By limiting the contractual arrangements with firms that have low ratings, managers can limit the credit risk they take on.

Each credit rating agency has its own terminology and codes, but the codes for Standard & Poor's are as follows:

Investment grade

AAA: The best quality borrowers, reliable and stable (many of them governments).

AA: Very strong capacity to meet financial commitments, a slightly higher risk than AAA.

A: Strong capacity to meet financial commitments, but could be susceptible to bad economic conditions.

BBB: Medium-class borrowers, with an adequate capacity to meet commitments. Satisfactory at the moment.

Non-Investment grade

BB: Not vulnerable in the near term but facing major ongoing uncertainties.

B: Currently has the capacity to meet commitments, but vulnerable to adverse conditions.

CCC: Currently vulnerable and dependent on favorable business and economic conditions.

CC: Currently highly vulnerable, very speculative bonds.

C: Virtually bankrupt, but payments of financial commitments are continued.

Moody's system includes codes Aaa, Aa, A, Baa etc., but is broadly similar. The credit ratings agencies failed spectacularly at the start of the global financial crisis, where they continued to rate certain CDOs (collateralized debt obligations) at the highest level shortly before they were announced to be 'toxic'. However, given that none of the agencies did conspicuously better than the others, and given the important role that the agencies play in the financial system, they have continued much as before.

10.2 Using credit scores for credit risk

Credit scores are drawn up with the express intention of giving guidance on the risk of a company defaulting on debt. The precise methodology that is used will vary: when assessing the risk associated with government bonds the agency will take a different approach than when making a corporate rating. For corporate risk the process is roughly as follows. A company that wishes to borrow money by issuing bonds will ask one of the ratings agencies to give a rating, and it will pay for this service. In fact, the two largest agencies, Standard & Poor's and Moody's will rate all large corporate bonds issued in the U.S., whether asked to or not. The issuer may, in any case, wish to pay the fee and embark on a more serious engagement with the rating agency to avoid a situation where the rating takes place without complete knowledge of the facts. In fact, institutional investors will prefer to have the bond issue rated by more than one agency.

The rating agency will consider two things: business risk and financial risk. Assessing business risk involves considering trading conditions, the outlook for the industry and the quality of the management team. Financial risk is assessed in a more quantitative way using accounting data looking at profitability, debt levels, and the financial strength and flexibility of the firm. Following a visit to the company by the analysts involved, a committee will meet to consider the analysts' recommendations and vote on the final grade assigned. The company is informed of the recommendation and the commentary on it that will be published and then given a final chance to provide additional information (or to ask for company confidential information to be removed from the commentary) before the rating is confirmed and published. If a company is issuing bonds with different characteristics then these are rated with reference to the overall company rating; though senior debt, with priority rights for payment, is likely to be given a notch higher rating than subsidiary debt. In the case of highly structured products like CDOs, where debt from various sources has been combined and sliced up, then the process is more complex, since the overall level of risk involves looking at the constituent parts of the structured product.

Once a rating has been given, the process moves to a surveillance mode where a separate team within the ratings agency is tasked with keeping an eye on the developments within the company in order that there can be a timely change in the rating (either up or down) if circumstances warrant it. The actual ratings are issued with plus or minus signs attached for all the grades between AA and CCC. Moreover, Standard & Poor's give what they describe as an outlook statement indicating the direction of change if they anticipate a change as likely in the next one to two years (with a status of 'developing' if a change is likely but it could be in either direction). If the company enters a phase where the agency believes there is a significant chance of change in a shorter time frame of three months or so, then the company is placed on 'credit watch'. We can see how this played out in the Liz Claiborne example in the timeline shown in Figure 10.1. Notice that there were announcements in May 2007, September 2007 and May

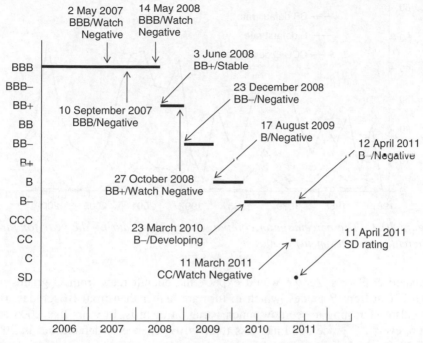

Figure 10.1   Timeline of announcements by Standard & Poor's of credit ratings for Liz Claiborne Inc.

2008 that reflected the possibility of a drop in rating as problems piled up for the company, but the first downgrade did not occur till June 2008.

Credit rating agencies perform a vital task in the market for capital. They have expertise in the task of evaluating risk and are independent of individual companies. The additional information that they provide to investors and market participants will ultimately mean that companies can raise money at lower costs, since the ratings provide an efficient way for investors to reduce their uncertainty. At the same time, ratings agencies have an important function for regulators who want to limit the risk that financial companies can take and can use ratings within the rules they set up.

The probability of a default obviously varies with overall macro-economic factors: the global credit crunch of 2008–2009 led to many more defaults occurring. The ratings agencies do not set out to give ratings aligned with particular probabilities of default, since this would require wholesale re-ratings as the economic climate changes from year to year. Instead, ratings agencies are concerned with relative probabilities: an investment grade company is less likely to default than one with a grade of BB, which in turn is less likely to default than a company with grade B and so on. Figure 10.2 shows how the percentage of defaults per year varies over time for the three non-investment grades given by

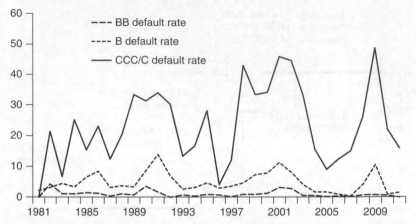

Figure 10.2   The percentage of companies defaulting during the year for three different (non-investment) grades.

Standard & Poor's. As we would expect, the default rates from C grades are higher than from B grades, which in turn are higher than from BB grades. But it is also interesting to see how much variation there is. In years like 2005 and 1996, even a C grade bond had less than a 10% chance of default, but in 2001 and 2009, entering the year with a C grade would mean more than a 45% chance of default before the end of the year.

10.2.1   A Markov chain analysis of defaults

In order to understand the risks associated with a particular grade we must specify the time horizon involved. There is very little chance that an AA-rated company will default this year, but in five years' time that same company may have slipped to a BB rating and with that the chance of default will have increased substantially. Standard & Poor's publish a report that includes the probabilities of making the various transitions that are possible. Table 10.1 shows the average transition

Table 10.1   Global corporate average transition rates for the period 1981–2011 (%).

	AAA	AA	A	BBB	BB	B	CCC/C	D	NR
AAA	87.2	8.7	0.5	0.1	0.1	0	0.1	0	3.4
AA	0.6	86.3	8.3	0.5	0.1	0.1	0	0	4.1
A	0	1.9	87.3	5.4	0.4	0.2	0	0.1	4.7
BBB	0	0	3.6	84.9	3.9	0.6	0.2	0.2	6.4
BB	0	0	0.2	5.2	75.9	7.2	0.8	0.9	9.8
B	0	0	0.1	0.2	5.6	73.4	4.4	4.5	11.7
CCC/C	0	0	0.2	0.3	0.8	13.7	43.9	26.8	14.4

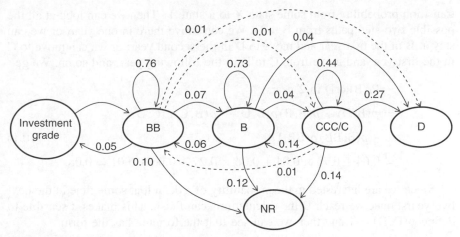

Figure 10.3   Annual probabilities of transitions between non-investment grades. Transitions with probability less than 0.01 are not shown, and transitions with probability less than 0.02 are shown dashed.

rates over a 30-year period. Thus, for example, on average 8.7% of firms rated AAA drop to AA during the course of a year. The table includes a column NR for not rated. Sometimes a firm will drop out of the ratings shortly before defaulting, but equally it may simply be a decision by the company that a rating is no longer necessary. The rating agency will track companies that drop out mid year and then default during that year, so that the figures on annual defaults are correct.

It is confusing to draw a diagram showing all the transitions that are possible – in Figure 10.3 we have shown just the annual transitions between the 'non-investment' grades. We have converted percentages to probabilities and the transitions with a probability of less than 0.01 have been omitted and those with a probability of less than 0.02 are shown dashed.

Given information on the chances of having a default or making a transition in one year, what does this imply about the chance of default over a three- or five-year time horizon? One simple way to approach this problem is to analyze a Markov chain model of the process. The Markov assumption is that changes in rating only depend on the current score. A company that has been AA for 15 years is no more or less likely to move to A than a company that only achieved AA status last year. And a company that, in successive years, has dropped from A to BBB to BB is no more or less likely to move down again to B than a company that has moved in the opposite direction, having been rated B last year but just moved up to BB.

One of the basic facts of a Markov chain is that the nth power of the transition matrix gives the probability of making a transition in n steps. To show why this happens we will analyze the probability that a company starting at state B will have a default over a two-year period using the information in Figure 10.3. (To simplify the calculations we use these approximate figures rather than the more accurate numbers given in Table 10.1). We will write $p(X, Y)$ for the transition probability from some state X to a state Y. Then we can look at all the possible two-step paths from B to D. We can move there in one year; or we can stay at B in the first year and move to D in the second year; or we can move to C in the first year and move from C to D in the following year, and so on. We get

$$\Pr(\text{B to D in 2 years})$$

$$= p(\text{B},\text{D}) + p(\text{B},\text{B})p(\text{B},\text{D}) + p(\text{B},\text{C})p(\text{C},\text{D})$$

$$+ p(\text{B},\text{BB})p(\text{BB},\text{D}) \qquad (10.1)$$

$$= 0.04 + 0.73 \times 0.04 + 0.04 \times 0.27 + 0.06 \times 0.01 = 0.08.$$

Since we are interested in the probability of a default at some time in the next two years, once we reach D the calculations can finish. This makes it sensible to define $p(\text{D},\text{D}) = 1$ and then we can see that the formula has the form

$$\Pr(x \text{ to } y \text{ in 2 steps}) = \sum_i p(x,i)p(i,y), \qquad (10.2)$$

where we sum over i being any of the states we can get to from x. In Equation (10.1) we have left out the terms like $p(\text{B},\text{NR})$ where the probability on the second step is zero. Suppose we take P to be the matrix shown in Table 10.1 augmented by two rows for D and NR, where in each case the probability of staying at the same state is 100%. Thus, P is a square matrix and it will be convenient to divide each element by 100 to express the transition probabilities directly rather than as percentages. Then we can write the element in the ith row and jth column as p_{ij}: it is the probability of making a jump from state i to state j. The rules of matrix multiplication tell us that the element in the ith row and jth column of $P \times P = P^2$ is given by $\sum_k p_{ik} p_{kj}$ which exactly matches Equation (10.2). Hence, P^2 simply gives all the two-step transition probabilities, and P^3 gives the three-step transition probabilities and so on.

Since Standard & Poor's also report on the three-year transitions, we can compare our default rate predictions using the Markov chain model with actual behavior. The full Markov chain comparisons are given in the spreadsheet BRMch10-Markov.xlsx with the array function MMULT used to carry out the matrix product. Figure 10.4 compares the actual and predicted three- and five-year default rates. We can see from this that the Markov model does not do a great job of predicting these rates. For example, starting at a BB grade, the Markov assumption predicts three- and five-year rates of 3.5% and 4.9% respectively, while the actual figures reported by Standard & Poor's are much higher, at 5.0% and 9.2% respectively.

The Markov chain model is a good way to think about credit rating movements, but is a big simplification of the real behavior. We can identify a number of possible reasons for the poor three- and five-year predictions from the Markov model.

1. *Grouping together of states.* Actual ratings are given with plus and minus signs, giving more states in total. When a Markov chain has states grouped

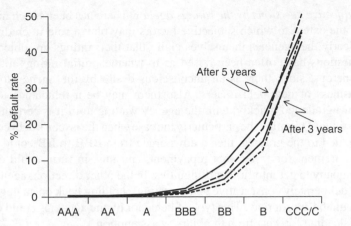

Figure 10.4   Actual (solid line) and predicted (dashed line) default rates from different starting grades over three and five years.

together, it no longer behaves as a Markov chain. For example, if companies typically move slowly through ratings from, say, BB+ to BB to BB− then knowing that the company has been in the grouping BB for some time may increase the chance of it being at the lower level BB−, thus increasing the chance of a jump to level B and breaking the Markov assumption.

2. *Different types of companies behave differently.* Suppose that different types of companies exist and they follow different Markov chains. This doesn't necessarily mean that one type of company is more risky than another (at the same rating); perhaps it is characteristic of companies in the financial sector to have a higher risk of immediate default and that is balanced by companies in the non-financial sector having a higher risk of moving to the next notch down. In any case, the existence of different types of company can mess up the Markov assumption. Exercise 10.2 gives an example of this.

3. *Bad years cluster together.* We have already observed that default rates (and more generally downgrades) vary substantially from year to year. This means that a more appropriate model might be one in which the probability of any transition depends on the year in question. We should then replace equations like Equation (10.2) with

$$\Pr(x \text{ at time } n \text{ to } y \text{ at time } n + 2) = \sum_i p_n(x, i) p_{n+1}(i, y)$$

where the subscript represents the time period. If the overall probability of a downgrade in one year is positively correlated with the probability of a downgrade in the next, then this can increase the overall probability of two successive downgrades when compared with the case when probabilities in one year are independent of those in the next year.

4. *Subjective decisions by the ratings agency.* There has been much discussion of the extent to which subjective factors may play a role in credit ratings. Clearly the agencies themselves claim that their ratings are objective, but questions have often been raised as to whether initial ratings may be too generous, since there is an unconscious desire by the agency to win the business of new debt issuers. Also there may be a reluctance to issue a downgrade too quickly, with the agency waiting until it is certain that it is justified. This might be particularly the case when the issuer's debt conditions are tied to the grading; then a downgrade from BBB to BB could trigger a requirement for faster debt repayment, and this, in turn, could cause the company to get into further difficulties. In the other direction, agencies may be deliberately conservative in their ratings, holding back on an upgrade that would be appropriate. This type of behavior by the agencies could explain a misleading forecast from the Markov assumption.

We can at least partially address the first two issues by using Standard & Poor's data broken down into more exact ratings and distinguishing between non-financial, insurance and financial companies. The second spreadsheet in the workbook BRMch10-Markov.xlsx compares the results for three-year default rates arising from a Markov model restricted to non-financial firms and the actually observed rates. The results are shown in Figure 10.5 and it seems that the Markov predictions are a little more accurate, but still involve substantial errors; for example, the Markov model predicts a three-year default rate starting at BB as 2.8% whereas the actual observed figure is 4.7%. Notice that we have not included in this analysis any recognition of the outlook statement that

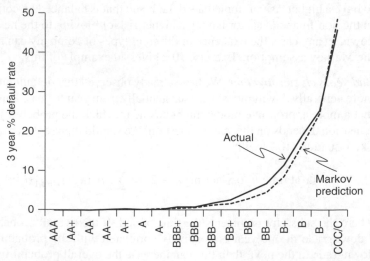

Figure 10.5   Three-year default predictions for non-financial companies compared with actual average figures.

accompanies a rating. One explanation of the relatively higher rates of default occurring (than are predicted by a Markov chain analysis) is that companies where ratings have just been cut are more likely to be given a negative outlook (in other words, they have a greater risk of default than other companies at the same rating grade).

10.3   Consumer credit

In the remainder of this chapter we will focus on consumer credit. Here the rating is made for an individual rather than a company. The first large-scale applications of automated ways of assessing credit were driven by the credit card explosion that happened in the 1960s and 70s. This led to lenders looking at credit histories and credit bureaus were set up with the aim of pooling data from different lenders so that a consumer who failed to pay off a store card, for example, would find that information was made available to other potential lenders (perhaps a car finance company).

Different credit bureaus (sometimes called credit reference agencies) operate in different countries. Credit bureaus collect together credit information from various sources and can provide a credit report on an individual when requested. Around the world, millions of these are issued every day (in the US alone more than two million a day) and the process is simply an automated interrogation of a database. The data held by credit bureaus vary from country to country and are affected by data protection laws. In the US, a huge amount of information is kept, while in some European countries it is limited to mainly publicly available information (for example, court records of bankruptcy). Usually a debt remains on the record only until it is repaid, or until a specified time limit has passed (perhaps 10 years). In many countries, consumers have the right to receive a free copy of their own credit record.

One widely used technique is to look at applicants for loans and try to judge on the basis of their characteristics how likely they are to default. The measure of default that is traditionally used is the likelihood that an applicant will go 90 days overdue on their payments within the next 12 months. However, it does not matter so much what this measure is: in the end a ranking of individuals occurs and the least attractive (that is, the most likely to default) are refused credit or funneled into a different type of credit arrangement.

The credit scoring method is straightforward. A lender is interested in whether or not to extend credit to an individual and in order to make this decision a number of variables are checked (such as the number of credit cards held, and whether the individual rents or owns their home). These variables are used to produce a 'score' and this is used to predict the likelihood that the individual defaults on the loan. Scoring occurs through a scorecard which simply adds together score components for various attributes that the individual possesses. For example, the fact that the credit applicant has lived at the same address for more than five years might be worth 15 points. The development of this scorecard is based on the credit histories of many thousands of other people. If you are interested in how your own score might come out, you can look at the website: www.myfico.com/ficocreditscoreestimator. The US credit score is commonly called a FICO score. In the UK, an equivalent website to let you estimate a credit score is www.checkmyfile.com (and the equivalent in Australia is checkmyfile.com.au). The exact scoring methods are not revealed, but some information is available (for example, 35% of the FICO score is related to payment history).

The objective of the lender is simply to make this decision as accurately as possible. Any information which can legally be used and which has a bearing on the creditworthiness of an individual will come into play. What is illegal? It is not permitted to discriminate on the basis of race, gender, sexuality or religion, so these questions cannot be asked. However, it is fine to consider an individual's postcode when carrying out the check. The rules on age and marital status are more complex and vary in different countries. In the US, card issuers cannot deny credit or offer less favorable terms on the basis of marital status.

10.3.1   Probability, odds and log odds

For a particular type of individual we can use previous data to predict the probability that they are *good*, i.e. that they will repay the loan. Write p_i for this probability for an individual of type i. An alternative is to look at the odds of being good as opposed to *bad*. This might be familiar from a betting context: if we say that the odds of a horse winning are 2 to 1, then we mean that the horse is twice as likely to win as not. The odds o_i are simply the probability of being good divided by the probability of being bad, i.e.

$$o_i = \frac{p_i}{1 - p_i}.$$

The probability of being good varies between 0 and 1. But the odds of being good can vary from 0 to any positive value. It can also be useful to look at the log odds, defined as

$$\log(o_i) = \log\left(\frac{p_i}{1 - p_i}\right).$$

These are natural logs taken to base e. Log odds can take any value, both positive and negative. Notice that both odds and log odds are increasing functions of the probability.

Suppose that we are considering extending credit to an individual. By doing this we will make a profit, perhaps from the interest that we charge or perhaps because we make a profit on a product that will not be sold unless we offer credit. There is, however, the possibility of a loss, usually much larger than the profit, that will occur if the individual does not repay the debt. Suppose we write L for the loss and R for the profit. Then the expected value to us of the loan is

$$p_i R - (1 - p_i)L.$$

This is positive (meaning we should go ahead with the loan) if

$$p_i > \frac{L}{L+R},$$

or equivalently if $o_i > (L/R)$, a condition that can also be written in terms of the log odds:

$$\log(o_i) > \log\left(\frac{L}{R}\right).$$

Based on all the data that we have, we can estimate the odds for an arbitrarily chosen individual. We define the *population odds* as

$$o_{\text{Pop}} = \frac{\Pr(G)}{\Pr(B)}.$$

These are the odds of an individual being good if they are chosen at random from the whole population. We estimate the probabilities $\Pr(G)$ and $\Pr(B)$ simply by looking at the proportions of good and bad in the entire population for which we have data.

We now begin the discussion of an example that we will continue to use in illustrating our discussion through the rest of this chapter.

Example 10.1   Bank of Sydney

Suppose that the Bank of Sydney has 1200 customers who have been loaned money on a short-term basis. Most have kept up with loan repayments: these are called *good* or G. Some have not kept up repayments, and anyone who falls a total of more than 90 days behind in repayments is classified as *bad* or B. Besides data on age at time of loan agreement, classified as under 30, 30–39, 40–49 and over 50, Bank of Sydney keeps data on whether the individuals are owners, renters or some other classification in respect to their home and also whether they have a credit card or not. The data for this example are all given in the spreadsheet BRMch10-BankofSydney.xlsx.

The Bank of Sydney data can be presented in various ways. Table 10.2 shows the number of good and bad individuals in each of the 24 different sub-categories obtained from 'credit card status (2)' × 'housing status (3)' × 'age bracket (4)'. In these data there are a total of 1070 goods and 130 bads. Thus, for the Bank of Sydney the population odds are $o_{\text{Pop}} = 1070/130 = 8.23$.

If we look at a single category of individual we find that the odds vary from this. For example, given that the individual is aged under 30, the first column in the Bank of Sydney data shows that the odds are

$$\frac{59 + 47 + 63 + 19 + 18 + 12}{5 + 10 + 6 + 2 + 9 + 2} = \frac{218}{34} = 6.412.$$

So, borrowers in this category are much more likely to be bad than the population as a whole. We can also carry out calculations of the odds for each of the sub-categories in Table 10.2. This gives the numbers shown in Table 10.3.   □

Table 10.2   Bank of Sydney data: number of good and bad individuals.

	Under 30	30–39	40–49	Over 50
Owner with credit card	$G = 59$ $B = 5$	$G = 111$ $B = 9$	$G = 118$ $B = 5$	$G = 232$ $B = 11$
Renter with credit card	$G = 47$ $B = 10$	$G = 16$ $B = 5$	$G = 22$ $B = 4$	$G = 64$ $B = 18$
Other with credit card	$G = 63$ $B = 6$	$G = 21$ $B = 2$	$G = 16$ $B = 3$	$G = 91$ $B = 5$
Owner without credit card	$G = 19$ $B = 2$	$G = 13$ $B = 3$	$G = 44$ $B = 4$	$G = 31$ $B = 2$
Renter without credit card	$G = 18$ $B = 9$	$G = 14$ $B = 10$	$G = 5$ $B = 1$	$G = 10$ $B = 3$
Other without credit card	$G = 12$ $B = 2$	$G = 26$ $B = 8$	$G = 6$ $B = 2$	$G = 12$ $B = 1$

Table 10.3   Bank of Sydney data: odds.

	Under 30	30–39	40–49	Over 50
Owner with credit card	11.80	12.33	23.60	21.09
Renter with credit card	4.70	3.20	5.50	3.56
Other with credit card	10.50	10.50	5.33	18.20
Owner without credit card	9.50	4.33	11.00	15.50
Renter without credit card	2.00	1.40	5.00	3.33
Other without credit card	6.00	3.25	3.00	12.00

Now we want to focus in on a particular type of individual, say an individual of category A. The odds for category A are given by

$$o_A = \frac{\Pr(G \mid A)}{\Pr(B \mid A)}.$$

These odds can be estimated by looking at the data, but now considering only the category A individuals: we simply look at the proportions of good and bad for this subset of the entire population.

The odds for a category A are linked to the odds for the population as a whole through something called the *information odds*, I_A, for a category A, which are defined as

$$I_A = \frac{\Pr(A \mid G)}{\Pr(A \mid B)}.$$

Because of Bayes' rule: $\Pr(G \mid A) = \Pr(A \mid G) \Pr(G) / \Pr(A)$ and similarly for $\Pr(B \mid A)$. Then we can rewrite the odds for category A as

$$o_A = \frac{\Pr(G \mid A)}{\Pr(B \mid A)} = \frac{\Pr(A \mid G)}{\Pr(A \mid B)} \times \frac{\Pr(G)}{\Pr(B)} = I_A \times o_{\text{Pop}}.$$

Here we have canceled the two terms $\Pr(A)$ to get the formula. So, we have shown that multiplying the information odds for a category by the population odds gives the odds for the category. The information odds provide a kind of modifier for the population odds to get to the odds in a particular category.

The *Weight of Evidence* (WoE) for a category is just the natural logarithm of the information odds for that category:

$$w_A = \log(I_A) = \log\left(\frac{\Pr(A \mid G)}{\Pr(A \mid B)}\right),$$

It is natural to make this definition because it enables us to find the log odds for a category simply by adding the weight of evidence and the log odds for the population:

$$\log(o_A) = \log(I_A \times o_{\text{Pop}}) = \log(I_A) + \log(o_{\text{Pop}})$$
$$= w_A + \log(o_{\text{Pop}}).$$

Example 10.1   (continued) Bank of Sydney

We will continue to look at the individuals who are under 30, corresponding to the first column in Table 10.2. As we have already seen, the odds for this category are $218/34 = 6.412$. We can also calculate the information odds for this category. Note that out of 1070 good individuals, 218 are in this age bracket, and out of 130 bad individuals, 34 are in this age bracket. In other words

$$\Pr(\text{age} < 30 \mid G) = 218/1070,$$

$$\Pr(\text{age} < 30 \mid B) = 34/130.$$

Hence, the information odds for 'age under 30' are

$$I_{\text{age}<30} = \frac{218/1070}{34/130} = \frac{218}{34} \times \frac{130}{1070} = 0.779.$$

We can calculate the weight of evidence for 'age under 30' as

$$w_{\text{age}<30} = \log\left(\frac{218}{34} \times \frac{130}{1070}\right) = \log(0.779) = -0.2497.$$

The rules for logs imply that the log odds for this category are

$$\log\left(\frac{218}{34}\right) = \log\left(\frac{218}{34} \times \frac{130}{1070}\right) + \log\left(\frac{1070}{130}\right)$$

$$= w_{\text{age}<30} + \log(o_{\text{Pop}})$$

$$= -0.2497 + 2.1079 = 1.8582. \qquad \square$$

Now suppose we are interested in the odds for an individual in a small category formed by the intersection of two or more other categories. Rather than just looking at the previous experience we have had with exactly matching individuals, it makes sense to try to make deductions from what we have observed in the larger categories that this person belongs to. It turns out that if the behavior under different attributes is independent, then the weights of evidence can be added together to find the log odds for an individual with a number of different attributes.

To see why this is true, consider an individual with two attributes A_1 and A_2. We wish to calculate the odds that this individual is good, i.e. we want to find $\Pr(G \mid A_1, A_2) / \Pr(B \mid A_1, A_2)$. We know from Bayes' rule that,

$$\Pr(G \mid A_1, A_2) = \Pr(A_1, A_2 \mid G) \frac{\Pr(G)}{\Pr(A_1, A_2)}.$$

If A_1 and A_2 are independent (so that information about one attribute does not tell us anything about the other) then we have $\Pr(A_1, A_2 \mid B) = \Pr(A_1 \mid B) \times \Pr(A_2 \mid B)$ and $\Pr(A_1, A_2 \mid G) = \Pr(A_1 \mid G) \times \Pr(A_2 \mid G)$. If this holds then the odds for an individual in the category of A_1 and A_2 are

$$\frac{\Pr(G \mid A_1, A_2)}{\Pr(B \mid A_1, A_2)} = \frac{\Pr(A_1, A_2 \mid G)}{\Pr(A_1, A_2 \mid B)} \frac{\Pr(G)}{\Pr(B)}$$

$$= \frac{\Pr(A_1 \mid G)}{\Pr(A_1 \mid B)} \times \frac{\Pr(A_2 \mid G)}{\Pr(A_2 \mid B)} \times \frac{\Pr(G)}{\Pr(B)}.$$

The same expression can be extended to any number of terms. So the log odds given $A_1, A_2, \ldots, A_n$ are

$$\log \left(\frac{\Pr(G \mid A_1, A_2, \ldots, A_n)}{\Pr(B \mid A_1, A_2, \ldots, A_n)} \right) = w_1 + w_2 + \ldots + w_n + \log(o_{\text{Pop}})$$

where o_{Pop} represents the population odds and w_i is the weight of evidence for the characteristic A_i.

This brings us to the idea of a scorecard. If we know the weights of evidence for each major category then, for an individual who sits in the intersection of a number of different categories, we can just add up the relevant w_j numbers, together with a constant term given by the log of population odds, to get a prediction for the log odds. This is called a *naive Bayes' scorecard*. More generally, a scorecard has a very simple structure: it associates every category with a score and then adds the scores for an individual together to get a final score. Now we return to our Bank of Sydney example.

Example 10.1   (continued) Bank of Sydney

We take the weights of evidence for the Bank of Sydney data. We have already seen that $w_{\text{age}<30} = -0.2497$. We can calculate the other weights of evidence in the same way. For example, there are 196 goods amongst the renters, and 60 bads, so

$$w_{\text{renter}} = \log \left(\frac{196}{60} \times \frac{130}{1070} \right) = \log(0.3969) = -0.9241.$$

Table 10.4   Scorecard derived from weight of evidence.

Attribute	Score
Age <30	−25
Age 30–39	−42
Age 40–49	30
Age ≥50	29
Owns home	62
Rents home	−92
Other	3
Has credit card	23
No credit card	−61
Constant	211

We also have a constant term given by $\log(o_{\text{Pop}}) = 2.1079$. These w values give us a scorecard, but it is convenient to first multiply all the numbers by 100 and round to the nearest integer. This gives the scorecard of Table 10.4. Using this we see that, for example, the score for an individual who is a homeowner, with a credit card and is age 42 is $30 + 62 + 23 + 211 = 326$, whereas a 28-year-old who lives at home and has no credit card has a score $-25 + 3 - 61 + 211 = 128$.

We can check how effective this is by using the scorecard to predict the odds for various categories of individual. The actual log odds for the data are given in Table 10.5.

In the example of the 42-year-old homeowner with a credit card, the scorecard gives 326, so the predicted log odds are 3.26. Hence, the odds of this individual being good are predicted to be

$$e^{3.26} = 26.05.$$

In this category there are 118 good individuals and 5 bad, so the actual odds are 23.6 to 1 (with corresponding log odds of 3.16). In the same way, we can look at the odds of a 28-year-old who lives at home and has no credit card being a 'good'. The score is 128, corresponding to log odds of 1.28, which means that the odds are predicted to be

$$e^{1.28} = 3.60.$$

Table 10.5   Bank of Sydney data: log odds.

	Under 30	30–39	40–49	Over 50
Owner with credit card	2.47	2.51	3.16	3.05
Renter with credit card	1.55	1.16	1.70	1.27
Other with credit card	2.35	2.35	1.67	2.90
Owner without credit card	2.25	1.47	2.40	2.74
Renter without credit card	0.69	0.34	1.61	1.20
Other without credit card	1.79	1.18	1.10	2.48

308 BUSINESS RISK MANAGEMENT

The actual results in this category are 12 good individuals and 2 bad individuals, giving odds of 6 to 1. Thus, the prediction in the first case is reasonable, but in the second case the prediction is less satisfactory. However, even in the second case things are not so bad, since if just one of the existing individuals was to be recategorized as bad, then this would turn the odds to 4 to 1, which is close to the estimation of 3.60. □

10.4 Logistic regression

A better way to predict probabilities for individuals is to use *logistic regression*, which is the subject of this section. This is the most common way to construct a scorecard. But we start with the fundamental question of how we should predict probabilities. Ordinary regression predicts a dependent variable y on the basis of the observation of a set of explanatory variables x_j, $j = 1, 2, \ldots, m$. The linear form of this prediction can be written

$$y = \beta_0 + \sum \beta_i x_i$$
$$= \beta \cdot \mathbf{x}$$

where we use a boldface letter to indicate a vector: $\mathbf{x} = (x_0, x_1, \ldots, x_m)$, $\beta = (\beta_0, \beta_1, \ldots, \beta_m)$ and we set $x_0 = 1$. We then estimate the βs by looking at previous data to find a good fit. We cannot use the same approach when dealing with probabilities, since we want to ensure that the predictions y are all between 0 and 1.

The logistic approach is to use a nonlinear transformation to get from x_i values to a probability prediction p. We set

$$p = \frac{e^{\beta \cdot \mathbf{x}}}{1 + e^{\beta \cdot \mathbf{x}}}. \tag{10.3}$$

If $\beta \cdot \mathbf{x}$ gets very large, then p approaches 1, if $\beta \cdot \mathbf{x} = 0$ then $p = 1/2$; and if $\beta \cdot \mathbf{x}$ is a large negative number, then p approaches 0. Figure 10.6 shows what this function looks like (and there are other functions we might choose which would produce similar results).

One advantage of the logistic function is that it fits beautifully with our previous definition of log odds. We have:

$$1 - p = 1 - \frac{e^{\beta \cdot \mathbf{x}}}{1 + e^{\beta \cdot \mathbf{x}}} = \frac{1}{1 + e^{\beta \cdot \mathbf{x}}}.$$

So

$$\text{odds} = \frac{p}{1 - p} = \frac{e^{\beta \cdot \mathbf{x}}}{1 + e^{\beta \cdot \mathbf{x}}} \frac{1 + e^{\beta \cdot \mathbf{x}}}{1} = e^{\beta \cdot \mathbf{x}},$$

and thus

$$\log_e \left(\frac{p}{1 - p} \right) = \log_e(\text{odds}) = \log_e(e^{\beta \cdot \mathbf{x}}) = \beta \cdot \mathbf{x}. \tag{10.4}$$

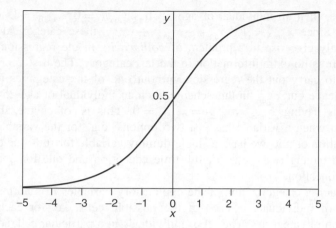

Figure 10.6   A graph of the logistic function $y = e^x/(1 + e^x)$.

Thus, if the actual probabilities of good or bad are given by a logistic model derived from an underlying linear function $\beta \cdot \mathbf{x}$, then the log odds will be a linear function of the observations x_i.

How can we estimate the values $\beta_0, \beta_1, \ldots, \beta_m$? In an ordinary regression we try to minimize the error term looking at the prediction against what actually happened. But for our logistic model, the prediction is a probability and what actually happened is either a good result or a bad result (which we can think of as either a 1 or a 0). There are two options: a simple approach using ordinary least squares regression or a more sophisticated logistic regression method. We deal with the simpler approach first.

The least squares regression works directly with Equation (10.4). Since the log odds are a linear combination of the explanatory variables, we can estimate the coefficients directly using an ordinary least squares regression based on the log odds for each category. In practice, this method works pretty well and has the advantage that it can be used without access to a data analysis tool that includes a logistic regression component (for example, we can use this ordinary least squares regression approach by applying the 'Data Analysis' add-in that comes with Excel).

We need to say more about the form of the explanatory variables x. In practice, the most important case is when the explanatory variables are categorical. Sometimes this follows from the nature of the data (for example, does an individual rent their home or not?). But even when this is not true, most credit scoring creates categorical data by assigning individuals to categories. The most obvious example is a variable like age: this is most naturally treated as a continuous variable, but credit scoring would normally determine certain age brackets and assign individuals to just one of these. The same holds true for income levels.

In this situation we use dummy variables for each category. So, for example, if we are using age brackets: (A) less than 30, (B) 30 to 39, (C) 40 to 49, (D)

50 or more, then an individual of age 33 has $x_{ageA} = 0$, $x_{ageB} = 1$, $x_{ageC} = 0$, $x_{ageD} = 0$. Since $x_{ageA} = 1 - x_{ageB} - x_{ageC} - x_{ageD}$, these categorical variables automatically give rise to a problem of *collinearity* in the regression; in other words, there is no extra information in the last category. The best approach here is simply to carry out the regression without one of the categories appearing. So, if we leave out x_{ageA} in this scheme, then an individual of age 25 is simply recorded as having $x_{ageB} = x_{ageC} = x_{ageD} = 0$. This is, of course, the natural thing to do when a variable has just two options: e.g. for the variable 'in full-time education or not' we have a single dummy variable for full-time education, rather than having two – one for full-time education and one for not being in full-time education.

With such a categorization of the explanatory variables it is normal to have a number of individuals in each of the possible cells, i.e. for each possible combination of categories. The set of individuals in a particular cell, some good, some bad, enables us to calculate the log odds value for the combination of dummy variables corresponding to that cell. These log odds values then become the dependent variable in our regression analysis.

There are two points to be borne in mind, however, in using this approach:

- The method gives equal weight to all categories no matter how many individuals are involved. In fact, we should be much more worried by errors in the log odds predictions when there are a large number of individuals in the category than when there are only a handful. (The more complex logistic regression automatically correctly weights the evidence from categories containing different numbers of individuals.)

- The least squares regression approach will fail if the log odds cannot be defined for a category because there are no 'bad' individuals in that category. In this case the log odds become infinite. The recommended procedure is to add 0.5 to both good and bad numbers to get an adjusted log odds value. So, for example, if a certain combination has only 10 individuals, all of them 'good', we might replace an infinite log odds value with $\log(10.5/0.5) = 3.0445$ (Note that this is still substantially higher than the log odds value if 9 out of 10 individuals were 'good', which is $\log(9/1) = 2.1972$).

Now we turn to the more sophisticated (and recommended) approach of logistic regression. The idea here is to use a maximum likelihood estimator for the coefficients β. In other words, we look at the data on individual good or bad results and assume that the probability of getting a good is given by Equation (10.3). Then we want to choose the values of β which maximize the probability of our set of observations (which are all 0 or 1).

To see how to work this out, suppose that there are two individuals who we think of as providing two observations, y_1 and y_2. Suppose that for a particular set of β values we calculate the probability that $y_1 = 1$ (good) and find it to be p_1. In the same way, we calculate the probability that $y_2 = 1$ and find this to be p_2. Then we can calculate from this the probability of observing the particular y_1 and y_2 values we have observed. This is called a likelihood associated with the β values. For example, the likelihood of observing $y_1 = 1$ and $y_2 = 0$ is given by $p_1(1 - p_2)$. If the x values associated with the first observation are the vector $\mathbf{x}^{(1)}$ and for the second observation the vector $\mathbf{x}^{(2)}$ we have

$$p_1(1 - p_2) = \frac{e^{\beta \cdot \mathbf{x}^{(1)}}}{1 + e^{\beta \cdot \mathbf{x}^{(1)}}} \left(1 - \frac{e^{\beta \cdot \mathbf{x}^{(2)}}}{1 + e^{\beta \cdot \mathbf{x}^{(2)}}} \right). \qquad (10.5)$$

The next stage is to make a maximum likelihood estimate of the vector β by choosing the value which maximizes this expression. The idea is that the observations we have made make some values of β unlikely. For example, if a particular β value led to p_1 near zero and p_2 near 1, then the observations $y_1 = 1$ and $y_2 = 0$ would have a very low chance of occurring (reflected in the low value of $p_1(1 - p_2)$) and this is evidence against this choice of β. Maximum likelihood turns this logic around and searches for the β values which give the highest likelihood for the observations made.

In general we will have n observations not just two, but the principle is the same: there will be a long product expression involving a single value of β and different values of the category variables x. A logistic regression involves maximizing this expression over all possible choices of β. We will give more details about this in the next section. But in any case a logistic regression requires the use of special purpose software. The logistic regression carried out for the Bank of Sydney example below has been obtained using the free software *Gretl*.

The beauty of logistic regression is that it works well when there are many more types of characteristic than the three in the Bank of Sydney example (credit card, housing type, age bracket). If we measure individuals through their answers to, say, 10 different questions, then even if each question is a simple yes/no we will still end up with more than 1000 cells. This means that there will be many cells with no individual and many more cells where there are no 'bad' individuals. But we still need a way to evaluate the credit risk from a new customer who may belong to one of the cells where we don't have many previous customers to compare to.

Example 10.1   (continued) Bank of Sydney

For the Bank of Sydney example, a logistic regression considers the 1200 individual observations and estimates the β values associated with the six variables $x_{\text{own}} = $ 'owner', $x_{\text{rent}} = $ 'renter', $x_{\text{cc}} = $ 'credit card holder', $x_{\text{ageB}} = $ 'age 30–39', $x_{\text{ageC}} = $ 'age 40–49' and $x_{\text{ageD}} = $ 'age over 50'. The three other variables 'other', 'no credit card' and 'age under 30' are linear combinations of other explanatory variables and do not add anything to the regression, so they are left out.

Thus, for example, the individuals in the top left-hand cell of Table 10.2 have $x_{\text{own}} = 1$, $x_{\text{rent}} = 0$, $x_{\text{cc}} = 1$, $x_{\text{ageB}} = 0$, $x_{\text{ageC}} = 0$ and $x_{\text{ageD}} = 0$. Thus, for each of these individuals the scalar product $\beta \cdot \mathbf{x} = \beta_{\text{own}} + \beta_{\text{cc}}$ which are also

Table 10.6   Output from logistic regression.

	Coefficient	Std. error
const	1.72256	0.287793
credit card	0.628808	0.211980
owner	0.493278	0.261849
renter	−0.983520	0.249570
age 30 to 39	−0.369558	0.271993
age 40 to 49	0.134251	0.319367
age 50 or more	0.204348	0.259001

the log odds. The predicted probability of being good for these 64 individuals is

$$p = \frac{e^{\beta_{own}+\beta_{cc}}}{1 + e^{\beta_{own}+\beta_{cc}}}.$$

This is the value that we try to match with the actual values through the right choice of the β values.

The data given in the spreadsheet BRMch10-BankofSydney.xlsx (sheet 2) have been used to run a logistic regression (using the free software Gretl) and part of the output is shown in Table 10.6.

Thus, we have the values of the β variables as follows:

$$\begin{aligned}
\beta_{cc} &= 0.629 \\
\beta_{own} &= 0.493 \\
\beta_{rent} &= -0.984 \\
\beta_{ageB} &= -0.370 \\
\beta_{ageC} &= 0.134 \\
\beta_{ageD} &= 0.204 \\
\beta_0 &= 1.723
\end{aligned}$$

where β_0 is the constant term. Hence, we obtain the following maximum likelihood estimator:

$$\log_e \left(\frac{p}{1-p} \right) = 1.723 + 0.629x_{cc} + 0.493x_{own} - 0.984x_{rent}$$

$$- 0.370x_{ageB} + 0.134x_{ageC} + 0.204x_{ageD}. \qquad (10.6)$$

We can use these values to produce a log odds table for each of the 24 categories. This is shown in Table 10.7, and can be compared with the data given in Table 10.5.

We can also compare these results with what happens if we run an ordinary least squares regression (the 'quick and dirty' approach). If this is done with the Bank of Sydney data, we get

$$\log_e \left(\frac{p}{1-p} \right) = 1.650 + 0.575x_{cc} + 0.527x_{own}$$

$$- 0.788x_{rent} - 0.349x_{ageB} + 0.090x_{ageC} + 0.424x_{ageD}.$$

Table 10.7   Bank of Sydney data: predicted log odds from logistic regression.

	Under 30	30–39	40–49	Over 50
Owner with credit card	2.84	2.48	2.98	3.05
Renter with credit card	1.37	1.00	1.50	1.57
Other with credit card	2.35	1.98	2.49	2.56
Owner without credit card	2.22	1.85	2.35	2.42
Renter without credit card	0.74	0.37	0.87	0.94
Other without credit card	1.72	1.35	1.86	1.93

For sub-categories in which there are a large number of individuals, we expect to see that the log odds predictions from a logistic regression are better than from an ordinary least squares approach. This is true for the Bank of Sydney data, where the log odds prediction from the logistic regression gives 3.05 for the most common type of customer (over 50, home owner with a credit card), and this is exactly right. The alternative least squares regression gives a log odds prediction of

$$1.650 + 0.575 + 0.527 + 0.424 = 3.18$$

which is a little high.   □

10.4.1   *More details on logistic regression

We have seen how with two observations y_1 and y_2, and with a probability that $y_i = 1$ given by p_i, the probability that $y_1 = 1$ and $y_2 = 0$ is $p_1(1 - p_2)$. We can generalize this expression to any combination of values for y_1 and y_2 by noting that the probability is given by

$$(p_1)^{y_1}(1 - p_1)^{(1-y_1)}(p_2)^{y_2}(1 - p_2)^{(1-y_2)}.$$

More generally, we can take n observations, $y_i, i = 1, 2, \ldots, n$ (with all y_i taking the value 1 or 0) and build up a big product having $2n$ terms:

$$\prod_{i=1}^{n} (p_i)^{y_i}(1 - p_i)^{(1-y_i)}. \qquad (10.7)$$

This is the probability of observing $y_1, y_2, \ldots y_n$ if the probabilities are really $p_1, p_2, \ldots p_n$. If the p_i come from a logistic model then

$$p_i = \frac{e^{\beta \cdot \mathbf{x}^{(i)}}}{1 + e^{\beta \cdot \mathbf{x}^{(i)}}}, \qquad (10.8)$$

where we have written $\mathbf{x}^{(i)} = (x_0^i, x_1^i, \ldots, x_m^i)$ for the independent variables associated with the ith observation. So the problem of finding a maximum likelihood estimator from a given set of data $(y_1, y_2, \ldots y_n)$ is equivalent to finding a set of β values that maximizes the product expression (10.7) where the p_i are given by Equation (10.8).

But if we want to maximize an expression, we can also maximize the log of that expression. So, rather than maximize the likelihood, we maximize the log likelihood

$$L = \log\left(\prod_{i=1}^{n} (p_i)^{y_i} (1 - p_i)^{(1-y_i)}\right)$$

$$= \sum_{i=1}^{n} y_i \log(p_i) + \sum_{i=1}^{n}(1 - y_i) \log(1 - p_i).$$

Since the p_i values are given by logistic functions, this can be simplified considerably:

$$L = \sum_{i=1}^{n} y_i \log\left(\frac{e^{\beta \cdot \mathbf{x}^i}}{1 + e^{\beta \cdot \mathbf{x}^i}}\right) + \sum_{i=1}^{n}(1 - y_i) \log\left(\frac{1}{1 + e^{\beta \cdot \mathbf{x}^i}}\right)$$

$$= \sum_{i=1}^{n} y_i (\log(e^{\beta \cdot \mathbf{x}^i}) - \log(1 + e^{\beta \cdot \mathbf{x}^i})) - \sum_{i=1}^{n}(1 - y_i) \log(1 + e^{\beta \cdot \mathbf{x}^i})$$

$$= \sum_{i=1}^{n} y_i \log(e^{\beta \cdot \mathbf{x}^i}) - \sum_{i=1}^{n} \log(1 + e^{\beta \cdot \mathbf{x}^i})$$

$$= \sum_{i=1}^{n} y_i (\beta \cdot \mathbf{x}^i) - \sum_{i=1}^{n} \log(1 + e^{\beta \cdot \mathbf{x}^i}).$$

To maximize L we take derivatives with respect to β_j. Now, since

$$\frac{d}{dz} \log(f(z)) = \frac{f'(z)}{f(z)} \quad \text{and} \quad \frac{d}{dz} e^{za+b} = a e^{za+b},$$

we get

$$\frac{\partial L}{\partial \beta_j} = \sum_{i=1}^{n} y_i x_j^i - \sum_{i=1}^{n} \frac{x_j^i e^{\beta \cdot \mathbf{x}^i}}{1 + e^{\beta \cdot \mathbf{x}^i}}, \quad j = 1, 2, \ldots, m,$$

$$\frac{\partial L}{\partial \beta_0} = \sum_{i=1}^{n} y_i - \sum_{i=1}^{n} \frac{e^{\beta \cdot \mathbf{x}^i}}{1 + e^{\beta \cdot \mathbf{x}^i}}.$$

To find the maximum, we set all these derivatives to zero and solve the simultaneous equations

$$\sum_{i=1}^{n} x_j^i \left(y_i - \frac{e^{\beta \cdot \mathbf{x}^i}}{1 + e^{\beta \cdot \mathbf{x}^i}}\right) = 0, \quad j = 1, 2, \ldots, m,$$

$$\sum_{i=1}^{n} \left(y_i - \frac{e^{\beta \cdot \mathbf{x}^i}}{1 + e^{\beta \cdot \mathbf{x}^i}}\right) = 0.$$

These are $m+1$ nonlinear equations to be solved for the $m+1$ values $\beta_0, \beta_1, \ldots, \beta_m$. It is not difficult to solve the equations using some iterative method (like Newton–Raphson). It is this solution that is provided by software carrying out logistic regression: the estimate of the β values is the one that maximizes the chance of observing the pattern of 1s and 0s that are observed.

10.4.2   Building a scorecard

A scorecard gives a way of calculating a score for an individual, but the scores themselves are not so critical, provided we can translate back from them to a probability of default. We have already met a scorecard that translates directly into (log odds)×100 in Table 10.4. For logistic regression we have

$$\log(\text{odds}) = \beta \cdot \mathbf{x} = \beta_0 + \beta_1 x_1 + \ldots + \beta_n x_n.$$

Since the x_i values are either 0 or 1, the β values are exactly what we require for a scorecard. So we simply take the regression estimates β, multiply by 100 and round to get the scorecard.

Notice that a scorecard will include each category even if the logistic regression produces estimates only for variables not implied by other variables. Thus, for example, in the naive Bayes' scorecard of Table 10.4 there is a score of 23 for 'has credit card' and -61 for 'no credit card'. The logistic regression would produce a score for just one of these variables and the other would be zero.

It is clear that adding the same number to each of a mutually exclusive and exhaustive set of variables is the same as adding a number to the constant term; each individual, no matter how they are classified, gets the same additional score. This means that we can carry out manipulations that will leave all the scores the same if we take off the same amount from the constant term as we add to each of a set of exclusive variables. Usually this is done in order to avoid any negative numbers in the scorecard.

Notice that for the standard (log odds)×100 scorecard, an increment of 100 in the score means an increment of 1 in the log odds and this corresponds to multiplying the odds by the fixed amount $e = 2.7183$. We can look at this another way and say that multiplying the odds by 2 will increase the score by $100\log(2) = 69$. Sometimes we want to construct a scorecard with the property that a particular base score corresponds to given odds and each doubling of the odds corresponds to an increment in the score by a particular amount. To do this we need to define the score as a linear transformation of the log odds. If the score for a category i is given by

$$s_i = a + b\log(o_i)$$

for constants a and b, then b can be determined by the fact that doubling the odds corresponds to increasing the score by $b\log 2$ and a can be determined from the required base score. We see how this works out in the Bank of Sydney example below.

Example 10.1   (continued) Bank of Sydney

Use the logistic regression results to construct a scorecard for the Bank of Sydney data where all the numbers are positive and a score of 200 corresponds to odds of 10 to 1 and each increment of 50 in the score corresponds to multiplying the odds by a factor of 2.

Solution

We begin with the logistic regression results of Equation (10.6). The β values can be multiplied by 100 and rounded to give the basic (log odds)$\times$100 scorecard labeled as Scorecard 1 in Table 10.8. Next we make adjustments to Scorecard 1 in order to obtain all positive numbers. We take a fixed amount off the constant term and add it to each category in a mutually exclusive and exhaustive set. We have made the minimum change to achieve this: we take 135 away from the constant term and add 98 to each of owner, renter and other, and in addition we add 37 to each age category. This gives Scorecard 2. Notice that any individual will get precisely the same score from Scorecard 2 as was obtained from Scorecard 1.

The final step is to use a linear transformation in order to achieve a score of 200 for odds of 10 to 1, and an increment of 50 when the odds are doubled. The new scorecard will have $s = a + b \log(\text{odds})$ instead of $s = 100 \log(\text{odds})$. The requirement is that

$$a + b \log(10) = 200$$

$$b \log(2) = 50.$$

Thus $b = 50/0.693 = 72.135$. And hence

$$a = 200 - 72.135 \log(10) = 33.903.$$

To translate the (log odds)$\times$100 scorecard, we need to first multiply all the terms (including the constant) by $72.135/100$ and then add 33.9 to the constant

Table 10.8   Adjusting a scorecard to have desirable characteristics.

Attribute	Scorecard 1	Scorecard 2	Scorecard 3
Age <30	0	37	87
Age 30–39	−37	0	61
Age 40–49	13	50	97
Age ≥50	20	57	102
Owns home	49	147	106
Rents home	−98	0	0
Other	0	98	71
Has credit card	63	63	45
No credit card	0	0	0
Constant	172	37	0

term and round. But to make the scorecard easier to apply, we take the additional step of adding the constant term, which is now $37 \times 0.721 + 33.9 = 60.59$, to each of the age categories so that we can drop the constant term. After rounding, this gives Scorecard 3 as shown in Table 10.8.   □

10.4.3   Other scoring applications

Having done all this work to determine credit scoring procedures, it is worth pointing out that exactly the same techniques can be used in another common management problem, which is the targeting of promotions. This is a kind of reverse to the credit scoring problem. We no longer want to identify the people who are likely to be bad in order to avoid giving them credit. Instead, we want to pick out the people who are more likely to respond positively to a promotion in order to justify the costs of a mailout targeted to them.

The idea of modeling the behavior in terms of a scorecard built up from different categories is still valuable. Both the credit scoring problem and the promotion targeting problem share the characteristic that quite a small proportion of the sample are in the category of responding (or in the category of a bad debt). From an estimation point of view, the consequence is that a small (absolute) number of individuals in any particular combination of categories will end up responding. This is the reason for using an indirect logistic regression procedure rather than simply taking the odds we observe in a single cell in the table and using this to predict the log odds for this cell. Because of the small numbers of responding individuals in some cells, just treating each cell on its own is a poor way to proceed. We are likely to get a better result by using the logistic regression model.

In order to avoid a negative log odds score, we need to deal with odds that are greater than 1. So, in this promotion targeting problem, we define the odds as:

$$\frac{\text{probability of not responding}}{\text{probability of responding}}.$$

With this change, everything goes through as before, except that in applying the scorecard we select individuals to mail with low values of the score corresponding to a relatively high probability of responding. Exercise 10.5 is an example of this kind of problem.

Notes

In the discussion of credit ratings agencies I have made extensive use of the information provided by Standard & Poor's. Table 10.1 is taken from Table 21 in Standard & Poor's 2011 Annual Global Corporate Default Study And Rating Transitions (publication date: 21 March 2012). Figure 10.2 graphs the changes over time shown in Table 3 of that report. Moreover, the spreadsheet BRMch10-Markov.xlsx contains material taken from Tables 21, 59, 61, 62 and 64 of the S&P 2011 Annual Global Corporate Default Study. This information is used to compare the predictions from a Markov assumption and the actual behavior seen, which forms the basis for Figures 10.4 and 10.5. The timeline for Liz Claiborne given in Figure 10.1 is taken from information given in Table 8 in the Standard & Poor's 2011 Default Synopses (publication date: 21 March 2012).

There is much to be said about the way that accounting data can be used to infer credit risk. This work goes back to a well-known model proposed by Altman in 1968 and a review of this literature is given by Altman and Saunders (1998). The use of techniques like logistic regression in this context provides a link between the corporate and consumer level credit risk. Another strand in the assessment of corporate risk relies on seeing the value of the firm evolving according to a stochastic process. Then, information on the volatility of the process and the upward drift in value can be translated into a statement about probability of default in a period of T years. This approach was originally proposed by Merton (1974) and since then has been adapted by KMV Corporation, acquired by Moody's in 2002. For more on this approach, see Chapter 16 in Culp (2001) or the paper by Bharath and Shumway (2008).

The data provided for the Bank of Sydney example are based, in part, on a similar set of simplified hypothetical data given by Thomas (2009) (he calls it Bank of Southampton data). The book by Thomas gives a much more detailed treatment of the way that consumer credit models operate and is a good place to start if you want to go deeper into this material. In the discussion we mention the problems of carrying out an ordinary least squares regression on log odds if categories have zero bad individuals. It may be overly optimistic to say that logistic regression avoids this problem. There is quite a literature on the way that the logistic regression maximum likelihood estimates are biased for small samples (see, for example, Firth, 1993 and the references there). One simple approach to reducing this bias is to add 0.5 to both the 'good' and 'bad' cells. The software Gretl that we have used for logistic regression is available at http://gretl.sourceforge.net.

References

Altman, E. and Saunders, A. (1998) Credit risk measurement: developments over the last 20 years. *Journal of Banking and Finance*, **21**, 1721–1742.

Bharath, S. and Shumway, T. (2008) Forecasting default with the Merton distance to default model. *Review of Financial Studies*, **21**, 1339–1369.

Culp, C. (2001) *The Risk Management Process: Business strategy and tactics*. John Wiley & Sons.

Firth, D. (1993) Bias reduction of maximum likelihood estimates. *Biometrika*, **80**, 27–38.

Merton, R. (1974) On the pricing of corporate debt: The risk structure of interest rates. *Journal of Finance*, **29**, 449–470.

Thomas, L. (2009) *Consumer Credit Models*. Oxford University Press, Oxford.

Exercises

10.1 Two-year default or NR probability

Using the information in Figure 10.3 (rather than the detailed figures in Table 10.1), calculate the probability that a company rated as CCC/C will have either ceased to be rated or defaulted within a two-year period.

10.2 Markov groupings

Verify the claim that grouping states together can destroy the Markov assumption. Suppose that there are four states A, B, C, D. Once either A or D is reached, there is no change possible. From B there is a 10% chance of moving to A and a 20% chance of moving to C, and otherwise there is no change. Similarly, from C there is a 10% chance of moving to B and a 20% chance of moving to D, and otherwise there is no change. New companies arrive at B in such a way that we expect the same number of companies in B as there are companies in C. Calculate the three-year probability of reaching D knowing that we are equally likely to start in either B or C, and compare this with the estimate made if we group together the states B and C.

10.3 Markov types

Verify the claim that different types of firms, each following a Markov chain, can produce non-Markov behavior in aggregate. Suppose that there are four states A, B, C, D. Once either A or D is reached, there is no change possible. Type X firms behave as in Exercise 10.2, i.e. starting from B, after one year there is a 10% chance of moving to A and a 20% chance of moving to C. Similarly, from C there is a 10% chance of moving to B and a 20% chance of moving to D. Type Y companies are the same except that they change state twice as often, i.e. from B there is a 20% chance of moving to A and a 40% chance of moving to C. Similarly, from C there is a 20% chance of moving to B and a 40% chance of moving to D. New companies arrive at B in such a way that we expect to have N companies of type X in B, N companies of type Y in B, N companies of type X in C, and N companies of type Y in C. Calculate the probability of moving to D in two steps from B for a Markov chain which matches the observed annual transitions and compare this with the true probability.

10.4 Octophone

Octophone is a mobile phone company that keeps data on its customers and rates them according to whether they fail to make scheduled contract payments or not in the first year of the contract term. The data available on application are age bracket, whether they have a credit card and whether they have had a mobile phone contract before (either with Octophone or another company). The results from 2000 of last year's applicants living in a single city are given in Table 10.9. In this table there are 1850 goods and 150 bads. Calculate the weights of evidence for the different attributes involved and use these to construct a (naive Bayes') scorecard.

Table 10.9   Octophone data: Goods and bad.

	age 18–21	age 22–29	age 30–45	46 and over
Previous phone, credit card	$G = 150$ $B = 6$	$G = 256$ $B = 8$	$G = 312$ $B = 9$	$G = 250$ $B = 7$
Previous phone, no credit card	$G = 114$ $B = 10$	$G = 123$ $B = 13$	$G = 92$ $B = 9$	$G = 91$ $B = 12$
No previous phone, credit card	$G = 99$ $B = 12$	$G = 182$ $B = 6$	$G = 59$ $B = 11$	$G = 45$ $B = 8$
No previous phone, no credit card	$G = 22$ $B = 9$	$G = 26$ $B = 14$	$G = 13$ $B = 7$	$G = 16$ $B = 9$

10.5   Octophone with contract costs

Octophone sells a variety of contracts but they can be classified on a dollars per month basis into three categories: low cost – less than \$30; medium cost – between \$30 and \$40; and high cost – more than \$40.

(a) Of the 2000 in the sample of Exercise 10.4 there are a total of 800 low-cost contracts, 600 medium-cost contracts and 600 high-cost contracts. The 150 bads are distributed with 40 on low-cost contracts, 40 on medium-cost contracts and 70 on high-cost contracts. Calculate the new weights of evidence including this additional information.

(b) A logistic regression is carried out on these data and produces the coefficients shown in Table 10.10. Here, the attributes age 45+; no credit card; no previous phone; and high-cost contract have all been omitted because of collinearity. Calculate a scorecard using these data.

Table 10.10   Logistic regression for Octophone.

	Coefficient
const	0.194171
age18_21	0.135372
age22_29	0.401779
age30_45	0.00415712
credit_card	1.38715
prev_phone	1.45171
low_cost	0.913791
med_cost	0.427606

(c) Use scaling to adjust the scorecard from (b) so that it has the following properties: (a) there is no constant term; (b) a score of 500 represents odds of 100 to 1; and (c) an increase in the scaled score of 100 points represents the odds being multiplied by 10. (So a score of 600 represents odds of 1000 to 1.)

(d) It is calculated that a mobile phone contract achieves an average profit of $15 per year, while every customer who fails to make scheduled payments will, on average, cost $100 (including cost of follow up, the average uncollected debt, and the replacement value of those handsets that are not recovered, after allowing for the average payments already made). For the scorecard in (c), what is the cutoff score where Octophone would not go ahead with the contract?

(e) Suppose that further breakdown of the figures suggests that the profits and losses are both dependent on the size of the contract as follows:

	Low cost	Medium cost	High cost
Profit per year	$10	$15	$20
Average cost if bad	$80	$100	$160

What would you recommend to Octophone?

10.6 Cosal Skincare

Cosal Skincare sells beauty products over the internet. Every six months Cosal runs a campaign to advertise one of its new products and to try to increase its customer base. This involves sending a free sample through the mail to potential customers drawn from a mailing list. The mailing list, which Cosal purchases, includes certain attributes of the recipients. For example, Cosal could pick out just females below the age of 30 who have a driving licence and live alone. Cosal uses the data from the last promotion to guide it as to the kind of potential customers to include in the mail out. As is common with this sort of exercise, only a small proportion of the recipients of a free sample will go on to make an online purchase. However, Cosal has found that a single online purchase is usually followed by others. It estimates that the profits earned from a single new online customer are approximately $150 on average. The free sample costs a total of $2.50 per recipient including mailing costs.

Table 10.11 gives the results from the last mail out to 10 000 potential customers. From this, 310 new online customers were achieved, which was regarded as a good result by Cosal. In this table, the P results are those who purchased online, the N results are those who did not make an online purchase.

(a) Use these data (also available on the Excel spreadsheet BRMch10-CosalSkincare.xls) to work out the log odds for each category and

Table 10.11   Results from Cosal Skincare mail out.

	under 30	30–39	40–49	over 50
Household of 1, full-time work	$N = 456$	$N = 412$	$N = 386$	$N = 386$
	$P = 34$	$P = 16$	$P = 8$	$P = 15$
Household of 1, not full-time work	$N = 453$	$N = 389$	$N = 375$	$N = 395$
	$P = 24$	$P = 12$	$P = 6$	$P = 12$
Household of 2, full-time work	$N = 463$	$N = 409$	$N = 387$	$N = 363$
	$P = 24$	$P = 15$	$P = 10$	$P = 15$
Household of 2, not full-time work	$N = 460$	$N = 401$	$N = 373$	$N = 371$
	$P = 18$	$P = 13$	$P = 5$	$P = 18$
Household of > 3, full-time work	$N = 448$	$N = 407$	$N = 405$	$N = 375$
	$P = 12$	$P = 6$	$P = 6$	$P = 9$
Household of >3, not full-time work	$N = 412$	$N = 404$	$N = 377$	$N = 383$
	$P = 12$	$P = 8$	$P = 4$	$P = 8$

then run an ordinary least squares regression to determine a set of β values to use.

(b) Develop a scoring rule to enable Cosal Skincare to decide who to include in its mail out.

10.7   Consistent with log odds

Suppose that Figure 10.7 is produced by a website offering a free credit score evaluation. Are the numbers shown consistent with a log odds scoring scheme?

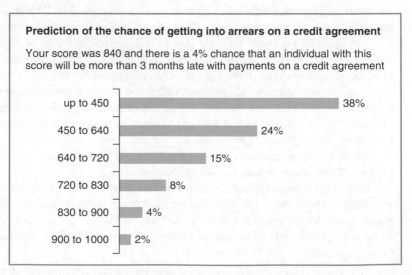

Figure 10.7   Graphic from website giving free credit scores.

Appendix A

Tutorial on probability theory

A.1 Random events

Probability theory is fundamental to much of our discussion in this book. It is
a topic that can be covered at many different levels. I have written this brief
tutorial with the specific aim of introducing the reader to the concepts that are
needed in the rest of the book. It may be useful as a single point to turn to for an
explanation of some probability concepts, or as a reminder of this material for
the reader who has been taught probability theory in the past, but has forgotten
the details.

Probability is the study of uncertain events; these are things that cannot be
precisely predicted in advance. For example, we might consider the event that
there is an earthquake in the San Francisco region next year of magnitude 7 or
above. This event either will or will not happen, but we cannot tell in advance. Or
we might consider tossing a coin and being interested in the event that it turns out
to be 'heads'. Again we cannot tell in advance what the result will be, but we can
talk about the *probability* of the event occurring. There are layers of philosophical
complications here about what exactly we mean by a probability, but it is not the
role of this tutorial to explore this area. We will assume that we are discussing
an event whose occurrence is determined by a '*random experiment*', the outcome
of which we cannot predict: for example, this might be tossing a coin.

We might argue that even the tossing of a coin is not truly random in reality;
if we knew the exact degree of force with which the spin was applied and we had
a sufficiently powerful computer, we could predict everything about the toss of
the coin; including which way up it would land. The same thing is true in theory
about the earthquake in California. At the moment this may be something that

Business Risk Management: Models and Analysis, First Edition. Edward J. Anderson.
© 2014 John Wiley & Sons, Ltd. Published 2014 by John Wiley & Sons, Ltd.
Companion website: www.wiley.com/go/business_risk_management

is impossible to predict, but with advances in geological science we may hope to know in advance when such events will occur, given enough data on what is happening in the rocks underground.

Since we want to develop the fundamental theory, we will sidestep these complications and assume that we do indeed have a random experiment and events that may occur but that cannot be predicted. If the events depend on specific circumstances, such as the force applied to flick a coin, then we will assume that we have no information about these factors. The random experiment may be an idealized concept, but it will be enough for us to develop the theory we want.

It is sometimes helpful to think of the random experiment producing one of a potentially large set of possible outcomes. We call the set of all possible outcomes the *sample space* (which is usually written Ω). With this definition we say that an event is a subset of Ω corresponding to a particular set of possible outcomes. So, for example, if we toss a coin 10 times we could define the event that exactly five heads occur.

Implicit within the terminology of a random experiment is the idea that it could (at least conceptually) be repeated. This is easy enough to see in relation to the tossing of coins, but we will extend the idea even to cases where there is no such possibility in reality. When we talk of the probability that the US economy is in recession during 2025, we know that this will or will not be the case, but we can carry out a thought experiment in which we imagine all sorts of different histories playing out between now and then, in some of which a recession occurs during that year and in others of which it does not.

One advantage of basing our discussion on a random experiment that can be repeated is that it allows us to use a frequentist definition of probability. Tossing a coin many times over should produce roughly equal numbers of heads and tails. So, the statement 'the probability of a head is equal to 0.5' can be interpreted as saying that with N tosses, the number of heads divided by N will approach 0.5 as N gets very large. More generally, we can consider for some event A (where A is a subset of the sample space Ω) the proportion of times that an outcome in A occurs out of N trials of the random experiment; then the probability of A is the limit of this proportion as N goes to infinity. We will not pursue this idea in detail or worry too much about why such limits will necessarily exist. Rather, we want to concentrate on the properties that probabilities must possess.

There are just three things that characterize probabilities: probabilities are non-negative; the probability of Ω is 1; and the probability of a union of disjoint events is the sum of the individual event probabilities. When there are only a finite number of outcomes possible in Ω, for example if the random experiment is the throwing of a dice, then everything is quite straightforward: we simply assign a probability to each possible outcome by making this the proportion of times we expect this outcome to occur, and then we define the probability of an event A as the sum of the probabilities of the outcomes within A.

Thus, if $\Omega = \{x_1, x_2, \ldots, x_n\}$ we let p_i be the probability that x_i occurs, and the probability of an event $A \subset \Omega$ is given by

$$\Pr(A) = \sum_{i \in J} p_i \text{ where } J = \{i : x_i \in A\}.$$

We need to have $p_i \geq 0$, $i = 1, 2, \ldots, n$ and $\sum_{i=1}^{n} p_i = 1$. The third property of probabilities (relating to the union of disjoint events) is automatic in this case.

The extra property of being able to add up the probabilities of disjoint events is important when we have an infinite number of possible outcomes in the sample space Ω. We would like to be able to describe probabilities associated with an infinite number of events. For example, consider a gambler who starts with a single dollar and repeatedly places one dollar bets on red at the roulette wheel, stopping when he has won \$20 or when he loses the \$1 he started with. His first bet is a dollar and he might lose straight away. If he doesn't, he will win an additional dollar. If his next two bets both lose, he will lose everything. More generally, we can see that the amount of money he has goes up or down by one dollar on each bet. We want to know the probability that he reaches zero before he reaches \$20. But this game can go on as long as we like, and the probability can be broken down into the probability of getting to zero (before \$20) after one step, three steps, five steps, etc. There are an infinite number of terms but they get smaller and smaller.

We need to get used to the terminology of sets in relation to events. The event that both A and B happen, when each of A and B is a subset of the sample space, is simply the intersection of the sets A and B. Similarly, the event that one or other of A and B happens is the union of the subsets A and B. This enables us to look at the probability of the union of A and B as the sum of the probability of three (disjoint) events: 'A without B', which we write as A\B and is the set of outcomes in A that are not in B; 'A intersection B', A ∩ B; and 'B without A', B\A.

Now, since A\B and A ∩ B are disjoint sets whose union is A we have, from our assumption on probabilities, that

$$\Pr(A) = \Pr(A\backslash B) + \Pr(A \cap B).$$

Similarly

$$\Pr(B) = \Pr(B\backslash A) + \Pr(A \cap B).$$

Thus

$$\Pr(A \cup B) = \Pr(A\backslash B) + \Pr(A \cap B) + \Pr(B\backslash A)$$
$$= \Pr(A) + \Pr(B) - \Pr(A \cap B).$$

A.2 Bayes' rule and independence

We say that two events are independent if one of them occurring makes no difference to the likelihood of the other occurring. This means that the probability of both A and B occurring is given by the product of the two individual probabilities, or in symbols

$$\Pr(A \cap B) = \Pr(A) \times \Pr(B).$$

To understand why this is true we need to go back to the idea of a conditional probability, which is a way of talking about the probability of an event given that we know some other event occurs. For example, if we know that two dice are thrown and they have different results, what is the probability that the sum of the two dice is an even number? It is simple enough to see that when two dice are thrown, the probability that the sum is an even number is exactly a half, since whatever the first throw, there will be a one half chance that the second is what is required to make the sum even (an even number if the first throw is even, odd otherwise). However, we want to find the probability *conditional* on the event that the dice show different numbers. One way to calculate this is to say that there are 36 equally likely outcomes from the throw of the two dice. Six of these outcomes have the two throws the same and the remaining 30 have the two numbers different. Of these 30, the 12 pairs listed below have a sum that is an even number.

Dice 1	Dice 2	Dice 1	Dice 2	Dice 1	Dice 2
1	3	3	1	5	1
1	5	3	5	5	3
2	4	4	2	6	2
2	6	4	6	6	4

Now, since all 36 original outcomes are equally likely, we can see that knowing the throws are different gives 30 equally likely outcomes of which 12 have the property we want. Thus, the conditional probability is $12/30 = 0.4$.

When dealing with conditional probabilities we use the notation $\Pr(A|B)$ to mean the probability that the event A occurs given that B occurs. Then the general principle that we have illustrated with the dice example is

$$\Pr(A|B) = \frac{\Pr(A \cap B)}{\Pr(B)}.$$

In words, we can say that the probability of A occurring given that B occurs is given by the probability of both A and B occurring divided by the probability that B occurs. This may be easier to grasp when we multiply through by $\Pr(B)$ to get

$$\Pr(A \cap B) = \Pr(B) \times \Pr(A|B),$$

which simply says that the probability of both A and B occurring together is the probability of B occurring multiplied by the probability that A occurs given that B occurs.

In the dice example, we have the probability that the throw is even given that the numbers are different is equal to the probability that the throw is even *and* the numbers are different (which is $12/36 = 1/3$) divided by the probability that the numbers are different (which is $30/36 = 5/6$).

The result can also be used to link $Pr(A|B)$ and $Pr(B|A)$. We have

$$Pr(A|B) = \frac{Pr(A \cap B)}{Pr(B)} = \frac{Pr(A \cap B)}{Pr(A)} \times \frac{Pr(A)}{Pr(B)} = \frac{Pr(B|A)\,Pr(A)}{Pr(B)}.$$

This is often called Bayes' Theorem.

Now we are ready to return to the question of independence. To say that the probability of A occurring is independent of whether or not B occurs is to say that the probability of A is the same as the probability of A given B. Thus

$$Pr(A) = Pr(A|B) = \frac{Pr(A \cap B)}{Pr(B)},$$

and we can rewrite this as

$$Pr(A \cap B) = Pr(A) \times Pr(B).$$

A.3   Random variables

Now we switch our focus from random events to *uncertain numbers* that are called *random variables*. Thus, we want to model a numerical value that we cannot accurately predict in advance. It could be the total rainfall next week, the price of oil six months from now, or the number we get when we throw a dice.

But even if we cannot predict the exact value in advance, we may be able to make statements about the likelihood of different values occurring. When we throw a dice we expect each of the six possible results to be equally likely to occur; when we need to plan for different amounts of rain we can go back to look at the records for this time of year to get an idea of the likelihood of a very large rainstorm; and if we want to predict the price of oil in the future we can turn to the financial markets to look at the pricing of derivatives and deduce from these the likelihood that traders in the market have guessed for different possible price outcomes.

The idea that different values of a random variable have different likelihoods is captured by assigning *probabilities* to different values. Here we need to distinguish between random variables that can take just one of a finite set of values (like the numbers 1 to 6 that might be thrown for a dice) and random variables that can take any value in a range (like the rainfall next week). The first type we call *discrete* random variables, and the second type are called *continuous* random variables. It is convenient to model these different types of random variable in different ways, but actually the difference between them is minor. After all, the price of oil is naturally modeled as a continuous random variable but in reality is determined only up to a certain accuracy (US cents per barrel). Usually we can translate fairly easily between continuous and discrete models by just taking a discrete random variable with a very fine increment in size.

For a discrete random variable we make the occurrence of one of the possible values an event, and then we see from our previous discussion that the probabilities for each possible outcome are non-negative numbers that add up to 1. More formally, we can say that when there are n possible values of the random variable: $x_1, x_2, \ldots, x_n$, then there are n associated probabilities $p_1, p_2, \ldots, p_n$ with $p_i \geq 0$, $i = 1, 2, \ldots n$ and $\sum_{i=1}^{n} p_i = 1$.

We can see immediately that a different approach will be required for continuous random variables. When dealing with a random variable that is continuous, for example the length of time that a call center agent requires to deal with a customer enquiry, the probability of any specific value occurring becomes infinitesimal. There is literally a zero probability of a call requiring exactly three minutes to deal with: if times are measured sufficiently accurately, that three-minute call will turn out to have been something like a 3.0001765 minute call.

When random variables are continuous it is helpful to look at events like $X \leq 3$ minutes, where X is the random variable that is the length of the call. For any continuous random variable X, there is an important function giving the probability that X takes a value less than or equal to t as t varies. This is called the *cumulative distribution function* (CDF) for X and is defined as

$$F(t) = \Pr(X \leq t).$$

When F is a continuous function then the small increment $F(t + \delta) - F(t)$ approaches zero as δ approaches zero. But

$$F(t + \delta) - F(t) = \Pr(X \leq t + \delta) - \Pr(X \leq t) = \Pr(t < X \leq t + \delta).$$

So, a continuous CDF corresponds exactly to the case where the probability that the random variable X falls into a small interval approaches zero as the length of the interval approaches zero. We can see this as equivalent to the statement that there is no single value of X that has a non-zero probability of occurring. Note that many people reserve the term 'continuous random variable' for a random variable that has a continuous CDF.

On the other hand, we may have a type of continuous random variable where there are certain values that are associated with a non-zero probability. For example, we may consider the time between calls received at a call center and record this as zero if calls are queued for attention. Then we have a continuous random variable that can only take a positive or zero value, but with a fixed probability of a zero. This type of behavior is linked to the CDF function F being discontinuous. Roughly speaking, we can say that a jump in the CDF function occurs at a value where there is an accumulation of possible outcomes.

We can formulate statements about the CDF that match the statements about probabilities. Since probabilities of events are non-negative, we can say that the CDF is increasing; i.e. $F(x) \geq F(y)$ when $X > Y$. Moreover, the probabilities of different values of X must add up to 1 (which is another way of saying that $\Pr(\Omega) = 1$). Thus, $\Pr(-t < X \leq t)$ approaches 1 as t approaches infinity. From this we deduce that

$$\lim_{t \to \infty} F(t) = 1 \text{ and } \lim_{t \to -\infty} F(t) = 0.$$

When the CDF function is well behaved, we can find its derivative, which we call the *probability density function*, $f(t)$. To define this formally we need to go back to the definition of a derivative: we have

$$f(t) = \lim_{\delta \to 0} \frac{F(t + \delta) - F(t)}{\delta} = \lim_{\delta \to 0} \frac{1}{\delta}(\Pr(t < X \leq t + \delta)).$$

Since the density function f is the derivative of the cumulative distribution function F, we can turn this around and say that the CDF is given by the integral of the probability density function.

It is a fact that continuous functions are differentiable almost everywhere, and when we take the integral of any function then its value at isolated points will not make a difference. Without getting into the fine detail needed for a theoretical treatment of this, we can say that *any* random variable with a continuous CDF will have a density function f with the property

$$F(t) = \int_{-\infty}^{t} f(x)dx.$$

There are many standard forms of continuous random variable distributions. We will introduce the normal distribution later. But other examples are the uniform distribution, which has a density function $f(x) = 1/(b - a)$ over an interval $x \in [a, b]$ and is zero outside this range, and the exponential distribution, which has a density function $f(x) = \lambda e^{-\lambda x}$ for $x \geq 0$ and is zero otherwise.

A.4   Means and variances

It is natural to ask about the average value of a random variable. Because the word average can have slightly different meanings depending on the context, we will usually use the term *mean value*, and we will also refer to this as the *expected value*. The expected value of a discrete random variable X, which takes values $x_1, x_2, \ldots, x_n$, and where $\Pr(X = x_i) = p_i$, $i = 1, 2, \ldots, n$ is given by

$$E(X) = \sum_{i=1}^{n} p_i x_i. \tag{A.1}$$

Thus, for example, the expected value of a fair dice which takes values from 1 to 6, each with probability 1/6 is given by

$$\frac{1}{6} \times 1 + \frac{1}{6} \times 2 + \frac{1}{6} \times 3 + \ldots + \frac{1}{6} \times 6 = \frac{21}{6} = 3.5.$$

The expected value can also be characterized as the average value obtained over N samples from the distribution. So, if we throw a dice 100 times we would expect the total value of all the dice throws to be $100 \times 3.5 = 350$. This follows from thinking about the number of times that different results are expected to occur: in 100 throws there should be $100p_1$ occasions when x_1 occurs, $100p_2$ occasions when x_2 occurs, and so on. Thus, the total of 100 draws of the random variable X will be approximately $\sum_{i=1}^{n} 100 p_i x_i = 100 E(X)$.

Equation (A.1) for a discrete random variable has its counterpart for a continuous random variable X with density function f_X. We have

$$E(X) = \int_{-\infty}^{\infty} x f_X(x) dx.$$

When the values that a random variable may take are restricted to a range $[a, b]$, then the density function is zero outside this range and we have $E(X) = \int_{a}^{b} x f_X(x) dx$.

Whether or not a random variable is discrete or continuous, if it is unbounded then it may not have a finite mean. For example, if we take a random variable X that takes values $2, 4, 8, \ldots, 2^n, \ldots$ where $n = 1, 2, \ldots$ and $\Pr(X = 2^n) = 1/2^n$ then the mean is not defined. The appropriate sum is

$$\sum_{i=1}^{\infty} p_i x_i = \sum_{i=1}^{\infty} \frac{1}{2^i} 2^i = \sum_{i=1}^{\infty} 1 = \infty.$$

On the other hand, if we consider the random variable Y that takes values $1, 2, 3, \ldots$ with $\Pr(Y = n) = 1/2^n$, then this has a mean given by

$$E(Y) = \sum_{i=1}^{\infty} \frac{1}{2^i} i = 2.$$

One thing we may want to do with a random variable is to take a function of it in order to create another random variable. For example, we might model the price of a particular stock in a year's time as a random variable X and be concerned to find the expected value of $\log(X)$. Then the log price is another random variable; price can only be positive but log price can take any value between $-\infty$ and ∞. It is important to realize that we cannot just say $E(\log(X)) = \log(E(X))$; this statement is false. Instead, we need to calculate the expectation using the following formula:

$$E(g(X)) = \int_{-\infty}^{\infty} g(x) f_X(x) dx.$$

One thing to watch out for is that we may have a random variable X where $E(X)$ exists, but there are functions g for which $E(g(X))$ does not exist. The only time when things are simple is when g is linear, i.e. $g(X) = aX + b$. Then

$$E(aX + b) = \int_{-\infty}^{\infty} ax \, f_X(x)dx + \int_{-\infty}^{\infty} b f_X(x)dx$$
$$= aE(X) + b. \tag{A.2}$$

Now we consider ways to measure the spread of possible values of a random variable. One straightforward way to measure this would be to look at the average of the difference between a value drawn from the random variable and the mean value of the random variable. Notice that we need to take the absolute value of the difference here, otherwise positive and negative differences will cancel each other out. Thus, if we write $\mu = E(X)$ then a measure of the spread of values in X would be $E(|X - \mu|)$ (whereas $E(X - \mu) = E(X) - \mu = 0$ from Equation (A.2)). Now even though this 'expected absolute difference to the mean' makes a good deal of sense, the modulus signs make it awkward to work with.

An alternative measure is to consider the square difference between a sample from X and the mean μ. Hence, we consider the variance of X which is defined as the *expected value of the square of the distance to mean*:

$$\text{var}(X) = E((X - \mu)^2).$$

Since the variance is the expectation of a random variable $(X - \mu)^2$ which is non-negative, then $\text{var}(X) \geq 0$. The expression for the variance can be simplified. As $(X - \mu)^2 = X^2 - 2\mu X + \mu^2$, we can write

$$\text{var}(X) = E(X^2) - 2\mu E(X) + \mu^2$$
$$= E(X^2) - \mu^2. \tag{A.3}$$

We can also define the standard deviation σ as the square root of the variance, so

$$\sigma = (E(X^2) - \mu^2)^{1/2}.$$

When a random variable is multiplied by a constant k, the standard deviation is multiplied by k and the variance is multiplied by k^2.

As an example, we can consider the random number X given by the number of heads in three tosses of a coin. We have $\Pr(X = 0) = 1/8$, $\Pr(X = 1) = 3/8$, $\Pr(X = 2) = 3/8$, $\Pr(X = 3) = 1/8$. The mean value of X is

$$\mu = \left(\frac{3}{8} \times 1\right) + \left(\frac{3}{8} \times 2\right) + \left(\frac{1}{8} \times 3\right) = \frac{3}{2}.$$

The variance of X can be calculated from looking at

$$E(X^2) = \left(\frac{3}{8} \times 1\right) + \left(\frac{3}{8} \times 4\right) + \left(\frac{1}{8} \times 9\right) = 3.$$

Thus, from Equation (A.3),

$$\text{var}(X) = E(X^2) - \mu^2 = 3 - \frac{9}{4} = \frac{3}{4},$$

and the standard deviation of X is $\sigma = \sqrt{3/4} = 0.866$.

A.5   Combinations of random variables

In the same way that we considered combinations of random events – looking, for example, at unions and intersections – we can also combine random variables. We need to think about an underlying random experiment that produces values for two or more different random variables. For example, we might throw a dice three times and let X be the random variable that is the sum of the first two throws minus the third and Y be the random variable that is the sum of the last two throws minus the first throw. This means that knowing information about the value of one random variable can give us information about the other. Because the random variables X and Y are defined from the same underlying random experiment, we can imagine plotting their possible values on the plane with X values on one axis and Y values on the other. If X and Y are discrete, then we will end up with points sitting on a grid, but if X and Y are continuous, then the possible values are spread over a region in the plane. For this example we have plotted the possible combinations of values for X and Y in Figure A.1. We can see that if we know that $X = -4$ then we can deduce that $Y = 6$. This is because $X = -4$ implies that both the first two throws are 1 and the third throw is 6.

The distribution of values for two (or more) random variables is called the *joint distribution*. Instead of a cumulative distribution function for one variable,

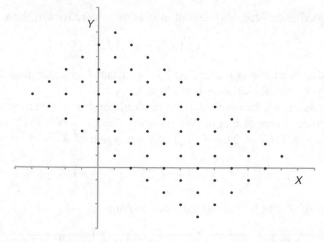

Figure A.1   Combinations of values for the random variables X and Y.

we now have a cumulative distribution function for the joint distribution. This is defined as

$$F_{X,Y}(x, y) = \Pr(X \leq x \text{ and } Y \leq y).$$

Whereas the simple CDF gives the probability that a random variable is less than a certain value, the joint CDF tells us the probability that the combined (X, Y) pair lies in a region (having a lower left-hand quadrant shape). If the underlying random experiment produces values for two continuous random variables, X and Y, then the properties of the joint distribution (giving the way that the combined values are spread out over the (X, Y) plane) are best described through the joint CDF. In many cases there is a joint probability density function that represents the distribution of joint values: in this case, the CDF $F_{X,Y}(x, y)$ is simply the integral of the density function $f_{X,Y}$ over the appropriate region. Thus

$$F_{X,Y}(x, y) = \int_{-\infty}^{x} \left(\int_{-\infty}^{y} f_{X,Y}(u, v) dv \right) du,$$

and

$$\Pr(a \leq X \leq b \text{ and } c \leq Y \leq d) = \int_{a}^{b} \left(\int_{c}^{d} f_{X,Y}(x, y) dy \right) dx.$$

If we are given the joint distribution then it will imply the behaviors for the individual random variables. Since

$$\Pr(X \leq x) = \Pr(X \leq x \text{ and } Y \leq \infty),$$

we can say that

$$F_X(x) = \lim_{y \to \infty} F_{X,Y}(x, y).$$

The distributions of X and Y derived from a joint distribution of (X, Y) are called the *marginal* distributions.

We say that two random variables are independent if the value of one of them gives no information about the value of the other. This can be converted into a statement about events: X and Y are independent random variables if for all sets A and B in R

$$\Pr(X \in A \text{ and } Y \in B) = \Pr(X \in A) \times \Pr(Y \in B).$$

This is very close to the definition we used earlier for random events: we have just replaced the event A (a subset of all possible outcomes Ω) by the event that the random variable X is in A (a subset of the real numbers R). So we could say that random variables X and Y are independent if every possible event defined by X alone is independent of every possible event defined by Y alone.

The next step is to think about the properties of the sum of two random variables; this is the simplest way to combine them. The formula for the mean of the sum of two or more random variables is easy, we simply add the individual means:

$$E(X + Y) = E(X) + E(Y).$$

This is always true – it doesn't require any properties like independence of the variables. One way to see this is by thinking of the average of N draws of the combination $X + Y$. The total is just the same as from N draws of X followed by N draws of Y.

If we are interested in other properties of the sum of random variables, like the variance or the standard deviation, then things are more complicated. If high values of one variable occur when there are low values of the other variable, then they will tend to cancel out in the sum. The end result is that the standard deviation of the sum of two random variables may be less than the standard deviation of either of them taken alone (and equally it might be more than this value).

However, when two random variables are independent then the expected value of their product is the product of their expected values and (derived from this) the variance of their sum is the sum of their variances. Algebraically we can write

$$E(XY) = E(X) \times E(Y), \tag{A.4}$$

$$\mathrm{var}(X + Y) = \mathrm{var}(X) + \mathrm{var}(Y). \tag{A.5}$$

To see why the first equation is true, we can consider X and Y as discrete random variables. In particular, we suppose that X takes values $x_1, x_2, \ldots, x_n$ with probabilities $p_1, p_2, \ldots, p_n$ and Y takes values $y_1, y_2, \ldots, y_m$ with probabilities $q_1, q_2, \ldots, q_m$. Then

$$E(XY) = \sum_{i=1}^{n} \sum_{j=1}^{m} \Pr(X = x_i \text{ and } Y = y_j) x_i y_j$$

$$= \sum_{i=1}^{n} \sum_{j=1}^{m} p_i q_j x_i y_j = \sum_{i=1}^{n} \left(p_i x_i \left(\sum_{j=1}^{m} q_j y_j \right) \right)$$

$$= E(X)E(Y).$$

The result about variances follows from the following argument:

$$\mathrm{var}(X + Y) = E[(X + Y)^2] - [E(X + Y)]^2$$

$$= E(X^2) + 2E(XY) + E(Y^2) - [E(X) + E(Y)]^2$$

$$= E(X^2) - (E(X))^2 + E(Y^2) - (E(Y))^2 + 2E(XY) - 2E(X)E(Y)$$

$$= \mathrm{var}(X) + \mathrm{var}(Y).$$

The standard deviation of a random variable X, which we write as σ_X, is just the square root of the variance, $\text{var}(X)$, so when X and Y are independent,

$$\sigma_{X+Y} = \sqrt{\text{var}(X) + \text{var}(Y)} = \sqrt{\sigma_X^2 + \sigma_Y^2}.$$

We can extend this formula to any number of random variables. The simplest case of all is where we have a set of random variables $X_1, X_2, \ldots, X_N$ which are all independent and also all have the same standard deviation, so we can write

$$\sigma_X = \sigma_{X_1} = \sigma_{X_2} = \ldots = \sigma_{X_N}.$$

Then

$$\sigma_{X_1+X_2+\ldots+X_N} = \sqrt{\sigma_{X_1}^2 + \sigma_{X_2}^2 + \ldots + \sigma_{X_N}^2} = \sqrt{N\sigma_X^2} = \sqrt{N}\sigma_X.$$

We define the covariance of X and Y as

$$\text{cov}(X, Y) = E(XY) - E(X)E(Y).$$

Thus

$$\text{var}(X + Y) = \text{var}(X) + \text{var}(Y) + 2\text{cov}(X, Y).$$

Now from Equation (A.5) we see that $\text{cov}(X, Y) = 0$ when X and Y are independent (though the implication doesn't go the other way around). Also

$$\text{cov}(X, X) = \text{var}(X).$$

If we multiply the random variable X by a constant, then the covariance between X and Y is also multiplied by the same constant, and similarly for the random variable Y. Thus, for any two constants w_X and w_Y we have

$$\text{cov}(w_X X, w_Y Y) = w_X w_Y \text{cov}(X, Y).$$

From this we can deduce

$$\text{var}(w_X X + w_Y Y) = w_X^2 \text{var}(X) + w_Y^2 \text{var}(Y) + 2w_X w_Y \text{cov}(X, Y).$$

If the covariance is positive then high values of X are associated with high values of Y. On the other hand, if the covariance is negative then high values of one variable are associated with low values of the other. The relationship shown in Figure A.1 above has a negative covariance between X and Y.

We can also normalize the size of the covariance by the standard deviation of the variables to produce the correlation coefficient

$$\rho(X, Y) = \frac{\text{cov}(X, Y)}{\sigma_X \sigma_Y},$$

where σ_X is the standard deviation for X and σ_Y is the standard deviation for Y. The effect of this scaling is to guarantee that the correlation lies between -1 (perfect negative correlation) and $+1$ (perfect positive correlation). The correlation is also independent of the scaling of the variables, so

$$\rho(w_X X, w_Y Y) = \rho(X, Y).$$

A.6   The normal distribution and the Central Limit Theorem

The most important continuous probability distribution is the normal distribution. The standard normal distribution can take values over the whole real line and has a density function

$$f(x) = \frac{1}{\sqrt{2\pi}} e^{-x^2/2}.$$

It turns out that the integral $\int_{-\infty}^{\infty} e^{-x^2/2} dx$ has the value $\sqrt{2\pi}$, so the constant term $1/\sqrt{2\pi}$ is inserted to ensure that the integral of the density function is 1. This random variable is symmetric around $x = 0$ so it has a zero mean. Moreover, the variance is also 1. To prove this we need to integrate by parts. Since $xe^{-x^2/2}$ is the derivative of $e^{-x^2/2}$ we have:

$$E(X^2) = \frac{1}{\sqrt{2\pi}} \int_{-\infty}^{\infty} x\left(xe^{-x^2/2}\right) dx$$

$$= \frac{1}{\sqrt{2\pi}} \int_{-\infty}^{\infty} e^{-x^2/2} dx - \left[xe^{-x^2/2}\right]_{-\infty}^{\infty} = 1.$$

We can use a simple transformation, scaling x and shifting it, to produce a normal distribution with any desired mean and standard deviation. To get a mean of μ and standard deviation of σ we use the density function

$$f_{\mu,\sigma}(x) = \frac{1}{\sigma\sqrt{2\pi}} e^{-\frac{1}{2}\left(\frac{x-\mu}{\sigma}\right)^2}.$$

The cumulative distribution function for the normal distribution function is written

$$\Phi_{\mu,\sigma}(x) = \int_{-\infty}^{x} f_{\mu,\sigma}(y) dy.$$

This is a useful function but does not have a closed-form formula. For many years people would look up values for Φ in tables. Nowadays it is better to use the inbuilt NORMDIST function in a spreadsheet.

The importance of the normal distribution arises from the way that it approximates the result of adding together a number of different independent random variables whatever their original distributions: this is the Central Limit Theorem.

There are a few different versions of this theorem and we will describe a version due to Lindeberg and Lévy. We suppose that we have a set of N random variables $X_1, X_2, \ldots, X_N$, with each X_i independent of all the others and each having the same distribution with mean μ and variance σ^2 (which is less than infinity). Then we know that

$$E \left(\sum_{i=1}^{N} X_i \right) = N\mu \text{ and var } \left(\sum_{i=1}^{N} X_i \right) = N\sigma^2.$$

Thus, if we consider the random variable

$$Y_N = \sqrt{N} \left(\frac{1}{N} \left(\sum_{i=1}^{N} X_i \right) - \mu \right)$$

then

$$E(Y_N) = \sqrt{N} \left(\frac{1}{N} (N\mu) - \mu \right) = 0.$$

Also, using the fact that $\text{var}(aX + b) = a^2 \text{var}(X)$,

$$\text{var}(Y_N) = \text{var} \left(\frac{1}{\sqrt{N}} \left(\sum_{i=1}^{N} X_i \right) \right)$$

$$= \frac{1}{N} \text{var} \left(\sum_{i=1}^{N} X_i \right) = \sigma^2.$$

Thus, the variable Y_N has mean 0 and variance σ^2. The Central Limit Theorem concerns the limit as we let the number of random variables being summed go to infinity. So, we suppose that there is an infinite set of independent identically distributed random variables, and we consider the random variable Y_N which involves the average of the first N of the random variables. Then, as N increases, the distribution of Y_N gets closer and closer to a normal distribution with mean 0 and variance σ^2. In particular, the Central Limit Theorem states:

$$\lim_{N \to \infty} \Pr(Y_N \leq \alpha) = \Phi_{0,\sigma}(\alpha).$$

Notice that a more natural way to express the Central Limit Theorem is to say that the distribution of a sum $\sum_{i=1}^{N} X_i$ gets closer and closer to a normal distribution with mean $N\mu$ and variance $N\sigma^2$. But this formulation, though capturing what happens, is unsatisfactory because the limiting distribution is expressed in terms of N and this is a number that goes to infinity. Hence, we have to rescale the sum $\sum_{i=1}^{N} X_i$, essentially to remove the dependence on N. This is what happens when we define the new random variable Y_N.

To illustrate the Central Limit Theorem, we show in Figure A.2 what happens when we consider a random variable that is obtained from the sum of nine individual random variables, each of which is uniformly distributed between 0 and 1. We have

$$Y_9 = \sqrt{9}\left(\frac{1}{9}\left(\sum_{i=1}^{9} X_i\right) - \frac{1}{2}\right)$$

$$= \frac{1}{3}\left(\sum_{i=1}^{9} X_i\right) - \frac{3}{2}$$

where each X_i has a uniform distribution on [0,1]. Thus, the density for X_i is simply $f(x) = 1$ for $0 \leq x \leq 1$. Each individual X_i has a mean of 0.5 but we also want to find the variance. This is given by

$$\text{var}(X_i) = E(X_i^2) - \left(\frac{1}{2}\right)^2$$

$$= \int_0^1 x^2 dx - \frac{1}{4}$$

$$= \left[\frac{x^3}{3}\right]_0^1 - \frac{1}{4} = \frac{1}{12}.$$

Thus, the Central Limit Theorem tells us that the variable Y_N is approximated by a normal distribution with mean 0 and variance 1/12 (i.e. the standard deviation is $\sqrt{1/12}$). Since

$$\sum_{i=1}^{9} X_i = 3Y_9 + \frac{9}{2},$$

this is equivalent to saying that the sum $\sum_{i=1}^{9} X_i$ approximates to a normal distribution with mean 9/2 and standard deviation $3\sqrt{1/12} = 0.86603$.

It is not possible to give any kind of formula for the distribution of $\sum X_i$, but we can generate random samples from its distribution and this has been done in Figure A.2. The circles are generated from a sample to find, for example, the probability that $\sum X_i$ takes a value between 4.25 and 4.5 – it is 0.1125. The solid line shows the corresponding normal distribution, which gives a predicted probability of 0.11358 for this number. The difference between the two shows up as the circle being just below the solid line at this value (the highest points in the diagram). Clearly the normal distribution provides an extremely good approximation for the sum of nine uniform distributions.

Besides arising naturally from the Central Limit Theorem, a normal distribution is also easy to work with because the sum of two or more independent random variables each with a normal distribution also has a normal distribution. Thus, if X has a normal distribution with mean μ_1 and variance σ_1^2 and Y has a normal distribution with mean μ_2 and variance σ_2^2, and X and Y are independent,

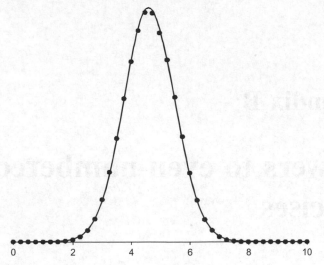

Figure A.2   Illustration of the Central Limit Theorem: sum of nine uniform distributions.

then $X + Y$ has a normal distribution with mean $\mu_1 + \mu_2$ and variance $\sigma_1^2 + \sigma_2^2$. Notice that this property does not hold for *any* other distribution: the sum of two uniform distributions is not a uniform distribution; the sum of two exponential distributions is not an exponential, etc.

Appendix B

Answers to even-numbered exercises

1.2 Connaught

(a) The executive accused of insider trading is rather a different problem than the underlying difficulties causing a 60% drop in market capitalization. The profit warning is related to business risk – a drying up of orders. But it also contains an element of operational risk arising from poor accounting systems (see part (b)).

(b) The unusual thing in this case is how quickly and dramatically the shares fell. This was because the firm had given no warning of such a sharp turnaround and had been claiming that the fundamentals were in good shape. The problem here is that the accounts have reflected a very optimistic view about what will happen. It is no surprise that the group's accounting practices have been under scrutiny. This is particularly a problem since the earnings that now look suspect were responsible for large bonus payments to executives. An appropriate risk management strategy would involve the company's accountants taking a more conservative view of future earnings.

1.4 Product form for heat map

It turns out to be surprisingly difficult to choose the numbers to make things work out, but here is one arrangement that works: Insignificant = $10 000;   Minor = $20 000;   Moderate = $42 000;   Major = $120 000;

Business Risk Management: Models and Analysis, First Edition. Edward J. Anderson.
© 2014 John Wiley & Sons, Ltd. Published 2014 by John Wiley & Sons, Ltd.
Companion website: www.wiley.com/go/business_risk_management

Catastrophic = \$300 000. Notice that we must have the \$ cost for moderate risk being greater than \$40 000 to avoid a 'medium' square switching to 'low' and less than \$45 000 to avoid a square that should be 'medium' switching to 'high'.

2.2 Changing a daily failure rate into a yearly one

(a) Let X_i be a random variable associated with day i that is 1 if there is a failure and 0 if there is not. The expected value of $X_i = 0.001 \times 1 + 0.999 \times 0 = 0.001$. Clearly the number of failures in 365 days is $N = \sum_{i=1}^{365} X_i$. Then

$$E(N) = \sum_{i=1}^{365} E(X_i) = \sum_{i=1}^{365} 0.001 = 0.365.$$

(b) The probability that there is no failure on any of the 365 days is the probability that each X_i is 0, $i = 1, 2, \ldots, 365$. If the X_i are independent, this probability is given by $0.999^{365} = 0.6941$. Thus, the probability that there is at least one failure over this period is $1 - 0.6941 = 0.3059$. (This is significantly smaller than the figure of 0.365 that might be our first guess of the probability.)

2.4 Combining union and intersection risk

(a) In the Venn diagram of Figure B.1 we have set A = poor weather, B = late delivery, and C = archaeological find. The shaded area in the diagram is the event of a greater than four-week delay.

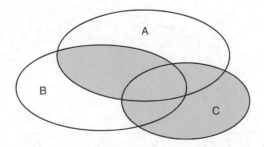

Figure B.1   Venn diagram for Exercise 2.4.

(b) This has probability (assuming independence) of

$$\Pr(C) + \Pr(A \cap B) - \Pr(A \cap B \cap C)$$
$$= 0.05 + 0.2 \times 0.1 - 0.2 \times 0.1 \times 0.05 = 0.069.$$

2.6 Portfolios with large numbers and different variances

We have $\sigma_i \leq \sigma_{\max}$, for each i, where $\sigma_{\max}$ is the largest standard deviation. Let $w_i = K/\sigma_i$ where we take

$$K = \frac{W}{\left(\frac{1}{\sigma_1} + \frac{1}{\sigma_2} + \ldots + \frac{1}{\sigma_N}\right)}$$

in order that $w_1 + w_2 + \ldots w_N = W$. Now each $1/\sigma_i$ is at least as big as $1/\sigma_{\max}$. Thus

$$K \leq \frac{W}{N(1/\sigma_{\max})} = \frac{W\sigma_{\max}}{N},$$

and the overall standard deviation is

$$\sigma_Z = \sqrt{w_1^2\sigma_1^2 + w_2^2\sigma_2^2 + \ldots + w_N^2\sigma_N^2}$$

$$= \sqrt{K^2 + K^2 + \ldots K^2} \leq \frac{W\sigma_{\max}}{N}\sqrt{N} = \frac{W\sigma_{\max}}{\sqrt{N}}.$$

2.8 Optimal portfolio with a risk-free asset

The optimal portfolios have standard deviations of 0.95, 0.47 and 0.16. Figure B.2 shows how these are arranged in a straight line.

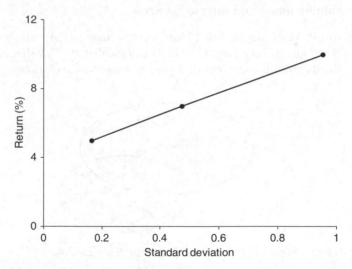

Figure B.2   Return is a linear function of standard deviation.

The portfolios are as follows:

Return	Weight on A	Weight on B	Weight on C	Weight on D
10	0.2236	0.4099	0.1491	0.2174
7	0.1118	0.2050	0.0745	0.6087
5	0.0373	0.0683	0.0248	0.8696

The underlying portfolio for A, B and C has weights 0.2857, 0.5238 and 0.1905 and these ratios remain the same throughout, but with a varying proportion on D.

2.10 Copula properties

We have the following values for $C(a, b)$ depending on the region:

$$= (2/3)ab \text{ when } a \le 3/4,\ b \le 3/4 \text{ (region } A)$$
$$= (1/2)a + 2a(b - (3/4)) \text{ when } a \le 3/4,\ b > 3/4 \text{ (region } B)$$
$$= (1/2)b + 2b(a - (3/4)) \text{ when } a > 3/4,\ b \le 3/4 \text{ (region} C)$$
$$= (3/8) + 2(3/4)(a - 3/4) + 2(3/4)(b - 3/4) - 2(a - 3/4)(b - 3/4)$$
$$\text{otherwise (region } D).$$

With these definitions we have

$$C(a, 1) = (1/2)a + 2a(1/4) = a \text{ when } a \le 3/4$$
$$C(1, b) = (1/2)b + 2b(1/4) = b \text{ when } b \le 3/4$$
$$C(a, 1) = (3/8) + (3/2)(a - 3/4) + (3/2)(1/4) - (a - 3/4)(1/2)$$
$$= a \text{ when } a > 3/4$$
$$C(1, b) = (3/8) + (3/2)(1/4) + (3/2)(b - 3/4) - (1/2)(b - 3/4)$$
$$= b \text{ when } b > 3/4.$$

Thus, C has uniform marginal distributions as required. We check the derivatives with respect to a (the b derivatives are similar).

$$\partial C/\partial a = (2/3)b > 0 \text{ in } A$$
$$= (1/2) + 2(b - (3/4)) > 0 \text{ in } B$$
$$= 2b > 0 \text{ in } C$$
$$= 2(3/4) - 2(b - 3/4) = 3 - 2b > 0 \text{ in } D.$$

Moreover, C is constructed to be continuous on the boundaries. So C is increasing in both arguments. However, the rectangle inequality does not hold since

$$C(3/4, 3/4) + C(1, 1) - C(3/4, 1) - C(1, 3/4)$$
$$= (3/8) + 1 - (3/4) - (3/4) = -1/8.$$

2.12 Upper tail dependence

Note

$$\Pr(X > F_X^{-1}(\alpha), Y > F_Y^{-1}(\alpha)) = 1 - \Pr(X \le F_X^{-1}(\alpha) \text{ or } Y \le F_Y^{-1}(\alpha))$$
$$= 1 - \Pr(X \le F_X^{-1}(\alpha)) - \Pr(Y \le F_Y^{-1}(\alpha))$$
$$+ \Pr(X \le F_X^{-1}(\alpha) \text{ and } Y \le F_Y^{-1}(\alpha))$$
$$= 1 - 2\alpha + C(\alpha, \alpha).$$

So, setting $\delta = 1 - \alpha$ the formula for λ_u becomes

$$\lambda_u = \lim_{\delta \to 0} \frac{1 - 2(1 - \delta) + C(1 - \delta, 1 - \delta)}{\delta}$$
$$= 2 + \lim_{\delta \to 0} \frac{C(1 - \delta, 1 - \delta) - 1}{\delta}.$$

Substituting for the Clayton copula we get

$$\lambda_u = 2 + \lim_{\delta \to 0} \frac{\left(\frac{2}{(1-\delta)^2} - 1\right)^{-\frac{1}{2}} - 1}{\delta}.$$

Now the limit here is equal to the negative of the derivative of $g(x) = (2x^{-2} - 1)^{-\frac{1}{2}} - 1$ evaluated at $x = 1$ (note $g(1) = 1$.) But

$$\frac{dg}{dx} = (1/2)(2x^{-2} - 1)^{-\frac{3}{2}}(4x^{-3}).$$

Since this takes the value 2 at $x = 1$ we obtain the result we want, that $\lambda_u = 0$.

3.2 VaR for a triangle distribution

We could calculate the CDF, but an alternative is to work directly with the area under a triangle. We will work in 1000s. We want to find the a such that $\int_a^{100} f(x)dx = 0.01$. Thus (using the half base times vertical height rule for the area of a triangle) we need

$$(1/2)(100 - a)f(a) = (1/2)(100 - a)^2/100^2 = 0.01.$$

Thus

$$(100 - a)^2 = 200$$

and $a = 100 - \sqrt{200} = 85.858$, giving a 99% VaR of \$85 858. Similarly, the 95% VaR is obtained from $(100 - a)^2 = 1000$, giving $a = 100 - \sqrt{1000} = 68.377$ and a 95% VaR of \$68 377.

3.4 Diversification reduces VaR

The 98% absolute VaR from investing in both A and B is $6000 – this follows from the table of possible results and their probabilities. There is a greater than 2% chance that one or other of the firms fails, but a less than 2% chance that they both fail. On the other hand, making a $20 000 investment in two bonds from A gives a 4% chance of collecting only 30% of the face value – and so the 98% absolute VaR is $14 000.

3.6 VaR estimates are a process

(a) Two totally different systems may have the same number of VaR breaches but those breaches occur at very different times. To achieve a process with the right number of breaches we could simply set up a VaR procedure which on 1 day out of 100 predicts a very large profit (so large that there is almost bound to be a breach) and on the other 99 days predicts a very large loss, large enough that the loss does not occur. This is a poor procedure because it tells us nothing about the real risks; which shows that we cannot judge the process just by the number of breaches.

(b) The answer is that they do not need to be equally good (see the discussion above). A good procedure is one that is informative, and this means that each number produced is the right number in the sense of predicting the right risk level for the day in question (rather than too high on some days and too low on others).

3.8 Expected shortfall is subadditive

(a) The expected shortfall is a risk measure that scales and shifts appropriately, so that we can say that the expected shortfall for a normal distribution with mean −5000 and standard deviation 3000 is

$$-5000 + 2.667 \times 3000 = 3001.$$

(b) The expected shortfall for the second project can be calculated in the same way. It is given by

$$-3000 + 2.667 \times 1500 = 1000.5.$$

The sum of the two projects has a normal distribution with mean −8000 and a variance that is given by the sum of the two variances. Hence, the standard deviation is given by

$$\sqrt{3000^2 + 1500^2} = 3354.$$

This means that the expected shortfall is

$$-8000 + 2.667 \times 3354 = 945.1$$

which is (much) less than the sum of the expected shortfalls from the individual projects, which is $3001 + 1000.5 = 4001.5$.

4.2 Fréchet distribution

With a tail index of 5 we know that the distribution of the maximum will approach a CDF of $F_{max}(a_N x) = \exp(-x^{-5})$. Assume that the number of times on any given day is consistent enough to allow them all to be modeled by the same distribution. We use the information on the mean of this distribution to estimate the scaling parameter a_N. The mean value of the Fréchet with parameter $\alpha = 5$ is given by $\Gamma(1 - (1/5)) = 1.1642$.

We have $F_{max}(x) \simeq \exp-(x/a_N)^{-5})$ which will have a mean of $1.1642 a_N$. (This needs careful thought – we might first think that the mean is $1.1642/a_N$.) Thus, $a_N = 12/1.1642 = 10.3$. Hence, we can estimate the distribution as

$$F_{max}(x) = \exp\left(-\left(\frac{x}{10.3}\right)^{-5}\right).$$

So the probability that x is less than 20 is $F_{max}(20) = \exp(-(20/10.3)^{-5}) = 0.964$. We have about a 3.6% chance of the guarantee being broken on any given day, and hence an expected value of $100/3.6 = 28$ days before this happens.

4.4 Calculating the mean of the GPD

The density function for the GPD is $(1/\beta)(1 + \xi x/\beta)^{-1/\xi-1}$. Hence, the mean of this distribution is given by the limit for large R of

$$\int_0^R (x/\beta)(1 + \xi x/\beta)^{-(1/\xi)-1} dx = \frac{1}{\xi - 1}\left[\frac{x + \beta}{(1 + \xi x/\beta)^{1/\xi}}\right]_0^R$$

$$= \frac{1}{\xi - 1}\left(\frac{R + \beta}{(1 + \xi R/\beta)^{1/\xi}} - \beta\right)$$

$$= \frac{\beta}{1 - \xi}\left(1 - \frac{1 + R/\beta}{(1 + \xi R/\beta)^{1/\xi}}\right).$$

If $\xi < 1$ then there is some $\delta > 0$ with $\xi = 1/(1 + \delta)$. Then

$$\frac{1 + R/\beta}{(1 + \xi R/\beta)^{1/\xi}} < (1 + \xi R/\beta)^{1-1/\xi} = (1 + \xi R/\beta)^{-\delta}.$$

But as R goes to infinity, $(1 + \xi R/\beta)^{-\delta}$ goes to zero. Hence, the integral takes the value $\beta/(1 - \xi)$ in the limit of large R.

4.6 Estimating parameters from mean excess figures

Write β_{10} for the β value for the $G_{\xi,\beta}$ distribution describing the excess over \$10 million, β_{20} for the equivalent for \$20 million, etc. Then the formula $\beta/(1-\xi)$ for the mean excess shows that

$$\beta_{10}/(1-\xi) = 9,$$
$$\beta_{20}/(1-\xi) = 12.$$

We also know that $\beta_{20} = \beta_{10} + 10\xi$. So, $10\xi/(1-\xi) = 3$ and hence $13\xi = 3$ and

$$\xi = 3/13 = 0.23$$

while

$$\beta_{20} = 12(1 - 0.23) = 9.24.$$

Finally we deduce that

$$\beta_{25} = \beta_{20} + 5\xi = 9.24 + 5 \times 0.23 = 10.39.$$

5.2 EUT and a business venture

(a) With a linear utility function, maximizing expected utility is the same as maximizing expected profit. Option A (invest) gives return of \$40 with certainty. If Kate pays back an amount x then the expected value of the investment from Option B (loan to Kate) is $0.7x$ (with probability 0.3 James ends with nothing). To make this attractive it has to at least match the certain profit, i.e.

$$x = 1040/0.7 = \$1485.71.$$

(b) The underlying equation for this problem is

$$u(1040) = 0.7u(x).$$

When $u(x) = \sqrt{x}$ we can rewrite this as

$$\sqrt{x} = \frac{\sqrt{1040}}{0.7} = 46.07.$$

So

$$x = 46.07^2 = \$2122.45$$

(which is an amount greater than Kate can repay.)

5.4 Stochastic dominance and negative prospects

We assume that $x_1 < x_2 < \ldots < x_n$ and everything is in dollar values. We may assume without loss of generality that the same outcomes appear in both A and B (by setting probabilities to zero if necessary). Notice that the worst outcomes in $-A$ and $-B$ are $-x_n$ and $-y_n$. The condition we need to show in order to establish that $-B$ stochastically dominates $-A$ involves sums for probabilities for outcome i and all better outcomes:

$$q_i + q_{i-1} + \ldots + q_1 \geq p_i + p_{i-1} + \ldots + p_1$$

for all i with at least one of these inequalities being strict. However, this condition is exactly the one given as an alternative to the first definition of stochastic dominance for A over B in Inequality (5.5).

5.6 Failure of stochastic dominance

We can sum over outcomes \$700 or better to find that q has the higher probability. But summing over outcomes \$1000 or better shows that r has the higher probability. Thus, neither stochastically dominates the other.

This also gives a clue as to how to arrange the utilities for the result we want. To get q preferred we can take utilities shown as Utility A in Table B.1 and to get r preferred take utilities shown as Utility B. The other columns sum up components to get an expected utility value (there are many different ways to achieve the same result).

Table B.1   Two sets of utility assignments for Exercise 5.6.

	Utility A	q	r	Utility B	q	r
\$100	1	0.1	0	1	0.1	0
\$300	2	0.4	0.6	2	0.4	0.6
\$400	3	0.6	0	3	0.6	0
\$500	4	0	1.2	4	0	1.2
\$700	20	6.0	4.0	5	1.5	1.0
\$900	21	4.2	2.1	6	1.2	0.6
\$1000	22	0	2.2	20	0	2.0
totals		11.3	10.1		3.8	5.4

6.2 Prospect Theory when gains turn to losses

Since all the x and y are positive and prospect A is preferred to B, we have

$$\sum_{i=1}^{n}[w^+(p_i + \ldots + p_n) - w^+(p_{i+1} + \ldots + p_n)]v(x_i)$$

$$> \sum_{i=1}^{n}[w^+(q_i + \ldots + q_n) - w^+(q_{i+1} + \ldots + q_n)]v(y_i).$$

Substituting for $v(x_i)$ and $v(y_i)$ using $v(-x_i) = -\lambda v(x_i)$ and $v(-y_i) = -\lambda v(y_i)$ shows that

$$-(1/\lambda)\sum_{i=1}^{n}[w^+(p_i + \ldots + p_n) - w^+(p_{i+1} + \ldots + p_n)]v(-x_i)$$

$$> -(1/\lambda)\sum_{i=1}^{n}[w^+(q_i + \ldots + q_n) - w^+(q_{i+1} + \ldots + q_n)]v(-y_i).$$

Multiplying through by $-\lambda$ changes the direction of the inequality, hence we have

$$\sum_{i=1}^{n}[w^-(p_i + \ldots + p_n) - w^-(p_{i+1} + \ldots + p_n)]v(-x_i)$$

$$< \sum_{i=1}^{n}[w^-(q_i + \ldots + q_n) - w^-(q_{i+1} + \ldots + q_n)]v(-y_i).$$

This is exactly the inequality we need to show that prospect $-B = (-y_1, q_1; -y_2, q_2; \ldots; -y_n, q_n)$ is preferred to $-A = (-x_1, p_1; -x_2, p_2; \ldots; -x_n, p_n)$.

6.4 Exponent in power law

The standard TK parameters have

$$v(x) = x^{0.88} \text{ and } w^+(p) = 0.132.$$

This means that values for the prospects are as follows

$$V(A) = v(1)(1 - w^+(0.05)) + v(381)w^+(0.05) = 25.52.$$
$$V(B) = v(20) = 13.96.$$
$$V(C) = v(301)(1 - w^+(0.05)) + v(681)w^+(0.05) = 172.81.$$
$$V(D) = v(320) = 160.15.$$

Prospect Theory predicts that both A and C are preferred.

Changing the exponent to 0.6 gives values

$$V(A) = 5.54, V(B) = 6.03, V(C) = 33.26, V(D) = 31.85,$$

which gives an accurate prediction of the experiment.

6.6 Splitting prospects can change choices

(a) For the four prospects the values are

$$V(A) = w^-(0.5)v(-100) + w^+(0.5)v(1000)$$

$$= 0.454 \times (-126.2) + 0.421 \times (501.2) = 153.71.$$

$$V(B) = w^+(0.4)v(1000) = 0.370 \times (501.2) = 185.44.$$

$$V(C) = (w^+(1) - w^+(0.7))v(200) + (w^+(0.7) - w^+(0.4))v(300)$$

$$+ (w^+(0.4))v(550)$$

$$= (1 - 0.534) \times 117.74 + (0.534 - 0.370) \times 169.59$$

$$+ 0.370 \times 292.64$$

$$= 190.96.$$

$$V(D) = v(340) = 189.82.$$

Hence C has the highest value and is preferred.

(b) The second stage gambles are $A' = (-\$440, 0.5; \$660, 0.5)$, $B' = (-340, 0.6; \$660, 0.4)$, $C' = (-140, 0.3; -40, 0.3; 210, 0.4)$.

$$V(A') = w^-(0.5)v(-440) + w^+(0.5)v(660)$$

$$= 0.454 \times (-478.8) + 0.421 \times (344.8) = -72.21.$$

$$V(B') = w^-(0.6)v(-340) + w^+(0.4)v(660)$$

$$= 0.518 \times (-379.6) + 0.370 \times (344.8) = -69.06.$$

$$V(C') = (w^-(0.6) - w^-(0.3))v(-40) + w^-(0.3)v(-140)$$

$$+ w^+(0.4))v(210)$$

$$= (0.474 - 0.328) \times (-55.3) + 0.328 \times (-170.8)$$

$$+ 0.370 \times 123.03$$

$$= -18.57.$$

All the values are negative, so none of these gambles would be accepted.

7.2 Ajax Lights

The firm believes that the monthly demand is $K - \alpha p$ where p is the price per bulb. So $K - 10\alpha = 50\,000$. And if S are sold next month at $8 then this will demonstrate that $K - 8\alpha = S$. Solving these two equations gives $\alpha = (S/2) - 25\,000$ and $K = 5S - 200\,000$.

Given Y and sales of S then the amount left to sell after one month is $100\,000 - S + Y$. The sales per month need to be half this and must match the demand per month. Thus (using the formulae for K and α that we have derived)

$$50\,000 - (S/2) + (Y/2) = (5S - 200\,000) - ((S/2) - 25\,000)p.$$

Solving for p gives

$$p = 10 - \frac{S}{5000} + \frac{Y}{50\,000}.$$

The stage 1 costs of the order (allowing for the previous cost of the existing stock assumed at \$5 per bulb) are

$$C_1(Y) = 500\,000 + 5Y.$$

Then write $Q(Y, S)$ for the total revenue arising from all sales (price times volume), so

$$Q(Y, S) = (100\,000 - S + Y)(10 - \frac{S}{5000} + \frac{Y}{50\,000}) + 8S.$$

Now suppose that $S = S_i$ with probability p_i (where $\sum_{i=1}^{N} p_i = 1$) then

$$\Pi(Y) = -C_1(Y) + \sum_{i=1}^{N} p_i Q(Y, S_i).$$

Taking derivatives, the optimal Y is given by

$$5 + \sum_{i=1}^{N} p_i \frac{(100\,000 - S_i + Y)}{50\,000} + \sum_{i=1}^{N} p_i (10 - \frac{S_i}{5000} + \frac{Y}{50\,000}) = 0$$

$$17 + \sum_{i=1}^{N} p_i \frac{-11S}{50\,000} + \frac{Y}{25\,000} = 0.$$

Hence

$$Y = \frac{11\overline{S}}{2} - 17 \times 25\,000$$

where $\overline{S} = \sum_{i=1}^{N} p_i S_i$ is the average value of S.

7.4 Non-anticipatory constraints in APS example

The demand in month 1 for scenario A and scenario C is different, and so normally this would allow different choices to be made for x_2 under these different scenarios. However, in this problem there is no correlation between demands in different periods. Moreover, the situation at the end of month 1 is the same for both scenarios – once demand is more than 50 there can be no extra sales. In both scenario A and scenario C, month 2 starts with just what was ordered at the start of month 1 (i.e. x_1), so a non-anticipatory solution (which cannot take account of demand information from the future) will order the same amount for these two scenarios.

7.6 VaR constraints on a portfolio with a bond

Suppose that the portfolio weights are w_A, w_B and w_C, where w_C is the proportion in the treasury bond. Then the expected profit after three years is proportional to $250w_A + 300w_B + 200w_C$. The distribution of profit is normal with this mean, and standard deviation

$$\sqrt{w_A 100^2 + w_B 150^2} = 10\sqrt{100w_A + 225w_B}.$$

To achieve a probability of less than 0.01 of having a negative profit implies a z value of 2.3263 or more. Thus the constraint is that

$$\frac{250w_A + 300w_B + 200w_C}{10\sqrt{100w_A + 225w_B}} \geq 2.3263.$$

Squaring this constraint we get an optimization problem

$$\text{maximize} \quad 250w_A + 300w_B + 200w_C$$

$$\text{subject to} \quad (25w_A + 30w_B + 20w_C)^2 \geq 5.4119(100w_A + 225w_B)$$
$$w_A + w_B + w_C = 1$$
$$w_A \geq 0, \, w_B \geq 0, \, w_C \geq 0.$$

This can be solved using a spreadsheet and Solver, or more directly. We can substitute for $w_C = 1 - w_A - w_B$ to get

$$\text{maximize} \quad 50w_A + 100w_B + 200$$

$$\text{subject to} \quad (5w_A + 10w_B + 20)^2 \geq 5.4119(100w_A + 225w_B)$$
$$w_A \geq 0, \, w_B \geq 0, \, w_A + w_B \leq 1.$$

Figure B.3 (where dashed lines are contours of the objective) shows that a solution occurs where the line $w_A + w_B = 1$ intersects the constraint $(5w_A + 10w_B + 20)^2 = 5.4119(100w_A + 225w_B)$, and this occurs when $w_A = 0.8012$ and $w_B = 0.1988$. Here, $w_C = 0$ and we do not invest in the bond in this case.

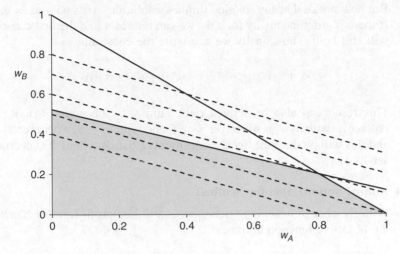

Figure B.3   Feasible region shaded for portfolio optimization.

8.2 Coefficients are unrelated

The formulation of this problem is similar to the budget of uncertainty arrangement, but with $B = 3$ so it does not provide an additional constraint. Thus, the constraint can be rewritten as

$$\overline{a}_1 x_1 + \overline{a}_2 x_2 + \overline{a}_3 x_3 + w_1 + w_2 + w_3 + 3t \leq b,$$

$$w_1 + t \geq \delta_1 |x_1|$$

$$w_2 + t \geq \delta_2 |x_2|$$

$$w_3 + t \geq \delta_3 |x_3|$$

$$w_i \geq 0, t \geq 0.$$

The question is: how can this be rewritten as a single constraint?

Observe that any solution (w_1, w_2, w_3, t) can be replaced with a new solution with the existing t added to each w_i and then the new t set to zero, i.e. we use the new solution $(w_1', w_2', w_3', t') = (w_1 + t, w_2 + t, w_3 + t, 0)$. This means that we can assume $t = 0$ without loss of generality to get constraints

$$\overline{a}_1 x_1 + \overline{a}_2 x_2 + \overline{a}_3 x_3 + w_1 + w_2 + w_3 \leq b$$

$$w_1 \geq \delta_1 |x_1|$$

$$w_2 \geq \delta_2 |x_2|$$

$$w_3 \geq \delta_3 |x_3|$$

$$w_i \geq 0.$$

But now notice that any solution implies a solution exists with $w_i = \delta_i |x_i|$. If there is strict inequality for a w_i we can reduce it and the first constraint will still hold. Thus, finally we can write the constraint as

$$\overline{a}_1 x_1 + \delta_1 |x_1| + \overline{a}_2 x_2 + \delta_2 |x_2| + \overline{a}_3 x_3 + \delta_3 |x_3| \leq b.$$

This result can also be derived quite simply by observing that we can replace a_i with $\overline{a}_i + \delta_i$ whenever $x_i \geq 0$ and with $\overline{a}_i - \delta_i$ whenever $x_i < 0$ and this will produce the tightest constraints existing within the uncertainty set A.

8.4 Robust optimization for Sentinel

With the change in selling price there is \$40 less profit from each sale, so the objective function becomes

$$120 z_L + 100 z_S - 70(x_L + x_S)$$

with the same constraints as before. The new final problem is

$$\text{maximize} \quad G$$

$$\text{subject to} \quad G \leq 120 x_L - 70(x_L + x_S)$$
$$G \leq 120 \times 5000 - 70(x_L + x_S)$$
$$G \leq 100 x_S - 70(x_L + x_S)$$
$$x_L \geq 0, x_S \geq 0.$$

The solution to this problem is just as claimed – everything zero. The implication is that with these data it is impossible to guarantee a profit, however x_L and x_S are chosen. The reason is that the loss from having too much (\$70) is more than the profits from sales \$50 and \$30 for the two sizes. So, if nature moves demand away from whichever is the larger order, the losses on that larger order are greater than the profits on the smaller order and Sentinel ends up making a loss.

8.6 Provence Rentals

(a) Let x be the number of GPS systems installed and p be the proportion of customers wanting a GPS system. Then Provence Rentals makes \$4 for each of $\min(x, 100p)$ cars and loses \$10 for each of $\max(100p - x, 0) = 100p - \min(x, 100p)$ cars. Thus, the profit made by Provence Rentals is

$$2000 \min(x, 100p) - 5000 \max(100p - x, 0) - 750x$$
$$= 7000 \min(x, 100p) - 500\,000p - 750x.$$

The robust optimization problem is

$$\max_{x} \left[\min_{p \in [0,1]} (7000 \min(x, 100p) - 500\,000p - 750x) \right].$$

(b) The objective function in the inner minimization is concave in p, in fact the p dependence is given by

$$7000 \min(x, 100p) - 500\,000p.$$

Thus, it is minimized at an end point of the interval $[0, 1]$.

(c) If $p = 0$ then the profit is $-750x$ and if $p = 1$ the profit is $7000x - 500\,000 - 750x$. So the problem becomes

$$\max_{x}[\min(-750x, 6250x - 500\,000)].$$

The maximization can never produce a guaranteed profit, but we do not want the loss to be too great. When taken as a function of x, the function $\min(-750x, 6250x - 500\,000)$ is maximized where the two lines cross, i.e. at

$$-750x = 6250x - 500\,000$$

$$x = 500\,000/7000 = 71.4.$$

Provence Rentals should fit 71 of its cars with GPS systems.

9.2 Pop concerts

(a) The expected value for the first investment is simple: \$12\,000 in expected ticket sales means that the expected profit is \$12\,000 − \$10\,000 = \$2000.

(b) We work in units of \$1000. Since investors are guaranteed to receive 9 back from their initial investment of 10, the net position is -1. On top of this, investors receive 5% of profit over 171. Write X for the expected ticket sales. X is a normal random variable with mean $\mu = 230$ and standard deviation 80. Then the expected profit is

$$E(\Pi) = -1 + (1/2)E[\max(X - 171, 0)]$$

$$= -1 + (1/20)[(230 - 171)(1 - \Phi_{\mu,\sigma}(171)) + 80^2 \varphi_{\mu,\sigma}(171)]$$

$$= -1 + (1/20)59(1 - 0.2304) + (1/20)80^2 \times 0.003799$$

$$= 2.486.$$

So the expected profit is \$2486, which is higher than (a) and so this is the preferred investment.

9.4 SambaPharm

A decision will be made on the basis of information on the effectiveness, which we measure as a fraction p. The cutoff value for p is given as $p = 0.4$. The profit for years 4 to 8 is given as \$50 k for each 0.1 increase in p above 0.4. Working in \$1000s this can be written as $500 \max(0, p - 0.4)$. Hence, we have a net present value that is given by

$$NPV = -60 - \frac{60}{1.08} - \frac{60}{1.08^2} + 500\max(0, p - 0.4)$$
$$\times \left(\frac{1}{1.08^3} + \frac{1}{1.08^4} + \frac{1}{1.08^5} + \frac{1}{1.08^6} + \frac{1}{1.08^7} \right).$$

We need to take the expectation of this for p uniformly distributed on $[0, 0.8]$. Using the formula we have

$$E[\max(0, p - 0.4)] = \frac{(0.8 - 0.4)^2}{2(0.8 - 0)} = 0.1.$$

So the expected NPV is

$$E(NPV) = -60 - \frac{60}{1.08} - \frac{60}{1.08^2}$$
$$+ 50 \left(\frac{1}{1.08^3} + \frac{1}{1.08^4} + \frac{1}{1.08^5} + \frac{1}{1.08^6} + \frac{1}{1.08^7} \right)$$
$$= 4.16.$$

9.6 Option prices imply distribution

A call option becomes more valuable as the price of the stock goes up. Assuming a normal distribution, a fair price for a call option with a strike price of a is

$$E(\max(X - a, 0)) = (\mu - a)(1 - \Phi_{\mu,\sigma}(a)) + \sigma^2 \varphi_{\mu,\sigma}(a).$$

We can set up a spreadsheet to search for the values of μ and σ that give the right values when plugged into this formula. We find

$$\mu = 452.35,$$
$$\sigma = 55.94.$$

A put option is more valuable as the stock price goes down; assuming a normal distribution its fair price is given by

$$E[\max(a - X, 0)] = (a - \mu)\Phi_{\mu,\sigma}(a) + \sigma^2 \varphi_{\mu,\sigma}(a).$$

Substituting $a = 450$ gives

$$E[\max(a - X, 0)] = (-2.35)\Phi_{\mu,\sigma}(450) + (55.94)^2\varphi_{\mu,\sigma}(450)$$
$$= 21.16.$$

It is interesting that, as mentioned in Section 9.4, the numbers here match the actual prices of American call options on Apple (for the dates mentioned) and the actual price for this put option is \$22.00 – a figure not dissimilar to the result of this calculation.

10.2 Markov groupings

The transition matrix is

$$P = \begin{bmatrix} 1 & 0 & 0 & 0 \\ 0.1 & 0.7 & 0.2 & 0 \\ 0 & 0.1 & 0.7 & 0.2 \\ 0 & 0 & 0 & 1 \end{bmatrix}.$$

We can take powers of this matrix to find the probabilities over three years (this can be done in Excel using the MMULT function, but you need to be familiar with array functions). We have

$$P^3 = \begin{bmatrix} 1 & 0 & 0 & 0 \\ 0.221 & 0.385 & 0.298 & 0.096 \\ 0.024 & 0.149 & 0.385 & 0.442 \\ 0 & 0 & 0 & 1 \end{bmatrix}.$$

If we are equally likely to be in B or C, then the probability of getting to D after three steps is $0.5(0.096 + 0.442) = 0.269$.

Now we consider grouping together the states B and C. Always assuming that we are equally likely to be in B or C, this gives a three-state chain with transition probabilities

$$\tilde{P} = \begin{bmatrix} 1 & 0 & 0 \\ 0.05 & 0.85 & 0.1 \\ 0 & 0 & 1 \end{bmatrix}.$$

The three-step transition probabilities are given by

$$\tilde{P}^3 = \begin{bmatrix} 1 & 0 & 0 \\ 0.129 & 0.614 & 0.257 \\ 0 & 0 & 1 \end{bmatrix}.$$

This gives a probability of getting to D after three steps of 0.257 (slightly smaller than without grouping).

10.4 Octophone

We can calculate the weights of evidence as follows:

	Good	Bad	Odds	WoE
18–21	385	37	10.41	−0.17
21–29	587	41	14.32	0.15
30–45	476	36	13.22	0.07
46+	402	36	11.17	−0.10
credit card	1353	67	20.19	0.49
no credit card	497	83	5.99	−0.72
previous phone	1388	74	18.76	0.42
no previous phone	462	76	6.08	−0.71

In the population there are 1850 goods and 150 bads. So $\log(o_{\text{Pop}}) = \log(1850/150) = 2.512$. The scorecard can be obtained simply by multiplying the weights of evidence by 100 to give:

	Scorecard
age 18–21	−17
age 21–29	15
age 30–45	7
age 46+	−10
credit card	49
no credit card	−72
previous phone	42
no previous phone	−71
constant	251

10.6 Cosal Skincare

(a) We get the following β values

$$\beta_0 = 3.89$$
$$\beta_{h1} = -0.58$$
$$\beta_{h2} = -0.57$$
$$\beta_{h3} = 0$$
$$\beta_{\text{full-time}} = -0.17$$
$$\beta_{\text{not-ft}} = 0$$
$$\beta_{<30} = -0.28$$
$$\beta_{30-39} = 0.17$$
$$\beta_{40-49} = 0.70$$
$$\beta_{>50} = 0$$

(b) To make it worthwhile we need the N/P odds to be less than $150/$2.50 = 60$ so we require log odds less than $\log(60) = 4.0943$. Using the normal procedure of multiplying by 100 and redistributing the constant term (189 to household attributes, 100 to work, and 100 to age) we get a scorecard as follows

Attribute	Score
Household of 1	131
Household of 2	132
Household of 3 or more	189
Full time work	83
Not full time work	100
Age under 30	72
Age 30 to 39	117
Age 40 to 49	170
Age over 50	100

It is only worthwhile sending the promotion to individuals with scores less than 409.

Index

Business Risk Management: Models and Analysis, First Edition. Edward J. Anderson.
© 2014 John Wiley & Sons, Ltd. Published 2014 by John Wiley & Sons, Ltd.
Companion website: www.wiley.com/go/business_risk_management